T0203920

Organic, Physical, and Materials Photochemistry

AND SUPRAMOLECULAR PHOTOCHEMISTRY

Series Editors

V. RAMAMURTHY

Professor
Department of Chemistry
Tulane University
New Orleans, Louisiana

KIRK S. SCHANZE

Professor
Department of Chemistry
University of Florida
Gainesville, Florida

ADDITIONAL VOLUMES IN PREPARATION

Organic, Physical, and Materials Photochemistry

edited by

V. Ramamurthy
Tulane University
New Orleans, Louisiana

Kirk S. Schanze
University of Florida
Gainesville, Florida

CRC Press
Taylor & Francis Group
Boca Raton London New York

CRC Press is an imprint of the
Taylor & Francis Group, an **informa** business

CRC Press
Taylor & Francis Group
6000 Broken Sound Parkway NW, Suite 300
Boca Raton, FL 33487-2742

First issued in paperback 2019

© 2000 by Taylor Francis Group, LLC
CRC Press is an imprint of Taylor & Francis Group, an Informa business

No claim to original U.S. Government works

ISBN-13: 978-0-8247-0404-9 (hbk)
ISBN-13: 978-0-367-39825-5 (pbk)

Visit the Taylor & Francis Web site at
http://www.taylorandfrancis.com

and the CRC Press Web site at
http://www.crcpress.com

Preface

This sixth volume of the Molecular and Supramolecular Photochemistry series provides an overview of scientific advances in the areas of organic, physical, and materials photochemistry. The contributions contained herein represent the breadth and diversity of the photochemical sciences at the dawn of the new millennium. Although the detailed aspects of these chapters differ, they share the common theme of examining the chemical and physical processes in molecular and supramolecular systems induced by the fundamental event of light absorption.

Within this volume, 13 active chemists working in the broad areas of organic, physical, and materials chemistry have summarized and critically evaluated the most important recent advances in their areas of expertise. In Chapter 1, Brousmiche, Briggs, and Wan provide an overview of recent advances in the area of photochemistry of hydroxyaromatic compounds. This review, written by one of the leading groups in the area of molecular organic photochemistry, gives a concise and authoritative description of hydroxyaromatic photochemistry. Chapter 2, written by Griesbeck and Fiege, and Chapter 3, written by Döpp, provide comprehensive overviews of organic photochemical transformations that afford cyclic products. These chapters are written with a mechanistic flair; however, the reactions described are of paramount significance to the field of synthetic organic chemistry. Chapter 4, written by Maggini and Guldi, provides an encompassing summary of the emerging field of fulleropyrrolidines—functionalized derivatives of the fullerenes, the very important class of carbon allotropes. This imaginatively written chapter describes the synthesis and photophysical characterization of a wide variety of supramolecular systems that are constructed by azomethine ylide cycloaddition to buckminsterfullerene (C_{60}). Maggani and Guldi's chapter is a nice follow-up to Chapter 9 of Volume 1 of this series, which described the photochemistry and photophysics of fullerene materials.

Chapters 5 and 6 of the current volume, authored by van Willigen and Falvey, respectively, consider applications of physical methods to study photochemically produced molecular reactive intermediates. In particular, in Chapter 5 van Willigen provides an authoritative primer and overview of the application of time-resolved EPR techniques to the study of free-radical intermediates. This chapter, written by one of the leaders in the field, covers the basic principles of the technique and then considers example applications to systems of interest to organic and organometallic chemists. The chapter is a valuable resource to practicing molecular photochemists who plan to use time-resolved EPR in their own work. Chapter 6 discusses the chemistry of nitrenium ions, important organic molecular reactive intermediates based on electron-deficient nitrogen. Falvey provides a comprehensive overview of this area, providing the reader with insight into methods of generation and reactivity to be expected of nitrenium ions.

Finally, Chapters 7 and 8 describe work being carried out by scientific leaders in the area of materials and polymer photochemistry. In Chapter 7, Murphy has authored a highly informative chapter that describes the properties of supramolecular assemblies produced by combining nucleic acids and inorganic nanoparticles. This emerging area is of interest not only from the fundamental standpoint but also because there are a number of applications of the nanoparticle-DNA assemblies to important problems in sensing biologically significant molecules. Chapter 8, authored by Lu and Winnik, closes the volume with an overview of the fundamental properties and applications of polymer-based sensors for dioxygen. These interesting materials, which are based on the photochemical quenching of luminescent molecular excited states by O_2, have very interesting properties and are finding widespread application in the fields of engineering and biology. Lu and Winnik provide a commanding discussion of the fundamentals of O_2 quenching in polymers, and then give examples of a number of important applications for the polymer-based sensors.

The chapters in this volume are a valuable asset to scientists and graduate students active in the areas of molecular and supramolecular photochemistry. Each chapter provides a wealth of information for specialists working in the areas of organic, physical, and materials photochemistry, and the volume should find a place on the shelves of all scientists interested in the broad field of molecular and supramolecular photochemistry.

Kirk S. Schanze
V. Ramamurthy

Contents

Contributors

Alexander G. Briggs, Ph.D. Department of Chemistry, University of Victoria, Victoria, British Columbia, Canada

Darryl W. Brousmiche Department of Chemistry, University of Victoria, Victoria, British Columbia, Canada

Dietrich Döpp, Dr. phil. Institut für Synthesechemie, Gerhard-Mercator-Universität Duisburg, Duisburg, Germany

Daniel E. Falvey, Ph.D. Department of Chemistry and Biochemistry, University of Maryland, College Park, Maryland

Maren Fiege, Ph.D. Department of Organic Chemistry, University of Cologne, Köln, Germany

Axel G. Griesbeck, Ph.D. Department of Organic Chemistry, University of Cologne, Köln, Germany

Dirk M. Guldi, Ph.D. Radiation Laboratory, University of Notre Dame, Notre Dame, Indiana

Xin Lu, M.Eng.   Department of Chemistry, University of Toronto, Toronto, Ontario, Canada

Michele Maggini, Ph.D.   Dipartimento di Chimica Organica, Università di Padova, Padova, Italy

Catherine J. Murphy, Ph.D.   Department of Chemistry and Biochemistry, University of South Carolina, Columbia, South Carolina

Hans van Willigen, Ph.D.   University of Massachusetts, Boston, Massachusetts

Peter Wan, Ph.D.   Department of Chemistry, University of Victoria, Victoria, British Columbia, Canada

Mitchell A. Winnik, Ph.D.   Department of Chemistry, University of Toronto, Ontario, Canada

Contents of Previous Volumes

1

Photochemistry of Hydroxyaromatic Compounds

Darryl W. Brousmiche, Alexander G. Briggs, and Peter Wan
University of Victoria, Victoria, British Columbia, Canada

I. INTRODUCTION

Although it is well known that phenols and hydroxyarenes are in general much stronger acids in the excited singlet state than in the ground or excited triplet state [1,2], the utility of this phenomenon for inducing new chemistry has not been extensively exploited. Much work continues in the study of the photoprototopic behavior of these compounds [3–5]. Of particular interest is intramolecular proton transfer, which may be rationalized as arising from the enhanced acidity of the phenolic (hydroxyarene) proton coupled with the enhanced basicity of another functional group on the same molecule. Such a broad topic is, however, beyond the scope of this chapter. Rather, the focus will be restricted to photochemical reactions of phenols and hydroxyarenes that may be loosely interpreted as resulting from their enhanced acidity in the singlet excited state. An attempt will be made to be inclusive rather than exclusive. For example, if replacement of the phenolic group by methoxy results in significant reduction in reaction efficiency (or changes the reaction path), then it is probable that the photoreaction concerned will be covered.

Several years ago, Wan and Shukla [6] reviewed organic photochemical reactions that may be mechanistically interpreted as resulting from a change in

1

acidity or basicity on photoexcitation. Several examples of chemical reactions which were claimed to be initiated by the enhanced acidity of phenols and naphthols in their excited state were presented. Reactions of this type will not be covered. Similarly, reactions that clearly involve radicals or radical cations (via direct or sensitized photolysis), those where the phenol (hydroxyarene) moiety plays an ancillary role, or those with very low quantum yields (Φ) will not be the focus of this chapter. Otherwise, the last several years have seen significant progress in the chemistry of phenols and hydroxyarenes where enhanced acidity in the excited state is probably vital for their observed photoreactivity.

II. HALOPHENOLS

A. Simple Halophenols

Extensive studies have been undertaken on the photochemical degradation of simple halophenols, due to their presence as pollutants in pulp mill effluents and in the commercial production of herbicides and pesticides [7,8]. This topic was recently reviewed in some detail [9], and as such only a discussion of selected mechanisms for degradation of monohalogenated phenols is presented here. The mechanisms for degradation of the higher halogenated phenols are extensively discussed in the above-mentioned review.

1. 2-Halophenols

Initial work by Omura and Matsuura [10] on the photolysis of 2-chlorophenol (1a) in aqueous base reported only a complex mixture of products. Later work by Boule and co-workers [8,11] showed that irradiation of 1b in neutral aqueous solution led to catechol (2). However, irradiation in aqueous base yielded a dimer of cyclopentadiene carboxylic acid. Formation of 2 presumably occurs through loss of Cl$^-$, to give an aryl cation, with subsequent attack by water. The Diels–Alder adduct arises via a complex mechanistic sequence involving loss of Cl$^-$, to give an α-ketocarbene, which undergoes a Wolff rearrangement to give ketene 3. Subsequent trapping by water gives fulvene 4, which leads to cyclopentadiene carboxylic acid (Scheme 1). Direct evidence for a carbene intermediate is at present not available. However, indirect evidence comes from laser flash photolysis (LFP) spectra, taken in neutral aqueous solution, which showed transients attributed to ketene 3 and fulvene 4, both of which require the intermediacy of the α-ketocarbene. More recent work [12,13] on 1b showed that carbene formation (leading to the Diels–Alder adduct) competes with formation of 2. LFP of 2-bromo and 2-fluorophenol gave identical results with the same transients being observed.

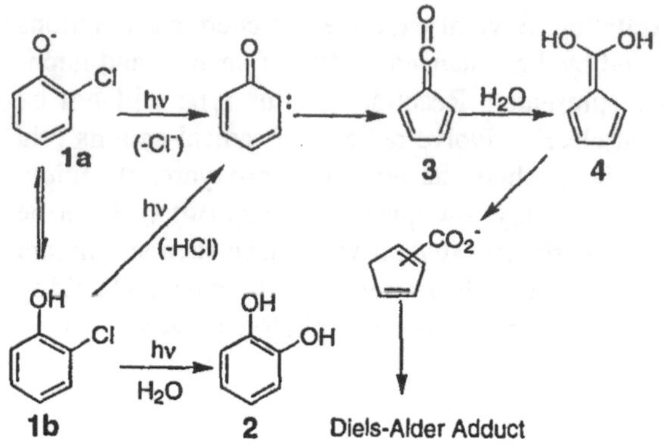

Scheme 1

2. 3-Halophenols

Photolysis of the 3-halo systems in either neutral or anionic form led to resorcinol (5) in greater than 80% yield, regardless of the halogen substituent [8,10,12,14].

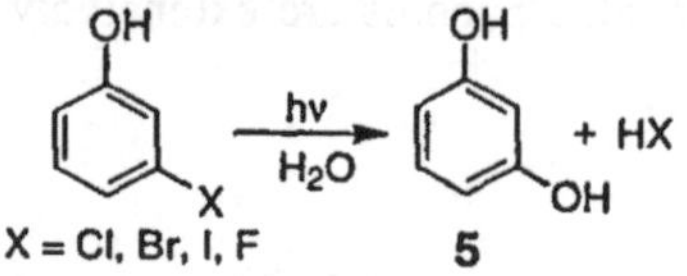

The Φ value for product formation was higher on irradiation of the phenolate than from the phenol, but no explanation has been proposed as to why this is so. The proposed mechanism of reaction was attack by water with concomitant departure of the chloride ion [8]. However, a more reasonable pathway is initial Ar–Cl bond heterolysis followed by attack of water on the aryl cation. Another possibility is a mechanism involving initial excited state proton transfer (ESPT) to solvent, to generate an excited state phenolate ion, which then undergoes Ar–Cl bond heterolysis.

3. 4-Halophenols

Omura and Matsuura [10] originally proposed that photolysis of 4-chlorophenol (6) resulted in initial homolysis of the C–Cl bond, since several coupling products (e.g., 7) along with hydroquinone (8) and p-benzoquinone (9) were observed. Later studies by Boule and co-workers [8] confirmed that 8 is indeed a product, although they proposed that its formation occurred through a mechanism involving initial bond heterolysis. Flash photolysis coupled with product studies by high-performance liquid chromatography (HPLC) [7,15] showed that **9** is the primary stable product from photolysis of **6** in aerated solutions. Electron spin resonance (ESR) spin trapping studies of the photolysis products of **6** [16] and 4-bromophenol [17] were identical, with the 4-hydroxyphenyl radical being identified in both cases. However, this proposed intermediate did not explain all of the products obtained from photolysis of the 4-halophenols. Other work by Oudjehani and Boule [18] proposed a triplet exciplex as an intermediate leading to **9**, while Brown and co-workers [19] proposed the intermediacy of a radical cation in its formation.

Using LFP studies, Grabner et al. [13] have shown that the initial photochemical intermediate is ketocarbene **10**, formed via formal loss of HCl (Scheme 2). This intermediate adequately explained the formation of all the observed photoproducts. Subsequent Fourier transform electron paramagnetic resonance (FT-EPR) [20] and LFP studies [21] have confirmed that the ketocarbene is indeed an intermediate in the photolysis. Studies of the corresponding bromo, fluoro, and iodo compounds have shown that they undergo a similar reaction pathway [14,22].

The initial step in the formation of ketocarbene **10** has been proposed by Grabner et al. [13] to involve heterolytic cleavage of the C–Cl bond followed by loss of the phenolic proton, rather than ESPT to solvent followed by loss of Cl⁻. However, it was noted by the authors that direct evidence for either pathway was not available. The justification for the proposed pathway is twofold: 1) The Φ value of ketocarbene formation is unaffected in the pH range 0–7, although

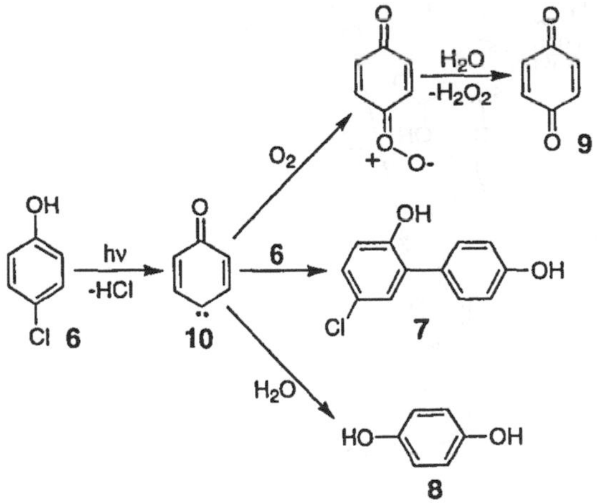

Scheme 2

the calculated pK_a of the excited singlet state halophenol is about 4 and 2) photolysis of 3-chlorophenol led cleanly to resorcinol (**5**), presumably via a substitution reaction and showed no LFP transients in the 50-ns time scale. As 3-halophenols cannot readily form ketocarbenes, it was proposed that only Ar–Cl heterolytic cleavage and not initial loss of a proton is required for substitution. An observation [13] in favor of the ESPT pathway is that carbene formation from **6** only took place in polar protic solvents, which are required for proton transfer.

B. Diflusinal Photochemistry

Several nonsteroidal anti-inflammatory drugs have been reported to have phototoxic side effects [23–26]; however, only diflusinal (**11**) contains a phenol moiety. Irradiation of **11** under anaerobic conditions yielded **13**, while similar experiments under O_2 gave several oxidation products which were not identified [26]. Fluorescence studies on **11** led to the proposal that an excited state intramolecular proton transfer (ESIPT) process was involved in the reaction pathway, due to the appearance of a red-shifted emission band attributed to **12**. Radical intermediates were also implicated in the reaction by observation of a decrease in the yield of **13** as scavengers were added, as well as by detection of the radical cation of methylviologen upon irradiation with **11** in deaerated solution. Based on these

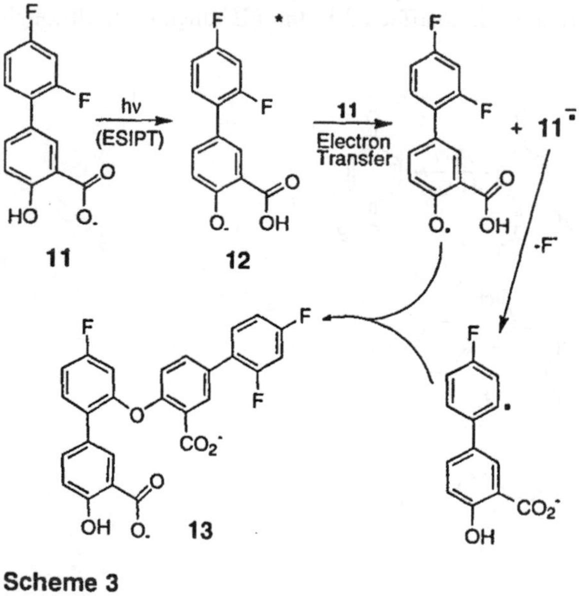

Scheme 3

results, the authors proposed that the reaction proceeded via an initial ESIPT followed by electron transfer to substrate (Scheme 3). Hemolytic activity was found to be due mainly to photolysis of **13**, although the radicals formed upon irradiation of **11** might also contribute. An alternative mechanism not involving electron transfer is direct Ar–F bond heterolysis from **12**, to generate the corresponding aryl cation, which is subsequently trapped by **11**.

III. HYDROXYSTYRENES, ACETYLENES, AND RELATED COMPOUNDS

A. *ortho*-Hydroxyaromatic Oximes and Imines

1. *o*-Hydroxy-Substituted Aromatic Oximes and Oxime Ethers

Grellmann and Tauer [27] reported that salicylaldoxime (**14**) undergoes photocyclization to 1,2-benzisoxazole (**15**) in hexanes or benzoxazole (**16**) and **15** in protic solvents. The analogous reaction of **17** to form benzisoxazole **18** and benzoxazole **19** was later mentioned by Ferris and Antonucci [28]. They proposed that both **18** and **19** were formed directly from irradiation of **17**. Subsequent work by this group [29,30] on **18** and by Haley and Yates [31] on **15** showed that **16** and **19** were actually formed as a secondary photoproduct from irradiation of **15** and **18**, respectively. A possible mechanism was proposed for formation of **16** (or **19**) [29] which involved homolysis of the N–O bond and formation of azirine **20**, followed by rearrangement and ring closure.

14 R, R$_1$ = H   15 R = H   16 R = H   20
17 R = CH$_3$; R$_1$ = H 18 R = CH$_3$ 19 R = CH$_3$

Haley and Yates [31] proposed that there were two possible pathways for the initial cyclization of **14** to **15**, based on fluorescence studies which showed that the phenol moiety could be either intra- or intermolecularly hydrogen-bonded, depending on the polarity of the solvent. In the mechanism proposed for the intermolecularly H-bonded system, deprotonation of the phenol to the solvent on excitation is followed by nucleophilic attack on the oxime nitrogen

14   H   15   H

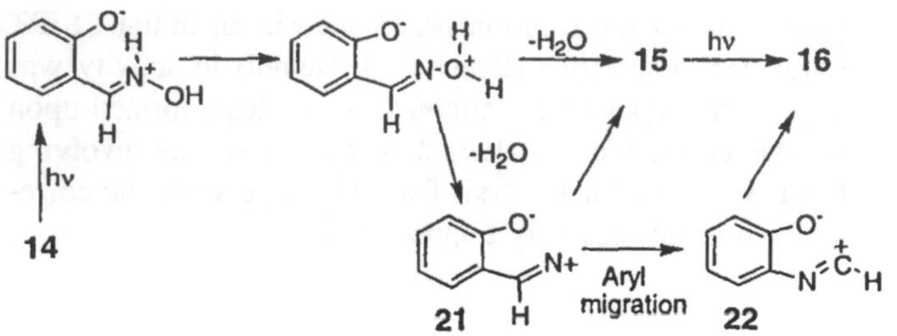

Scheme 4

atom with concomitant loss of OH⁻ to yield **15**. The proposed mechanism for the intramolecularly H-bonded system is more complex (Scheme 4). It involves an initial ESIPT from the phenol to nitrogen. This is followed by a proton shift to the oxime OH which loses water to give either **15** (if it is concomitant with nucleophilic attack by the phenolate) or **21**, which can subsequently collapse to give **15**. In addition, **21** may undergo an aryl migration to give **22**, which cyclizes to form **16**, thus bypassing **15**.

2. Photocyclization of *N,N'*-Diphenyl-1,5-dihydroxy-9,10-anthraquinone diimine

Kobayashi et al. [32] reported that ESIPT is necessary in the photocyclization of **23** to **24**. Irradiation of **25** and **26** gave no photoproduct, indicating that the phenol moiety is necessary for reaction. Time-resolved IR spectroscopic studies of **23** showed two transients which were assigned as *o*-quinone methide **27** ($\tau \approx$ 1.3 ms) and **28** ($\tau \approx 90$ μs). The proposed mechanism involved ESIPT of the phenol proton to the imine to give electronically excited **27**, followed by conrotatory electrocyclic ring closure to **28** (note stereochemistry). Subsequent oxidation of **28** (by O₂) gave **24**. Further irradiation of **24** gave **29** in low yield, presumably via a similar mechanism involving initial ESIPT from the enamine to the ketone.

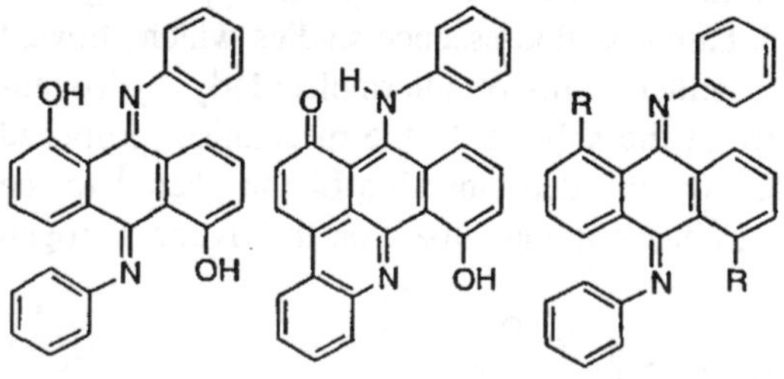

23　　　　**24**　　　　**25** R = OCH₃

　　　　　　　　　　　　　　　26 R = H

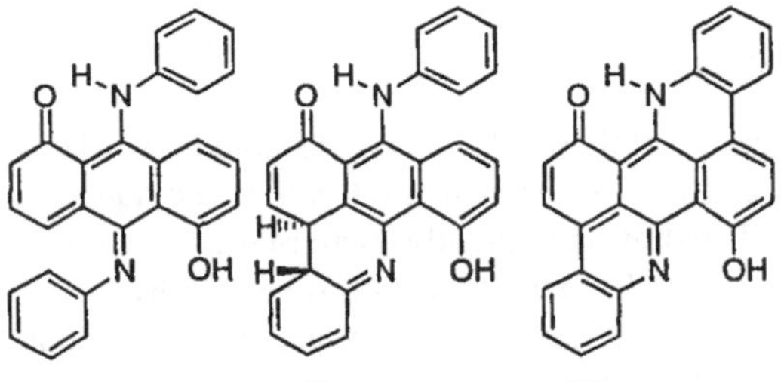

27   **28**   **29**

B. Photoisomerization of 2-Hydroxybenzonitrile

Photolysis studies by Ferris et al. of o-hydroxybenzonitrile (**30**) [28,33] showed formation of **16** via a singlet pathway with no trace of **15**, although they showed that photolysis of **31** (the amino analog of **30**) formed **33** (an analog of **16**) via photoisomerization of **32**. LFP studies of **15** and **30** under O_2 by Richard and

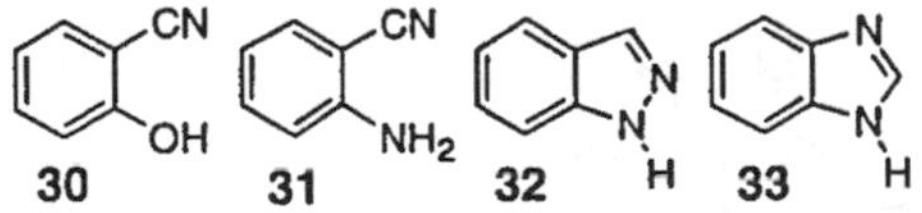

30   **31**   **32**   H   **33**   H

co-workers [34] showed the same transient, which was assigned to **20**, due to the similarity of the observed UV-Vis spectrum to the 2,4-cyclohexadienone chromophore. Thus, they proposed that **20** may be a common intermediate in the formation of **16** from both **15** and **30**. Formation of **16** was found [28] to decrease in protic solvents. As these solvents favor ESPT (to solvent), it can be inferred that ESIPT is required for formation of **16** from **30**. Although no mechanism was presented, **16** may arise from an initial ESIPT process where the proton is transferred from the phenol to the cyano group followed by rearrangements (Scheme 5) to yield **20** and, ultimately, **16**. Photolysis of 4-cyanophenol has also been proposed to result in formation of an azirine [34] with the final product being an isonitrile. However, triplet-quenching experiments showed it was not formed via the singlet manifold.

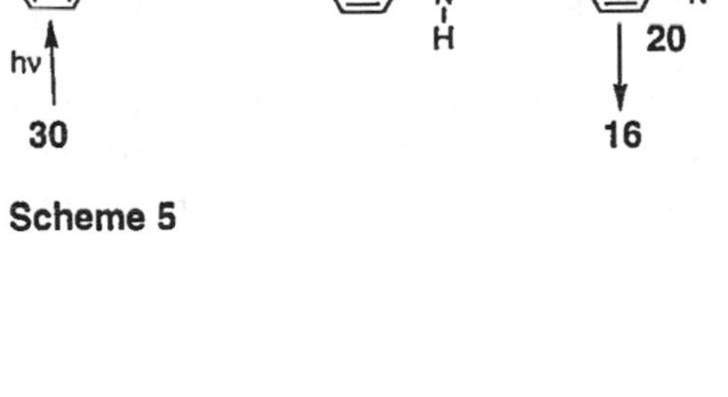

Scheme 5

C. Alkenyl Phenols and Naphthols

1. Photohydration and Photoamination

Photochemical studies by Woolridge and Roberts [35] and Yates and co-workers [36,37] of various phenyl alkenes and acetylenes in neutral and acidic aqueous solution have shown that the rate of hydration is greatly increased upon excitation. In an extension of these studies [38,39] experiments on several o-hydroxy-substituted styrenes and acetylenes were undertaken in neutral and acidic aqueous solution to determine if ESIPT played a role in the anticipated photohydration reaction. Results indicated that this was the case, as 34 and 35 were both efficiently photohydrated at neutral pH to give 36 and 37, respectively.

Plots of the product Φ vs. pH for 34 and 35 indicated that there was no pH dependence in the range of 7–0. However, there was a strong dependence observed for the methoxy derivatives. These results are indicative that ESIPT is rate limiting for the hydroxy compounds. Moreover, the methoxy analogs and the parent hydrocarbons were found to have a much lower Φ value for photohydration in neutral solution compared to 34 and 35, indicating that another mechanistic pathway is operating for the photohydration of these systems. In addition, Φ was observed to drop on irradiation of 34 and 35 at high pH. This was attributed to formation of the phenolate in the ground state, which cannot undergo ESIPT. Based on these results, a mechanism for photohydration was proposed for 34, involving initial ESIPT from the singlet excited state, to generate an o-quinone methide, which is subsequently trapped by water to give the Markovnikov photohydration product. Photohydration of 35 proceeds in a similar manner. Subsequent work by Kalanderopoulos and Yates [39] expanded on the initial studies by investigating the effects of ring and alkene substituents in the o-hydroxystyrene (34) system. Rotation of the phenyl–alkene double bond due to substitution of an α-H (e.g., 38) allowed better overlap between the phenylhydroxy proton and the alkene π electrons which resulted in a higher Φ for product formation. That such a hydrogen bond–like interaction existed was shown in the IR spectrum of **34**, which contained two OH bands, attributed to free and hydrogen-bonded phenols.

Recent studies undertaken by Fischer and Wan [40,41] explored the photo-hydration of *m*-and *p*-hydroxystyrenes **39** and **40** and indene **41**. Unlike the ortho systems, these molecules do not have the phenol and alkene moieties in close proximity. Even so, product studies indicated that efficient photohydration was occurring to yield the corresponding hydration products with significant Φ (~0.2) in neutral solution. Fluorescence studies in acetonitrile with water as quencher were found to fit a cubic dependence (modified Stern–Volmer analysis). This indicated that the formal ESIPT process is mediated by a water trimer, which effectively acts as a bridge between the hydroxy and alkene moieties in the ex-cited singlet state. LFP studies showed formation of the corresponding *m*-quinone methides as intermediates in the photohydration.

Irradiation of several substituted hydroxystyrene derivatives **42–45** in ace-tonitrile in the presence of amines [42,43] resulted in the respective Markovnikov addition products, while irradiation of the respective acetyl esters (of the phenol) gave no reaction. The initial report [42] proposed that the phenolate was deproto-nated by the amine in the ground state and that upon excitation the proton was transferred from the ammonium ion to the alkene. In subsequent work [43], a new mechanism was proposed (Scheme 6) in which adiabatic proton transfer occurs from the excited state phenol to the amine, with subsequent protonation of the alkene by the ammonium ion to form an *o*-quinone methide. This is fol-lowed by nucleophilic attack by amine to give the product. The authors ruled out direct proton transfer from the phenol to the alkene due to the fact that no amination products were observed in the related para system **46**. However, Fis-cher and Wan [40,41] (vide supra) have shown that a polar protic solvent (prefer-ably water) is required to mediate the ESIPT process in meta and para systems.

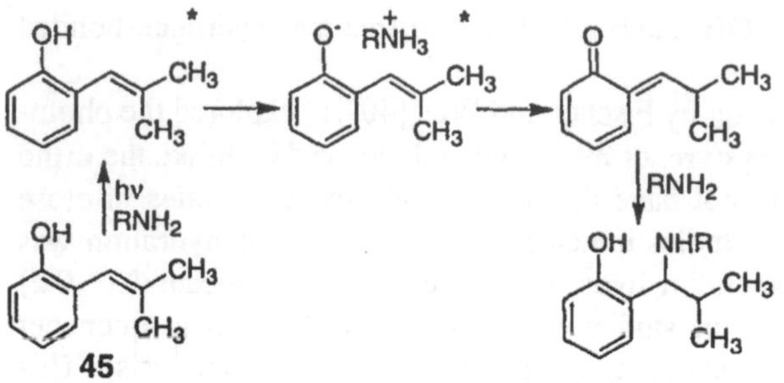

Scheme 6

It is likely that ESIPT occurs between the phenol and the alkene in the ortho systems, given that a hydrogen bond has been shown to exist between these moieties [39]. Fluorescence studies showed a decrease in phenol emission with added amine, with a concomitant increase in emission from the phenolate. This is most likely due to a nonreactive pathway involving ESPT to the amine, as a decrease in phenolate emission would be expected if the excited phenolate ion were to react with the ammonium ion.

2. Photocyclization of a Vinylnaphthol Derivative

Uchida and Irie [44] reported the interesting photocyclization of **47** to **48** in hexane. As the cyclization is reversible and **48** is colored, this reaction forms the basis of a photochromic system. The authors proposed initial ESIPT from the phenol to the alkene to generate the corresponding *o*-quinone methide (which was not detected), followed by electrocyclic ring closure to give chromene **48**. In general, electrocyclic ring closures of photogenerated *o*-quinone methides are not seen due to the formation of a highly strained four-membered ring. In this case, the π system of the thiophene ring allows for an electrocyclic ring closure involving a six-membered ring. No cyclized product was found upon irradiation in methanol, which is in agreement with the proposed mechanism, since methanol would readily quench the *o*-quinone methide by nucleophilic attack.

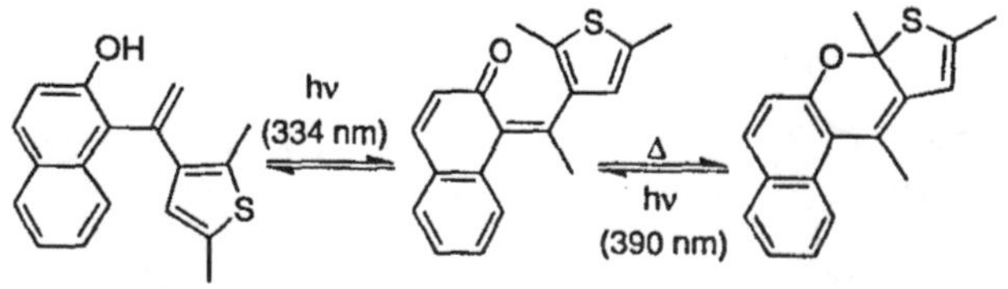

IV. HYDROXYAROMATIC KETONES AND RELATED COMPOUNDS

A. Photocyclization of Hydroxychalcones

1. Formation of Flavanones from 2′-Hydroxychalcones

Stermitz et al. [45] reported the formation of flavanone (50) from 2′-hydroxychalcone (49) on irradiation in benzene. Similarly, they found that irradiation of sorbophenone 53 yielded flavanone 54. In both cases, it was proposed that the reaction proceeded by initial ESPT from the phenol to solvent, followed by nucleophilic attack of the phenolate on the vinyl carbon β to the carbonyl. It was implied that proton transfer occurred between the phenol and solvent. However, ESIPT between the phenol and ketone seems more likely. Interestingly, the reverse reaction (photochemical transformation of 50 to 49) had previously been reported by Mack and Pinhey [46], who believed that the reaction followed a radical pathway. Similarly, Nakashima et al. [47] showed that chalcone 51 was formed on irradiation of flavanone 52. However, they proposed that the reaction occurred through electrocyclic ring opening.

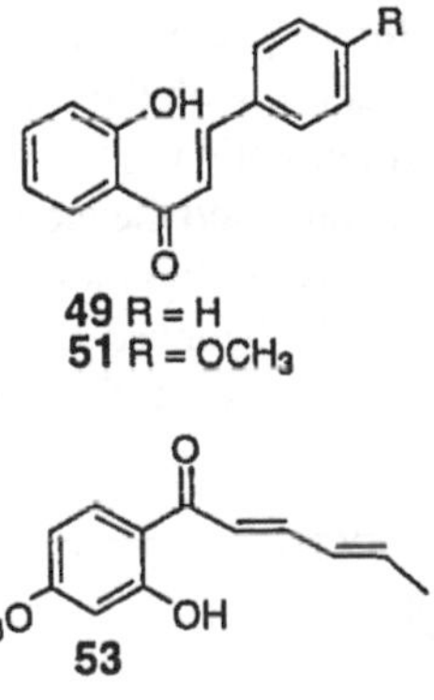

49 R = H
51 R = OCH₃

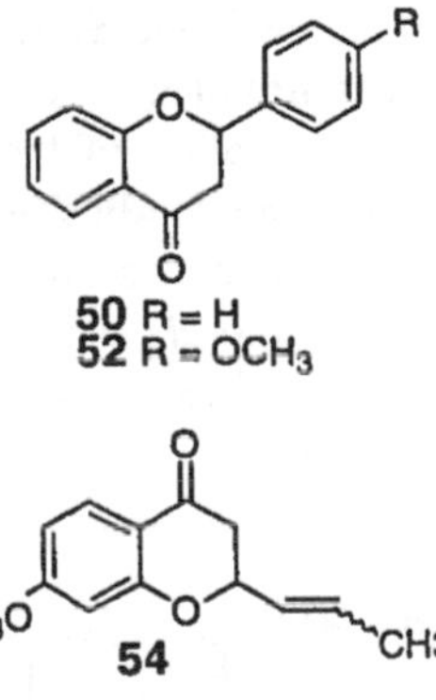

50 R = H
52 R = OCH₃

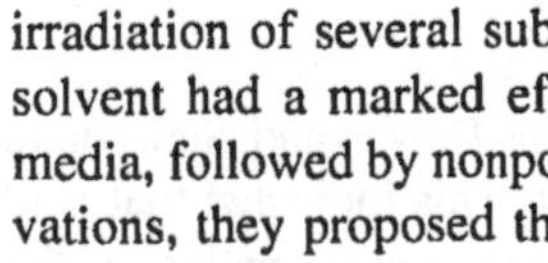

53

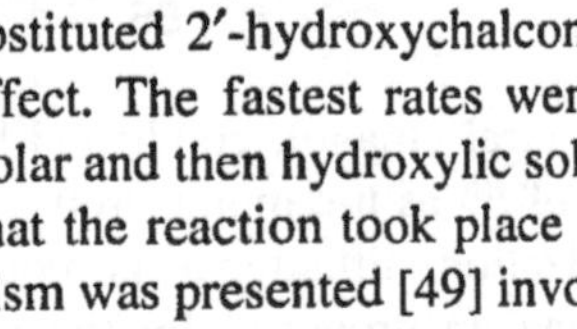

54

Matsushima and co-workers [48,49] showed that triplet quenchers and radical scavengers had no effect on the yield or the rate of flavanone formation upon irradiation of several substituted 2′-hydroxychalcones. However, the choice of solvent had a marked effect. The fastest rates were observed in polar aprotic media, followed by nonpolar and then hydroxylic solvents. Based on these observations, they proposed that the reaction took place via ionic rather than radical intermediates. A mechanism was presented [49] involving initial ESIPT from the phenol to the carbonyl (Scheme 7), to generate an o-quinone methide. Isomerization about the exocyclic alkene yields the intermediate required for electrocyclic ring closure to the flavone product. Quantum yield measurements showed a linear dependence on light intensity consistent with a one photon process. As such, the authors argued that rotation about the exocyclic alkene occurred adiabatically.

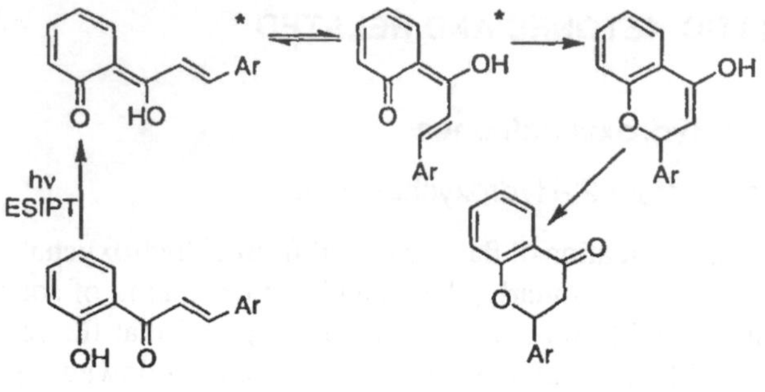

Scheme 7

The electrocyclic ring closure could also occur adiabatically, although evidence for this was not available.

2. Formation of Flavylium Cations from 2-Hydroxychalcones

Jurd [50] has shown that 2-hydroxychalcone **55** formed the respective flavylium cation **56** much more rapidly when irradiated (in aqueous acid), with the conclusion that photolysis aided in the formation of the flavylium cation. This work [50], as well as subsequent ground-state studies by McClelland and Gedge [51], showed that in solution 2-hydroxychalcones are in the *trans* configuration. As cyclization necessarily occurs from the *cis* isomer, Jurd [50] believed that the required *trans-cis* isomerization was aided by light, while the cyclization itself proceeded thermally, mediated by acid.

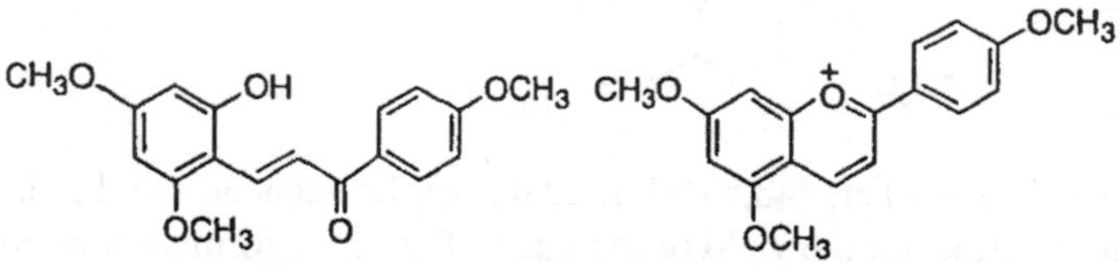

Photolysis of **57** in neutral ethanol, conducted by Dewar and Sutherland [52], showed the formation of **59** as the major product. This indicated that acid is not required for the cyclization, although it is required for formation of the flavylium cation. The mechanism proposed for the reaction was attack of the phenolate on the carbonyl following its deprotonation in the excited state, to give initially **58**, and then **59** on acetal exchange with ethanol. Although it can be assumed that the phenolate resulted from an ESPT to solvent, no indication was given as to how the required *cis* isomer was formed.

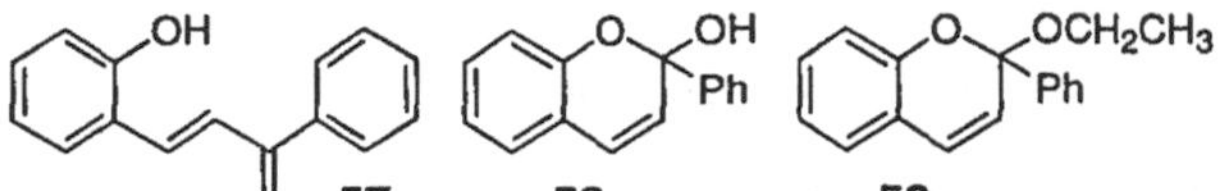

57   58   59

Matsushima and co-workers [53–57] have conducted extensive studies on the photochromic and actinometric properties of a variety of 2-hydroxychalcones. The initial mechanism proposed [53,54] for cyclization to the flavylium cations was a two-photon process, requiring one photon for the isomerization and one for the cyclization. Later work [55,56] indicated that the reaction involved only a single photon, based on the observed first-order relationship between the rate of reaction and light intensity. It was believed that trans–cis isomerization might proceed following an ESIPT from the phenol to the ketone (to generate an o-quinone methide), with subsequent electrocyclic ring closure (Scheme 8), similar to the mechanism they proposed for cyclization of the 2′-hydroxychalcones (Scheme 7).

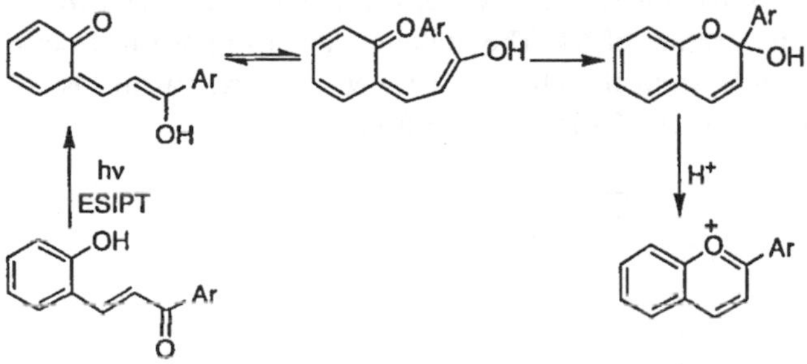

Scheme 8

B. Photosolvolytic Rearrangement of p-Hydroxyphenacyl Esters

Recent work by Givens and co-workers [58–60] has shown that the p-hydroxyphenacyl moiety can be utilized as a photocaging group due to its facile release of protected functionalities (e.g., **60**). Previous work on p-methoxyphenacyl derivatives **61** [61–64] for the photorelease of a number of functional groups have implicated reaction via T_1, with the mechanism being either 1) hydrogen abstraction by the ketone (from solvent) to give the ketyl radical, followed by fragmentation to release ˙OR and the enol of p-methoxyacetophenone or 2) direct homolysis of the C–OR bond. Photodecaging from **60** was found to occur most efficiently in water, with formation of **62** and released HOR. Although Givens and co-workers [58–60] have suggested a mechanism via T_1 involving radical intermediates,

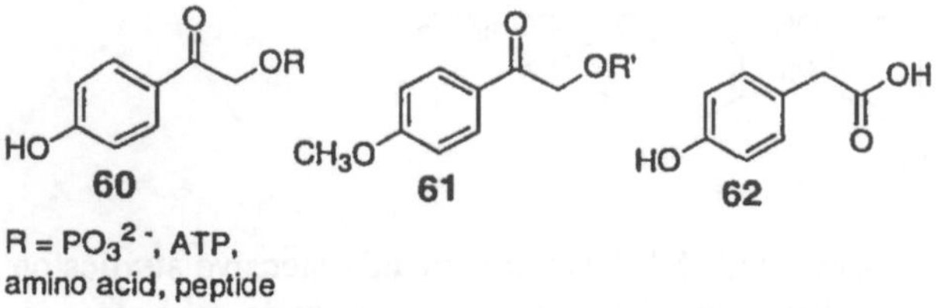

60 **61** **62**

R = PO$_3^{2-}$, ATP,
amino acid, peptide

recent work by Zhang et al. [65] has shown no decrease in formation of **62** from **63** in the presence of O$_2$ and triplet quenchers. In addition, irradiation of **63** in neat acetonitrile or the *p*-methoxy derivative in aqueous acetonitrile gave no reaction. These results indicate that reaction of **63** proceeds via S$_1$ and that both the *p*-hydroxy group and water are required for the decaging reaction. Based on this and other data, a new mechanism for reaction of **63** has been proposed (Scheme 9). Excitation results in water-mediated ESIPT from the phenol to the ketone to give *p*-quinone methide **64**, presumably still electronically excited. Loss of acetic acid with rearrangement (Favorskii-like process) gives cyclopropane **65**, which then leads to **62** on attack by water. The key steps in this new mechanism are ESIPT and the adiabatic reaction of **64**. The authors have also not ruled out a mechanism involving ESPT from the phenol to water, concerted with loss of acetate and rearrangement to give **65** in one step from S$_1$.

Scheme 9

V. ALLYLPHENOLS AND ALLYLNAPHTHOLS

A. Photocyclization of Allylphenols and Allylnaphthols

Horspool and Pauson [66] reported the Markovnikov photocyclizations of **66** and **67** to **68** and **69**, respectively, although no mechanism was presented. Further work by Fráter and Schmid [67] expanded on the allylphenols studied. Irradiation of unevenly substituted alkenes gave the Markovnikov cyclization product in

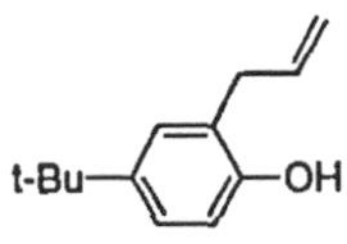

66 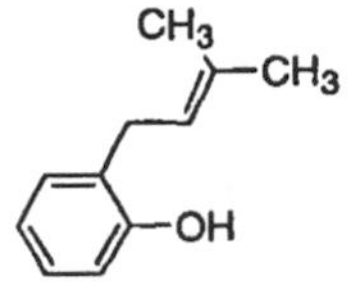   67

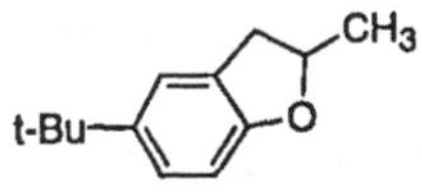

68 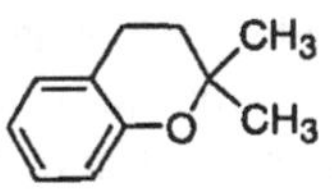   69

high yield, while in cases where the double bond was more evenly substituted, a mixture of the respective benzopyrans and benzofurans was obtained (e.g., **71** and **72** from **70**) [67]. Increasingly lower yields were obtained as the solvent polarity increased. It was proposed that the photocyclization proceeded via an ESIPT between the hydrogen-bonded phenol and alkene in a Markovnikov fashion, presumably followed by nucleophilic attack of the phenolate on the carbocation.

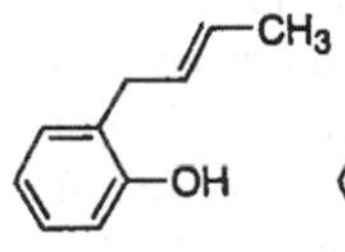

70 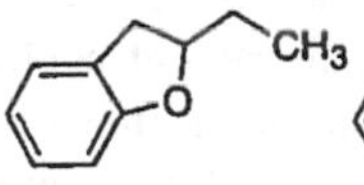   71 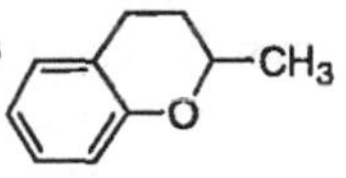   72

The photocyclization/rearrangement of **73** to yield **74** was reported by Shani and Mechoulam [68], but without a proposed mechanism. Later work [69] on cannabidiol (**75**) gave **76** and **77** as the major products in cyclohexane. The authors proposed a radical mechanism for their formation as both alkenes are attacked in essentially equal proportions, although they did acknowledge that ESIPT was another alternative. In hexanes it can be expected that both phenolic groups in **75** will be hydrogen-bonded to some extent to the relevant alkene moiety. As such, it is reasonable to expect that an ionic ESIPT pathway could also occur for both alkenes.

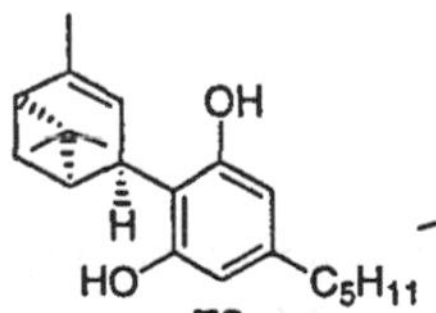

73 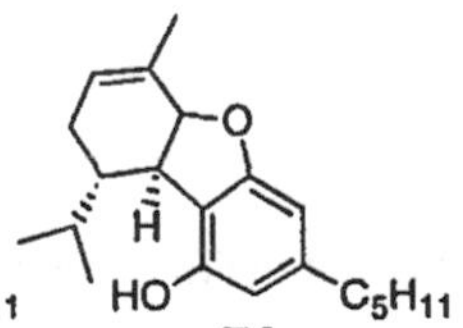   74

Shani and co-workers [70] have conducted experiments on the photocyclization of several substituted *o*-allylphenols in various solvents which show the importance of hydrogen bonding between the phenol and the alkene in these systems. Those 2-allylphenols that contained a polar substituent ortho to the phenol did not undergo photocyclization. This lack of reactivity was attributed to the formation of a strong intramolecular H-bond between the polar substituent and the phenol, resulting in a nonproductive ESIPT. In compounds with nonpolar substituents, cyclization was found to occur except in solvents that had the ability to hydrogen bond to the phenol. These results are in accordance with a photocyclization mechanism via an ESIPT. Related work by Kim et al. [71] involving *o*-allylphenol in water and ammonia clusters has shown that ESIPT does not occur when the solvent cluster size is greater than 3.

An alternative to ESIPT has been presented [72] to explain the photocylization in these systems. It involves initial electron transfer between the phenol and the alkene, followed by hydrogen transfer and then radical cyclization. In contrast, Chow et al. [73] have presented more recent work that favors ESIPT for the photocyclization of 2-allyl-1-naphthol (**78**) and 1-allyl-2-naphthol (**79**). Irradiation in benzene gave the expected mixture of benzopyrans and benzofurans, as well as secondary photoproducts. Product yields were significantly decreased upon photolysis in polar solvents, presumably due to an unproductive ESPT to solvent, and in accordance with reaction occurring via an ESIPT pathway. Decreases fluorescence intensity and product Φ with a quencher such as triethylamine (which hydrogen bonds to the phenol) were used to show that the reaction proceeds through a singlet pathway.

Miranda and Tormos [74] have shown that cyclization may not be the only reaction occurring upon excitation of *o*-allylphenols. They found formation of seven other photoproducts (<1%) from irradiation of **70**, in addition to the expected cyclic ethers. It was proposed that formation of these products occurred through a carbene and other reactive intermediates, including radical cations or anions (by electron transfer to or from solvent, respectively).

Cyclization in a manner analogous to that observed for the *o*-allylphenols has been reported for 1,1'-bis-2-naphthol (**80**) and its monomethoxy derivative **82** [75]. Irradiation in methanol or aqueous acetonitrile yielded **81** and **83**, respectively. No product formation was observed upon irradiation of the dimethoxy derivative **84**. The authors proposed that initial photolysis of **80** results in ESIPT from the naphthol to the 3-carbon of the other naphthalene system, to yield **85**, followed by ring closure. Presumably excitation results in intramolecular charge transfer to this position, making it more basic. Although no proposal of how the proton is transferred was presented, it is most likely via a solvent-mediated pathway, given that the reaction does not occur in nonprotic solvents.

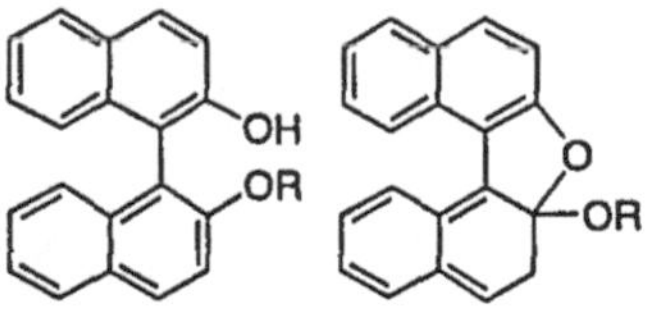

80 R = H
82 R = CH₃

81 R = H
83 R = CH₃

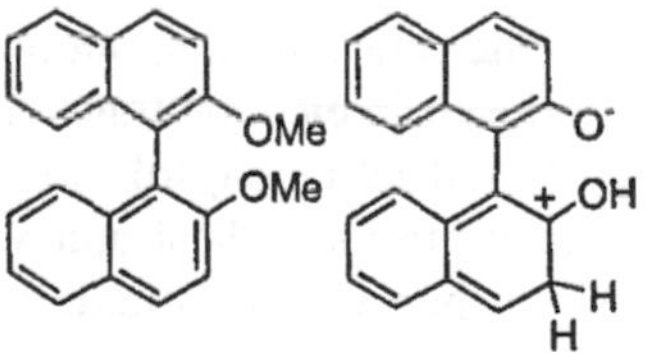

84

85

B. Photocyclization of Cinnamylphenols and Cinnamylnaphthols

Studies [76] on *o*-(*trans*-2-cinnamyl)phenol (**86**) showed that both the five- and six-membered cyclic ethers (**87** and **88**, respectively) were formed on irradiation, along with *o*-(*cis*-2-cinnamyl)phenol (**89**). Photolysis in deaerated hexanes resulted in **89** as the major product (48%), although **87** and **88** were also formed. In aerated solutions the yield of **89** dropped significantly, with a concomitant increase in the yields of **87** and **88**, while photosensitization with acetone gave only **89**. These observations indicated that cyclization occurred via the singlet manifold, while isomerization was via the triplet pathway.

As these systems are bichromophoric, they have the possibility of reaction occurring from excitation of either the phenol or the styrene chromophore. Irradiation of benzene, as singlet sensitizer, gave increased ratios of **87** to **88**, while use of other sensitizers with lower singlet energies gave more **88** than **87**. This indicates that formation of the benzopyran comes from the styrene singlet state (which has the lower singlet energy), while the benzofuran is formed via excitation of the phenol. ESIPT from each of these chromophores results in a different carbocation, which subsequently reacts in the ground state to give the respective product.

In addition to ESIPT, it was proposed [76,77] that formation of **88** could also occur from irradiation of **86** via an electron transfer pathway in the presence of an appropriate electron donor. Experiments with added trimethyl and triethylamine gave a decreased yield of **87** with increasing amine concentration, but an increase in the yield of **88**. Excitation was believed to result in electron transfer from the amine to excited state **86**, to yield the styrene radical anion, followed by intramolecular proton transfer from the phenol moiety to the radical anion. Reverse electron transfer from the amine radical cation then yields a diradical which can cyclize to form **88**.

Derivatives of the parent *o*-(2-cinnamyl)phenol with substituents on the styrene ring have been synthesized, as these were expected to lower the energy of the styrene chromophore and allow its selective excitation [78]. Irradiation of **90–92** in benzene resulted in enhanced ratios of the respective benzopyrans relative to the benzofurans, verifying that the six-membered cyclic ethers were the product of excitation of the styrene chromophore. The *p*-chloro and bromo derivatives of the parent *o*-(2-cinnamyl)phenol (on either the styrene or the phenol ring) have also been synthesized and their reactivity compared to the parent system [79]. Loss of the halogen atom occurred when the energy of the respective S_1 state (styrene or phenol) was above that of the C–X bond homolysis energy. However, if the S_1 energy was lower, then cyclization predominated.

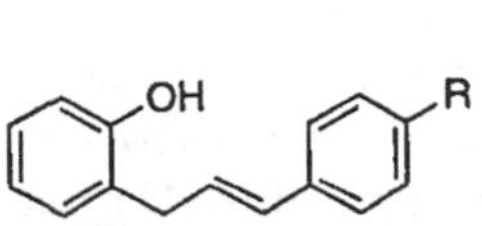

90 R = Ph
91 R = CH₃
92 R = OCH₃

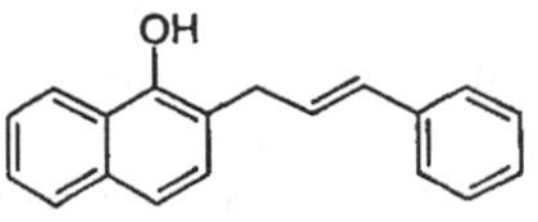

93

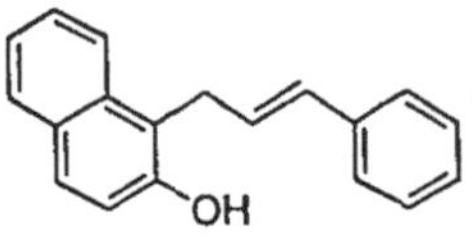

94

In related work [80], investigations were undertaken in the photochemistry of the cinnamylnaphthols **93** and **94**. Here the naphthol chromophore was expected to dictate the reactivity of the system. The Φ values of fluorescence for **93** and **94** were significantly lower than for the parent naphthols, indicating that the singlet state was being quenched, presumably by ESIPT to the alkene. However, quenching studies showed it was likely that electron transfer was also occurring in the cyclization of **94**.

VI. HYDROXYBENZYL ALCOHOLS AND RELATED COMPOUNDS

A. Chromenes and Hydroxyaromatics with Labile Benzylic Substituents

Early reports [81,82] of photogenerated *o*-quinone methides resulted from an investigation of the photochromism of synthetic and naturally occurring 2H-chromenes. The variously substituted synthetic chromenes (e.g., **95,96**) gave, upon UV irradiation over a range of temperatures and in liquid or solid (frozen solvent glass) solution, yellow, orange, or red photoproducts whose color faded within hours. Natural chromenes, which occur widely in plants, behaved similarly. Strong evidence for *o*-quinone methide intermediates such as **97** was provided by in situ hydride reduction to isolable phenolic derivatives such as **98**. No such product was detected upon treating 6-methoxy-2H-chromene **96** with LiAlH₄ in the absence of irradiation.

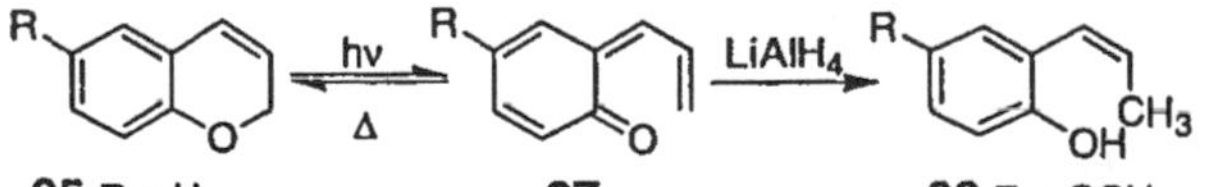

95 R = H
96 R = OCH₃

97

98 R = OCH₃

Seiler and Wirz [83] invoked quinone methide–type intermediates (e.g., **100**) to explain the photohydrolysis of a series of trifluoromethyl-substituted phenols and naphthols (e.g., **99**). They remarked on the dramatic (seven orders of magnitude) decrease in pK_a of such substrates upon excitation. Their proposal of heterolytic C–F cleavage from S_1 leading to quinone methide–type intermediates was strengthened by the observation of the corresponding hydroxybenzoic and hydroxynaphthoic acids as hydrolysis products, and by flash photolysis which gave long-wavelength transients (500–550 nm) attributable to **100**.

The variety of substrates that lead to o-quinone methides on photolysis was extended by Hamai and Kokubun [84,85] in a study of the photochromism of 2-hydroxytriphenylmethanols **101**. Irradiation in various aprotic solvents produced the o-fuchsones **102** (which are o-quinone methides), as supported by the new postirradiation λ_{max} at 340 and 440 nm (for R = H) and by comparison with a known fuchsone [85]. An investigation [86] into the solid-state photochemistry of a closely related 4-hydroxytriphenylmethanol system concluded that the favorable intermolecular alignment within the crystal lattice of H^--donating (phenolic) and H^--accepting (benzyl alcohol) groups facilitated photodehydration to fuchsones.

In another ring-opening example, Padwa et al. [87] established the intermediacy of o-quinone methide **104** in the photolysis of 3-phenylisocoumaranone (**103**). With methanol as solvent, the intermediate **104** was largely trapped to give the ether **105**, demonstrating the electrophilic nature of the methide carbon. Further, the formation of the byproduct xanthene **106** is fully consistent with the expected reactivity of **104**.

The body of work cited above suggests the synthetic utility of photo-induced benzylic C–X bond cleavage with substrates containing *o*- or *p*-phenolic groups. Turro and Wan [88] established the facile photogeneration of benzyl cations by the acid-catalyzed photodehydroxylation of *m*- and *o*-methoxy-substituted benzyl alcohols. *p*-Methoxybenzyl alcohol was much less reactive. A logical outcome of the various mechanistic components outlined above was achieved by Wan and Chak [89]. They showed that the enhanced acidity of excited-state phenols could be harnessed for intramolecular dissociation of benzylic C–OH bonds in *o*-hydroxybenzyl alcohol (**107**) to give the parent *o*-quinone methide intermediate **108**. This was trapped by methanol to yield the corresponding ether **109**. Interestingly, photo-induced benzylic C–OH heterolysis was also achieved [90] in alkaline solutions (pH ≥ 10) of **107**, resulting in the photochemical synthesis of phenol-formaldehyde resins. This presumably occurs with the intermediacy of the *o*-quinone methide **108**, which is trapped by the ground-state phenolate ion of **107**, followed by further oligomerization.

The heterolysis of benzylic C–OH bonds can be facilitated by an *o*-amino group. Yang and Wan [91] postulated the formation of the *o*-quinone methide imine **110** by photolysis of *o*-aminobenzyl alcohol in acidic aqueous alcohol solutions. Formation of the methyl ether **111** was taken as evidence for trapping of **110**. The question of intramolecular catalysis by the excited singlet-state protonated amino group remains unresolved. At pH values at which the starting amine remains largely unprotonated (pH ≥ 3.5), nucleophilic attack on **110** by the substrate amino group led to mixed oligomers.

An investigation of photo- and thermochromic properties of Mannich bases led Komissarov et al. [92] to propose an interesting equilibrium involving both *o*- and *p*-quinone methides. Irradiation or heating of the morpholine-substituted naphthalene derivative **112** gave the isolable yellow *p*-quinone methide **114**. They proposed that the less stable isomeric *o*-quinone methide of the naphthalene fragment (113) was the intermediate initially generated. Addition of morpholine moves the equilibrium to the left with concomitant decolorizing of the solution.

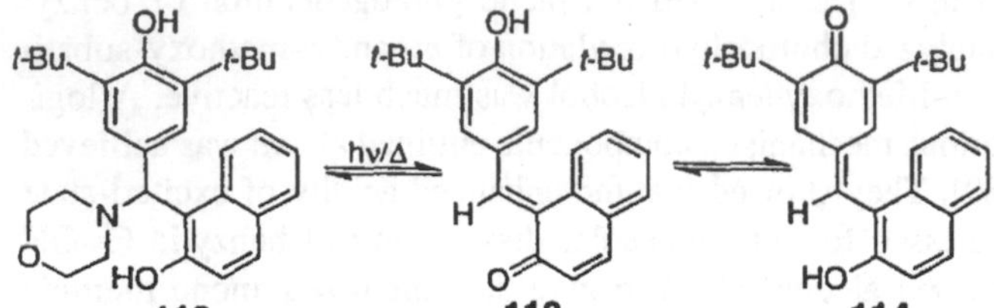

112   113   114

Recently, Nakatani et al. [93] trapped photogenerated *o*-quinone methide 116, formed efficiently by irradiation of Mannich bases such as 115 in neutral aqueous acetonitrile. Reaction with ethyl vinyl ether formed the expected Diels–Alder adduct 117. This was part of an ongoing investigation into low-energy ($\lambda > 300$ nm) generation of quinone methides and their role in alkylation of DNA. The authors noted that photolysis of 115 in basic solution (i.e., as the phenolate ion) does not lead to *o*-quinone methide formation, in contrast to the observation made by Wan and Henning [90] for a simpler phenolic system.

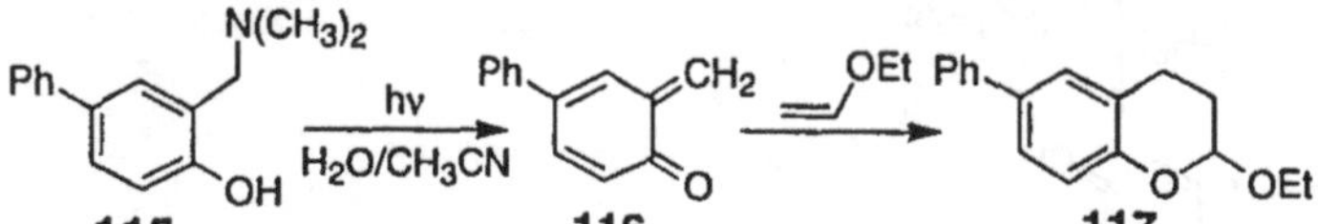

115   116   117

B. Hydroxybenzyl Alcohols

Almost all of the photolytically generated quinone methides and analogs discussed above have involved the ortho isomer. *o*- And *p*-quinone methides are believed to have widespread occurrence in biological transformations and have been used in synthesis (generated thermally), while *m*-quinone methides have been a subject of theoretical interest (94). A simple and general photolytic method for generating all three isomers of quinone methide has been reported by Wan and co-workers [94]. These authors have trapped the prototype *o*-quinone methide 104 by Diels–Alder cycloaddition with in situ ethyl vinyl ether to give the chroman derivative 119. Similar [4+2] adducts were formed on photolysis of 118 in the presence of dihydropyran and dihydrofuran. LFP studies gave a transient absorption spectrum with $\lambda_{max} = 340$, 450 nm attributable to 104.

Photolysis of the para analog 120 in aqueous methanol gave the expected ether product 122. LFP gave a spectrum with $\lambda_{max} = 360$ nm attributable to 121. A strongly absorbing transient ($\tau \leq 30$ ns, $\lambda_{max} = 430$ nm) was also observed on LFP of the meta isomer 123 and attributed to the corresponding *m*-quinone methide 124. Preparative photolysis of 123 in aqueous methanol gave high yields of the corresponding methyl ether product **125**, establishing that the *m*-quinone methides photogenerated using this approach are zwitterionic rather than diradical in character.

The increased understanding of photogeneration of quinone methides has led to a new breadth of opportunity in the synthesis of biologically related molecules. For example, the hexahydrocannabinol ring system **128** has been synthesized [95] in high yield by photolysis of the diol **126**. The *o*-quinone methide intermediate **127** undergoes intramolecular [4+2] cycloaddition with the electron-rich alkene portion to form **128** stereospecifically. Thermal conversion of **126** to **128** requires forcing conditions (180°C).

Quinone methides have become increasingly studied because of putative roles in biochemical processes, but the photochemical aspects of this study are only at the beginning phase. Vitamin B$_6$ (pyridoxine, **129**) has been photolyzed [96] to give the methanol-trapped product **131**. The *o*-quinone methide-type intermediate **130** formed on LFP gave λ_{max} = 370 nm and sufficient intensity for an investigation of the pH dependence of the reaction. This intermediate was longest lived at pH 7. It is interesting that the product from reaction with the corresponding *m*-quinone methide was not observed, although it is possible in principle. The obviously favored *o*-quinone methide was also trapped by cycloaddition with ethyl vinyl ether to give the corresponding adduct in high yield.

C. Biaryl Systems

The photochemistry of hydroxybenzyl alcohols seems always to lead to quinone methides as intermediates, and the following reports by Huang and Wan [97,98] introduced the additional feature of changing dihedral angle in biphenyl variants. They proposed that xanthene (**106**) should be the anticipated product upon photolysis of *o*-phenoxybenzyl alcohol. Intramolecular Friedel–Crafts alkylation by the photogenerated benzyl cation would lead to the desired product **106**. However, the major product was 6H-dibenzo[b,d]pyran (**134**). A detailed investigation [97–99] of the mechanism of this intriguing dehydration–rearrangement led to the proposal of photolytic homolysis followed by recombination to initially give the biphenyl benzyl alcohol **132**. Subsequent excitation results in the biphenyl quinone methide intermediate **133**, which then readily cyclizes to the dibenzopyran **134**. A key feature of this mechanism is the twisting of the biphenyl toward increased coplanarity upon excitation to S_1, which facilitates the reaction. A plot of Φ for formation of **134** on photolysis of **132** as a function of pH showed two sigmoid regions (pH ~ 1 and pH ~ 10). Above pH 10 the phenolate ion of **132** is the predominant form and excitation of this species gave the highest Φ. The sigmoid region at pH ~ 1 (the expected pK_a for S_1 of the phenol moiety) indicates that the reaction at intermediate pHs occurs via the excited-state phenolate ion. The authors also proposed that in very acidic media ($H_0 < -1$) in which the phenolic group would not be deprotonated, dibenzopyran **134** may be formed via a mechanism involving a photogenerated benzylic cation.

Interestingly, it was also shown [100] that xanthene (106) in fact undergoes photoisomerization to the dibenzopyran 134 in good yield. This process is favored in the presence of water, perhaps because polar hydroxylic solvents stabilize postulated (polar) keto intermediates. The procedure was extended to the photolysis of 9-methylxanthene to give the corresponding methyl-substituted pyran. However, the 9-alkyl substituent opens new reaction pathways, so that other products, including phenanthrene, become dominant [100].

Shi and Wan [101] have utilized the tendency of excited-state biphenyls to favor planarity (relative to the more twisted geometry of the ground state) in an investigation of the change in charge polarization for hydroxy-and alkoxy-substituted biphenyls upon excitation to S_1. For example, heating of 4-phenylphenol with D_2SO_4 (in acetic acid) gave electrophilic deuteration at position 2. In contrast, irradiation in D_2SO_4/acetonitrile gave mainly deuteration at position 2' and a lesser amount of the 4'-deuterated product (i.e., on the phenyl ring not attached to the hydroxyl group). This kind of photo-induced charge redistribution is important in the context of possible molecular conductors and switches. As part of the same study, the biphenyl alcohols 135 and 136 were studied and shown to react via loss of the benzylic hydroxide ion from S_1, giving rise to novel and highly conjugated biphenyl quinone methides 137 and 138.

Similar considerations apply to the photolysis of the biphenyl alcohol 139. Due to the favorable orientation of substituents, 139 cyclizes in high yield to pyran 141 in aqueous acetonitrile [102] (see also the reaction of 132 to give 134). The expected biphenyl quinone methide intermediate 140 could not be detected by LFP, presumably because of its very short lifetime. The reaction proceeds, albeit less efficiently, in anhydrous acetonitrile, suggesting a partial role for ESIPT from the phenol group in S_1. The same cyclization occurred when crystals of 139 were irradiated, showing that protic solvents are not required. However, the solid-state molecular configuration is such that *intermolecular* ESPT may be postulated to account for the reaction.

The use of light to induce physical motion (as well as chemical reactions) in molecules is of key interest in the design of molecular "machines." In order to further define the utility of photo-induced twisting of biaryls, Shi and Wan [103] synthesized several (e.g., binaphthyl **142**) having large ground-state dihedral twist angles (up to 90°). Upon irradiation (λ = 254 nm) these convert to the corresponding pyrans (e.g., **143**) with a much smaller dihedral twist angle (~37°). Moreover, the reaction is photoreversible at longer wavelengths (>350 nm). The fluorescence behavior of model compounds suggests that S_1 deprotonates *before* the twisting motion that ultimately leads to the more planar pyrans. Subsequently, Burnham and Schuster [104] studied this system in a search for chiral photochromic optical triggers for liquid crystals. These authors resolved the two enantiomers of **142** and **143** and measured the activation energy for thermal racemization of **143** to be 26 kcal/mol. Interestingly, irradiation of optically active **142** at 254 nm gave optically active **143**, and irradiation of optically active **143** at greater than 350 nm gave optically active **142**. LFP of **143** in cyclohexane gave a transient spectrum with λ_{max} = 560 nm attributed to the (naphtho)quinone methide **144**. This study concludes that ground-state **144** is the intermediate through which both thermal and photochemical interconversion and racemization of **142** and **143** occur. Unfortunately, this system proved unsuitable as an optical trigger because the enantiomeric excess achievable at any reasonable conversion of either compound was inadequate.

Finally, the finely tuned competition between cationic intermediates and quinone methides is illustrated by the recent work of Fischer et al. [105] involving hydroxy-9-fluorenols. Photolysis of **145** in aqueous methanol gave the ether **147** in good yield. Evidence for the quinone methide **146** was the observation of a long-lived transient (τ = 5–10 s, λ_{max} = 450 nm) from LFP in water throughout the pH range. Additionally, intermediate **146** was trapped by cycloaddition with in situ ethyl vinyl ether in high yield. Photolysis of the *m* isomer **148** in aqueous methanol also gave the corresponding methyl ether. In contrast, however, LFP of **148** in aqueous solution gave no observable transient, suggesting the intermediacy of the formally antiaromatic cation **149**. LFP of **151** in water gave a transient (τ = 47 ns, λ_{max} = 480 nm) assigned to the cation **150**. The authors suggested that a *m*-quinone methide–type intermediate may be formed upon irradiation of **151** under strongly alkaline conditions, but otherwise the cation seems to be preferred. Therefore, the simple presence of a phenolic OH group is not sufficient to force the preference for reaction via quinone methides under all conditions, though *ortho*-OH appears to strongly favor this, presumably because of better proximity of the functional groups for concerted loss of water.

VII. CONCLUSIONS

This chapter has focused attention on the photochemistry of phenols and naph-thols that may be interpretable as due to the enhanced acidity of the phenol (naph-thol) proton in S_1. A much larger body of knowledge on the general photochemis-try of these and related compounds exists, and the reader should be aware that this chapter has only presented a portion of this literature. It seems clear that the reactions reviewed show remarkable diversity in structure, reaction outcome, and mechanistic detail. One of the more general reaction types is that of quinone methide photogeneration from appropriately substituted phenols and related com-pounds. Although ortho-substituted systems have been the most studied and play the most prominent role, a number of meta and para isomers show similar reactiv-ity. Photolytic pathways to quinone methide intermediates offer mild conditions for the formation of a wide range of desirable organic and biochemical targets. The basic photochemistry of these systems is now understood and the door is open to exciting new possibilities. We can only conclude that much more remains to be learned in the photochemistry of phenols and naphthols.

ACKNOWLEDGMENTS

Support of this work came from the National Science and Engineering Research Council (NSERC) of Canada and the University of Victoria. D.W.B. thanks NSERC for a postgraduate scholarship.

REFERENCES

1. Ireland, J. F.; Wyatt, P. A. H. *Adv. Phys. Org. Chem.* **1976**, *12*, 131.
2. Vander Donckt, E. *Prog. React. Kinet.* **1970**, *5*, 273.
3. Brown, R. G.; Ormson, S. M.; Le Gourrierec, D. *Prog. React. Kinet.* **1994**, *19*, 211.
4. Arnaut, L. G.; Formosinho, S. J. *J. Photochem. Photobiol. A: Chem.* **1993**, *75*, 21.
5. Arnaut, L. G.; Formosinho, S. J. *J. Photochem. Photobiol. A: Chem.* **1993**, *75*, 1.
6. Wan, P.; Shukla, D. *Chem. Rev.* **1993**, *93*, 571.
7. Lipczynska-Kochany, E.; Bolton, J. R. *J. Chem. Soc., Chem. Commun.* **1990**, 1596.
8. Boule, P.; Guyon, C.; Lemaire, J. *Chemosphere* **1982**, *11*, 1179.
9. Burrows, H. D.; Ernestova, L. S.; Kemp, T. J.; Skurlatov, Y. I.; Purmal, A. P.; Yermakov, A. N. *Prog. React. Kinet.* **1998**, *23*, 145.
10. Omura, K.; Matsuura, T. *Tetrahedron* **1971**, *27*, 3101.
11. Guyon, C.; Boule, P.; Lemaire, J. *Tetrahedron Lett.* **1982**, *23*, 1581.
12. Boule, P.; Richard, C.; David-Oudjehani, K.; Grabner, G. *Proc. Indian Acad. Sci.* **1997**, *109*, 509.
13. Grabner, G.; Richard, C.; Köhler, G. *J. Am. Chem. Soc.* **1994**, *116*, 11470.
14. Lipczynska-Kochany, E. *Chemosphere* **1992**, *24*, 911.
15. Lipczynska-Kochany, E.; Bolton, J. R. *J. Photochem. Photobiol. A: Chem.* **1991**, *58*, 315.
16. Lipczynska-Kochany, E.; Kochany, J.; Bolton, J. R. *J. Photochem. Photobiol. A: Chem.* **1991**, *62*, 229.
17. Lipczynska-Kochany, E.; Kochany, J. *J. Photochem. Photobiol. A: Chem.* **1993**, *73*, 23.
18. Oudjehani, K.; Boule, P. *J. Photochem. Photobiol. A: Chem.* **1992**, *68*, 363.
19. Durand, A.-P. Y.; Brattan, D.; Brown, R. G. *Chemosphere* **1992**, *25*, 783.
20. Ouardaoui, A.; Steren, C. A.; van Willigen, H.; Yang, C. *J. Am. Chem. Soc.* **1995**, *117*, 6803.
21. Durand, A.-P.; Brown, R. G.; Worral, D.; Wilkinson, F. *J. Photochem. Photobiol. A: Chem.* **1996**, *96*, 35.
22. Durand, A.-P.; Brown, R. G.; Worrall, D.; Wilkinson, F. *J. Chem. Soc. Perkin Trans. 2* **1998**, 365.
23. Condorelli, G.; Costanzo, L. L.; De Guidi, G.; Giuffrida, S.; Miano, P.; Sortino, S.; Velardita, A. *EPA Newsletter* **1996**, *58*, 60.
24. Constanzo, L. L.; De Guidi, G.; Condorelli, G.; Cambria, A.; Famà, M. *J. Photochem. Photobiol. B: Biol.* **1989**, *3*, 223.
25. Constanzo, L. L.; De Guidi, G.; Condorelli, G.; Cambria, A.; Famà, M. *Photochem. Photobiol.* **1989**, *50*, 359.

26. De Guidi, G.; Chillemi, R.; Giuffrida, S.; Condorelli, G.; Cambria Famà, M. *J. Photochem. Photobiol B: Biol* **1991**, *10*, 221.
27. Grellmann, K. H.; Tauer, E. *Tetrahedron Lett.* **1967**, *20*, 1909.
28. Ferris, J. P.; Antonucci, F. R. *J. Am. Chem. Soc.* **1974**, *96*, 2010.
29. Ferris, J. P.; Antonucci, F. R. *J. Am. Chem. Soc.* **1974**, *96*, 2014.
30. Ferris, J. P.; Trimmer, R. W. *J. Org. Chem.* **1976**, *41*, 13.
31. Haley, M. F.; Yates, K. *J. Org. Chem.* **1987**, *52*, 1825.
32. Kobayashi, K.; Iguchi, M.; Imakubo, T.; Iwata, K.; Hamaguchi, H.-O. *J. Chem. Soc., Chem. Commun.* **1998**, 763.
33. Ferris, J. P.; Antonucci, F. R.; Trimmer, R. W. *J. Am. Chem. Soc.* **1973**, *95*, 919.
34. Scavarda, F.; Bonnichon, F.; Richard, C. *New J. Chem.* **1997**, *21*, 1119.
35. Wooldridge, T.; Roberts, T. D. *Tetrahedron Lett.* **1973**, *41*, 4007.
36. Wan, P.; Culshaw, S.; Yates, K. *J. Am. Chem. Soc.* **1982**, *104*, 2509.
37. Wan, P.; Yates, K. *J. Org. Chem.* **1983**, *48*, 869.
38. Isaks, M.; Yates, K.; Kalanderopoulos, P.*J. Am. Chem. Soc.* **1984**, *106*, 2728.
39. Kalanderopoulos. P.; Yates, K. *J. Am. Chem. Soc.* **1986**, *108*, 6290.
40. Fischer, M.; Wan, P. *J. Am. Chem. Soc.* **1998**, *120*, 2680.
41. Fischer, M.; Wan, P.*J. Am. Chem. Soc.* **1999**, *121*, 4555.
42. Yasuda, M.; Sone, T.; Tanabe, K.; Shima, K. *Chem. Lett.* **1994**, 453.
43. Yasuda, M.; Sone, T.; Tanabe, K.; Shima, K. *J. Chem. Soc., Perkin Trans. 1* **1995**, 459.
44. Uchida, M.; Irie, M. *Chem. Lett.* **1991**, 2159.
45. Stermitz, F. R.; Adamovics, J. A.; Geigert, J. *Tetrahedron* **1975**, *31*, 1593.
46. Mack, P. O. L.; Pinhey, J. T. *J. Chem. Soc., Chem. Commun.* **1972**, 451.
47. Nakashima, R.; Okamoto, K.; Matsuura, T. *Bull. Chem. Soc. Jpn.* **1976**, *49*, 3355.
48. Matsushima, R.; Hirao, I. *Bull. Chem. Soc. Jpn.* **1980**, *53*, 518.
49. Matsushima, R.; Kageyama, H. *J. Chem. Soc., Perkin Trans. 2* **1985**, 743.
50. Jurd, L.*Tetrahedron* **1969**, *25*, 2367.
51. McClelland, R. A.; Gedge, S. *J. Am. Chem. Soc.* **1980**, *102*, 5838.
52. Dewar, D.; Sutherland, R. G. *J. Chem. Soc., Chem. Commun.* **1970**, 272.
53. Matsushima, R.; Miyakawa, K.; Nishihata, M. *Chem. Lett.* **1988**, 1915.
54. Matsushima, R.; Suzuki, M. *Bull. Chem. Soc. Jpn.* **1992**, *65*, 39.
55. Matsushima, R.; Mizuno, H.; Kajiura, A. *Bull. Chem. Soc. Jpn.* **1994**, *67*, 1762.
56. Matsushima, R.; Mizuno, H., Itoh, H. *J. Photochem. Photobiol. A: Chem.* **1995**, *89*, 251.
57. Matsushima, R.; Suzuki, N.; Murakami, T.; Morioka, M. *J. Photochem. Photobiol. A: Chem.* **1997**, *109*, 91.
58. Givens, R. S.; Park, C.-H. *Tetrahedron Lett.* **1996**, *37*, 6259.
59. Park, C.-H.; Givens. R. S. *J. Am. Chem. Soc.* **1997**, *119*, 2453.
60. Givens, R. S.; Jung, A.; Park, C.-H.; Weber, J.; Bartlett, W. *J. Am. Chem. Soc.* **1997**, *119*, 8369.
61. Epstein, W. W.; Garrossian, M. *J. Chem. Soc., Chem. Commun.* **1987**, 532.
62. Baldwin, J. E.; McConnaughie, A. W.; Moloney, M. G.; Pratt, A. J.; Shim, S. B. *Tetrahedron* **1990**, *46*, 6879.

63. Givens, R. S.; Athey, P. S.; Kueper III, L. W.; Matuszewski, B.; Xue, J.-Y. *J. Am. Chem. Soc.* **1992**, *114*, 8708.
64. Givens, R. S.; Athey, P. S.; Matuszewski, B.; Kueper III, L. W.; Xue, J.-Y.; Fister, T. *J. Am. Chem. Soc.* **1993**, *115*, 6001.
65. Zhang, K.; Corrie, J. E. T.; Munasinghe, V. R. N.; Wan, P. *J. Am. Chem. Soc.* **1999**, *121*, 5625.
66. Horspool, W. M.; Pauson, P. L. *J. Chem. Soc., Chem. Commun.* **1967**. 195.
67. Fráter, G.; Schmid, H. *Helv. Chim. Acta* **1967**, *50*, 255.
68. Shani, A.; Mechoulam, R. *J. Chem. Soc., Chem. Commun.* **1970**, 273.
69. Shani, A.; Mechoulam, R. *Tetrahedron* **1971**, *27*, 601.
70. Geresh, S.; Levy, O.; Markovits, Y.; Shani, A. *Tetrahedron* **1975**, *31*, 2803.
71. Kim, S. K.; Hsu, S. C.; Li, S.; Bernstein, E. R. *J. Chem. Phys.* **1991**, *95*, 3290.
72. Morrison, H., *Organic Photochemistry*, Padwa, A. Ed.; Marcel Dekker, Inc.: New York, 1979; Vol. 4, pp 143.
73. Chow, Y. L.; Zhou, X.-M.; Gaitan, T. J.; Wu, Z.-Z. *J. Am. Chem. Soc.* **1989**, *111*, 3813.
74. Miranda, M. A.; Tormos, R. *J. Org. Chem.* **1993**, *58*, 3304.
75. Cavazza, M.; Zandomeneghi, M.; Ouchi, A.; Koga, Y. *J. Am. Chem. Soc.* **1996**, *118*, 9990.
76. Jiménez, M. C.; Márquez, F.; Miranda, M. A.; Tormos, R. *J. Org. Chem.* **1994**, *59*, 197.
77. Jiménez, M. C.; Leal, P.; Miranda, M. A.; Tormos, R. *J. Org. Chem.* **1995**, *60*, 3243.
78. Jiménez, M. C.; Miranda, M. A.; Tormos, R. *Tetrahedron* **1997**, *53*, 14729.
79. Jiménez, M. C.; Miranda, M. A.; Tormos, R. *J. Org. Chem.* **1998**, *63*, 1323.
80. Jiménez, M. C.; Leal, P.; Miranda, M. A.; Scaiano, J. C.; Tormos, R. *Tetrahedron* **1998**, *54*, 4337.
81. Becker, R. S.; Michl, J. *J. Am. Chem. Soc.* **1966**, *88*, 5931.
82. Kole, J.; Becker, R. S. *J. Phys. Chem.* **1967**, *71*, 4045.
83. Seiler, P.; Wirz, J. *Helv. Chim. Acta* **1972**, *55*, 2693.
84. Hamai, S.; Kokubun, H. *Bull. Chem. Soc. Jpn.* **1974**, *47*, 2085.
85. Hamai, S.; Kokubun, H. Z *Phys. Chem. (Frankfurt)* **1974**, *88*, 211.
86. Lewis, T. W.; Curtin, D. Y.; Paul, I. C. *J. Am. Chem. Soc.* **1979**, *101*, 5715.
87. Padwa, A.; Dehm, D.; Oine, T.; Lee, G. A. *J. Am. Chem. Soc.* **1975**, *97*, 1837.
88. Turro, N. J.; Wan, P. *J. Photochem.* **1985**, *28*, 93.
89. Wan, P.; Chak, B. *J. Chem. Soc., Perkin Trans. 2* **1986**, 1751.
90. Wan, P.; Hennig, D. *J. Chem. Soc., Chem. Commun.* **1987**. 939.
91. Yang, C.; Wan, P. *J. Photochem. Photobiol. A. Chem.* **1994**, *80*, 227.
92. Komissarov, V. N.; Ukhin, L. Y.; Kharlanov, V. A.; Lokshin, V. A.; Bulgarevich, E. Y.; Minkin, V. I.; Filipenko, O. S.; Novozhilova, M. A.; Aldoshin, S. M.; Atovmayn, L. O. *Akad. Nauk SSSR, Ser. Khim.* **1992**, *10*, 2389. (*Engl. Trans.*, **1993**, 1875)
93. Nakatani, K.; Higashida, N.; Saito, I. *Tetrahedron Lett.* **1997**, *28*, 5005.
94. Diao, L.; Yang, C.; Wan, P. *J. Am. Chem. Soc.* **1995**, *117*, 5369.
95. Barker, B.; Diao, L.; Wan, P. *J. Photochem. Photobiol. A. Chem.* **1997**, *104*, 91.
96. Brousmiche, D.; Wan, P. *J. Chem. Soc., Chem. Commun.* **1998**, 491.

97.   Huang, C.-G.; Wan, P. *J. Chem. Soc., Chem. Commun.* **1988**, 1193.
98.   Huang, C.-G.; Wan, P. *J. Org. Chem.* **1991**, *56*, 4846.
99.   Huang, C.-G.; Beveridge, K. A.; Wan, P. *J. Am. Chem. Soc.* **1991**, *113*, 7676.
100.   Huang, C.-G.; Shukla, D.; Wan, P. *J. Org. Chem.* **1991**, *56*, 5437.
101.   Shi, Y.; Wan, P. *J. Chem. Soc., Chem. Commun.* **1995**, 1217.
102.   Shi, Y.; MacKinnon, A.; Howard, J. A. K.; Wan, P. *J. Photochem. Photobiol. A: Chem.* **1998**, *113*, 271.
103.   Shi, Y.; Wan, P. *J. Chem. Soc., Chem. Commun.* **1997**, 273.
104.   Burnham, K. S.; Schuster, G. B. *J. Am. Chem. Soc.* **1998**, *120*, 12619.
105.   Fischer, M.; Shi, Y.; Zhao, B.; Snieckus, V.; Wan, P. *Can. J. Chem.* **1999**, *77*, 868.

2

Stereoselectivity of Photocycloadditions and Photocyclizations

Axel G. Griesbeck and Maren Fiege
University of Cologne, Köln, Germany

I. INTRODUCTION

In Volumes 1 and 3 of the series "Molecular and Supramolecular Photochemistry" from 1997 and 1999, respectively, two chapters were published that describe the regio- and stereoselectivity of photocycloadditions, photocyclizations, and other photochemical reactions [1]. The reader might wonder about the motivation to write another article with stronger emphasis on the stereoselectivity of photocyclizations and photocycloadditions. We feel that the fascination of organic photochemistry mainly originates from the combination of well-known ground-state effects (such as solvent or substituent effects, steric, electronic, and stereoelectronic control) and additional effects that originate from the excited-state nature of the substrates and unusual spin multiplicities of reagents or intermediates. This makes the analysis of many photochemical processes more difficult, the design of new reactions tricky, and the predictability of efficiency and selectivity problematic. On the other hand, new mechanistic aspects might lead to an important new dimension in control of stereoselectivity in photochemistry and, as recently stated by Peter Wagner, "stereochemistry is the essence of organic chemistry" [2]. Recent investigations of temperature, spin, and substituent effects on the stereochemistry of photochemical processes have shown that at least two concepts have to be taken seriously into account: reactive triplet biradical conformations and the complex kinetic interplay between product formation and deactivation to restore ground-state molecules from reactive intermediates.

Photocycloaddition and photocyclization reactions cover a large part of organic photochemistry and lead to carbo- or heterocyclic compounds with various ring sizes. Depending on the structure of the starting materials and the course of the reaction, several new stereogenic centers are produced. The analysis of the stereochemistry often allows detailed insight into the mechanism of these multistep procedures.

The easiest situation can be described by electronic excitation of the substrate (or one of several substrates) followed by product formation, in this case essentially ring formation. In between the primary excited singlet state and the ground-state product, however, there might be (and there actually exists for most examples) a complex world of photophysical processes, i.e., bond-forming and bond-cleaving steps and intermediates whose lifetimes span over several orders of magnitudes. If such a photochemical reaction involves solely singlet states and if it is perfectly controlled by orbital symmetry rules, the analysis of stereochemistry is straightforward and uses the rules originally formulated by Woodward and Hoffmann. Unfortunately, these rules are not valid for triplet state reactions, and the stereoselectivity-controlling factors become obscure and are a matter of constant debate.

In this chapter, the results of recent research in the fields of stereoselective photocycloaddition and photocyclization reactions are summarized and compared with theoretical models which have been discussed in recent years. The authors of this chapter understand themselves as synthetic organic chemists and thus apologize right from the beginning for possible simplifications in the adaptation of theoretical models. A concise presentation of the overwhelming number of photochemical ring formations described in the literature is also not the aim of this chapter and interested readers should contact relevant reviews written in the last decade (references given in the respective chapters).

II. DESCRIPTION OF PHENOMENOLOGICAL TRENDS

Singlet photocycloadditions, and especially singlet photocyclizations, often show high stereoselectivity concerning the transformation of a given substrate configuration into the product configuration as well as high "simple" diastereoselectivity originating from the formation of a new carbon–carbon bond. In the corresponding triplet reactions stereoselectivity is often diminished and the direction of stereocontrol is in many cases inverted. This phenomenon has been described in the older literature solely as a reason for the lifetime of the intermediates. Singlet biradicals have been detected only in the last two decades and are extremely short-lived if not highly stabilized by spin-diluting substituents. The spin-isomeric triplet biradicals, however, are known for a much longer time and have been detected by spectroscopy, trapping experiments, or radical clock experiments. The lifetime of these species is in the nano- and microsecond region, which gives these transients enough time for molecular motions, especially bond rotations, which are normally not available for their singlet spin isomers. Thus, the conservation of configuration when going from the starting material to the product is no longer expected for triplet photoreactions, and stereochemistry can be used as a simple tool for differentiating between these two reaction channels.

Photocycloadditions involving excited aromatic substrates, such as meta-cycloadditions, are not discussed in this chapter. The main part of this chapter deals with the Paternò–Büchi and the enone–ene photocycloaddition reaction, and the rules discussed there for explaining and predicting product regio- and stereoselectivity are subsequently applied in a condensed way for photocyclization reactions.

III. PHOTOCYCLOADDITION REACTIONS

A. Paternò–Büchi Reactions

The [2 + 2] photocycloaddition of a carbonyl compound with an alkene is called a Paternò–Büchi reaction. The process described in the original publication by Paternò and Chieffi in 1909 was the addition of benzaldehyde to 2-methyl-2-butene using solar irradiation [3]. Despite being more than 90 years old this specific experiment still raises the challenging question, What is the reason for the high regioselectivity and the moderate stereoselectivity? And as the new century dawns, it reminds us of the early days of solar chemistry. Many synthetic [4] and mechanistic [5] features of this reaction have been summarized in excellent reviews.

The "early" step of the reaction mechanism involves the addition of the triplet excited aldehyde to the alkene and the formation of a triplet 1,4 biradical [6] (Scheme 1). The constitution of this biradical is still a matter of debate. From a practical point of view and by using chemical intuition, it seems logical that 1) the electrophilic nature of the oxygen atom of the excited carbonyl group favors reaction at this center and formation of a new C–O bond, and 2) spin stabilization also favors a bis carbon-centered 1,4 biradical. Concerning the alkene component, spin stabilization and the local softness model were used to explain the regioselectivity of the carbonyl oxygen attack [7]. The structure of the products isolated from oxygen-trapping experiments also bolster the assumption of 2-oxatetramethylenes as decisive intermediates [8]. Furthermore, the incorporation of cyclopropyl rings either in the carbonyl or in the alkene component of the Paternò–Büchi reaction leads to the formation of ring-enlarged products via cyclopropylmethyl radical ring-opening reaction. The efficiency of these reac-

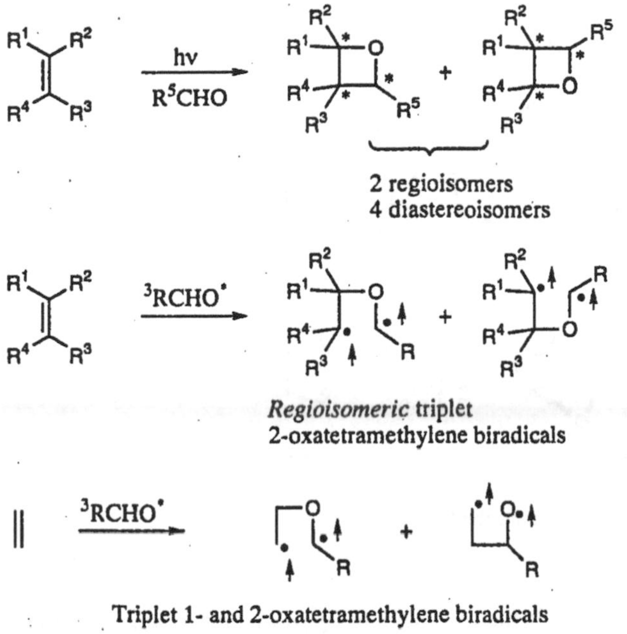

2 regioisomers
4 diastereoisomers

Regioisomeric triplet
2-oxatetramethylene biradicals

Triplet 1- and 2-oxatetramethylene biradicals

Scheme 1 The Paternò–Büchi reaction.

tions strongly depends on the proper substitution of the cyclopropyl ring in order to stabilize the homoallylic 1,7 biradical. This process has already been documented by several research groups [9]. It has been used for biradical lifetime determination [10] and recently also for the synthesis of tetrahydrooxepines [11] (Scheme 2).

The "rule of five" was also helpful in evaluating the structure of the primary intermediates [12]. When acetylnorbornene (1), the classical substrate for intramolecular photocycloaddition, is irradiated, exclusively the oxetane 2 bridging the tetrahydrofuran ring is formed [13]. The only reasonable approach to this structure is C–O bond formation prior to the second C–C bond–forming step. This high-yielding reaction, which leads after ring opening to the diquinane 3, has recently been extended for the synthesis of complex triquinanes [14].

Recently published calculations, however, questioned the assumption of the 2-oxatetramethylene as the exclusive intermediate ("indeed there is no experimental evidence that contradicts the conjecture that both mechanisms may operate" [15a]) to some extent; therefore, the following analysis should be considered only as the most likely avenue to oxetane formation [15]. Stereoselectivity already comes into the game at this early step, e.g., in the antique benzaldehyde

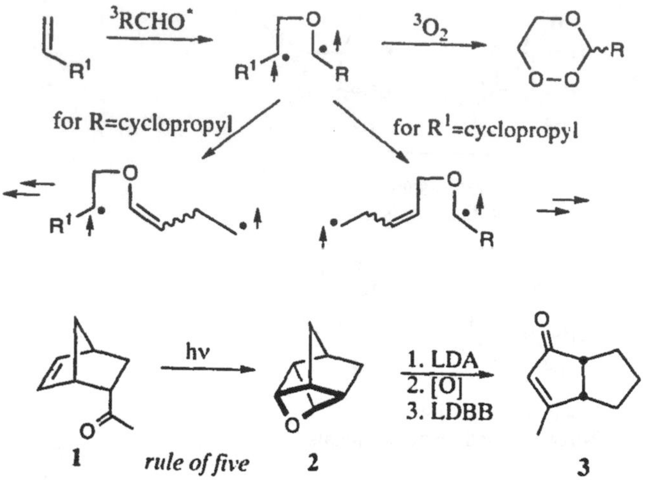

^{3}RCHO$^•$

^{3}O$_2$

for R=cyclopropyl

for R^1=cyclopropyl

hv

1. LDA
2. [O]
3. LDBB

1　　*rule of five*　　**2**　　　　　　**3**

Scheme 2　Arguments for 2-oxatetramethylenes.

addition to trimethylethylene. If the regioselectivity of this reaction is controlled by the "most stable biradical rule," the stereogenic center at C-2 of the product is already defined at this stage [16]. The "late" steps of the Paternò–Büchi reaction are connected to the triplet nature of the intermediary 1,4 biradical. The spin-imposed barrier for triplet to singlet intersystem crossing (ISC) separates primary and secondary bond formation in time. The rate of ISC is connected with several geometrical and energetic factors which make the exact analysis complex. Progress in this direction has been made in recent years by theoretical calculations for relatively simple model structures and will be discussed later.

A milestone in the field of stereoselective Paternò–Büchi reactions was the work on facial diastereoselectivity performed in the group of Hans-Dieter Scharf in the mid-1980s [17]. They used highly electron-rich cycloalkenes (e.g., 1,3-dioxolanes **4**, furans **5**) as π^2 components and phenylglyoxylates **6** as carbonyl compounds. The carbohydrate and menthol-derived chiral auxiliaries were localized in the ketoester part and thus relatively far from the reactive carbonyl group. The simple diastereoselectivity controls the relative configuration of the stereogenic center C-7. In all cases, independent of the structure of the cycloalkene and the ester substituent, exclusively the *endo*-phenyl diastereoisomers **7** and **8**, respectively, were formed [18] (Scheme 3).

The formation of this stereogenic center is decoupled from the facial diastereoselectivity, which rises with increasing size of the ester substituent. The Paternò–Büchi reactions of the phenylglyoxylic acid 8-phenylmenthyl ester with several cycloalkenes resulted in the formation of bicyclic oxetanes **7, 8** with

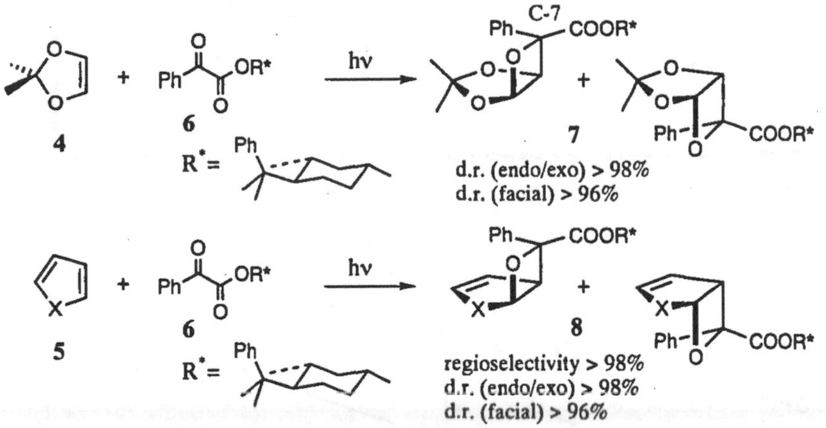

Scheme 3   Facial diastereoselectivity with phenylglyoxylates.

high simple and induced diastereoselectivities [19]. In addition to the remarkable and synthetically useful extent of facial diastereoselectivity, the Scharf group discovered a striking temperature dependence of the facial selectivity [20] which resembled the isoselectivity relations investigated by Giese [21]. In addition to isoselectivity points, however, inversion temperatures were discovered at which the influence of the reaction temperature on the degree of stereoselectivity was inverted [22]. This effect is a fundamental new phenomenon and has recently also been discovered in several nonphotochemical reactions [23] (Scheme 4).

A straightforward mechanistic interpretation of this phenomenon in Paternò–Büchi reactions is a two-stage process with one stage predominantly determined by the activation entropy term and the other by the activation enthalpy. At the inversion temperature, the selectivity-determining step changes from entropy- to enthalpy-determined and thus also the temperature/selectivity behavior. No inversion temperatures were detected for the simple diastereoselectivity. Actually, the diastereoselectivity of the second C–C bond formation is more than 98% in all cases investigated by Scharf et al. This clearly results from a process at stage 2 that cannot be influenced by temperature effects on stage 1. From the analysis of simple and induced diastereoselectivities, as well as from the temperature dependence of the facial diastereoselectivity, a reaction mechanism results which describes the origin of product stereochemistry as a subtle combination of facial approach, biradical conformational equilibration, retrocleavage, and C–C bond formation. A simple kinetic picture for this scenario is shown in Scheme 4. The endo-/exo- (simple) selectivity results from the last step in the reaction sequence, i.e., after spin-inversion from several possible biradical con-

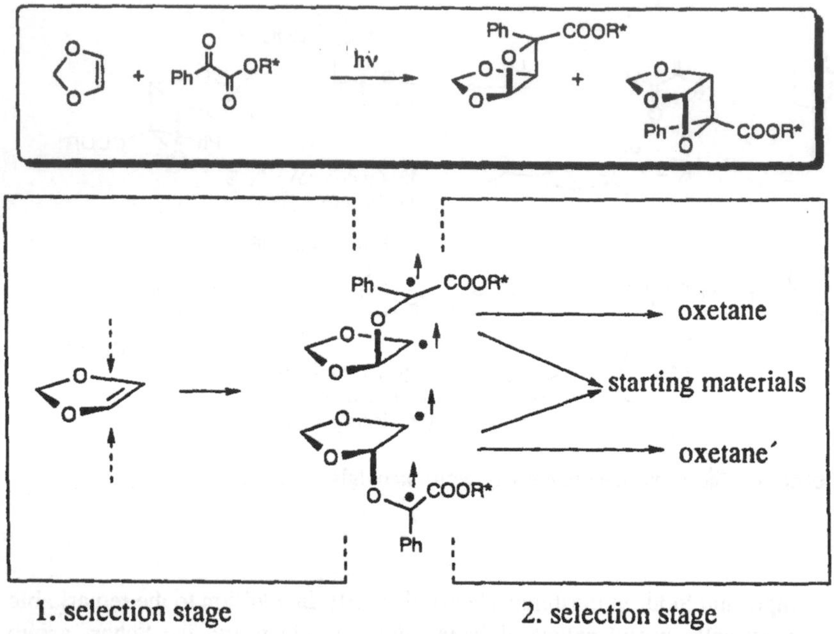

1. selection stage   **2. selection stage**

Scheme 4 The two-stage model for the Paternò–Büchi reaction.

formers, the exo- and the endo diastereoisomers (with already determined facial selectivity) are formed in competition with C–O bond cleavage. Whether or not singlet biradicals are involved in this process is still an open question. In any case, the lifetimes for these species are expected to be extremely short, and bond rotations cannot compete with C–C bond formation or C–O bond cleavage. Thus, the simple diastereoselectivity maps the ISC geometry of the triplet 1,4 biradicals [24].

In order to separately analyze facial and simple diastereoselectivity and the termination step of the Paternò–Büchi reaction, we investigated model reactions between cycloalkenes similar to those used by Scharf et al. As carbonyl addends prochiral aldehydes and ketones were applied. The most intensively investigated alkene was 2,3-dihydrofuran (**9** = DHF), which gave the Paternò–Büchi product **10** with high (98:2) regioselectivity (Scheme 5). The regiochemical aspect of the Paternò–Büchi reaction is often discussed in terms of the most stable biradical rule. However, this rule ignores the fact that the approach geometry between the electronically excited carbonyl and the alkene component does not necessarily have to result in the energetically more stable 1,4 biradical. Exciplexes with strong charge transfer character have yet not been detected in the Paternò–Büchi

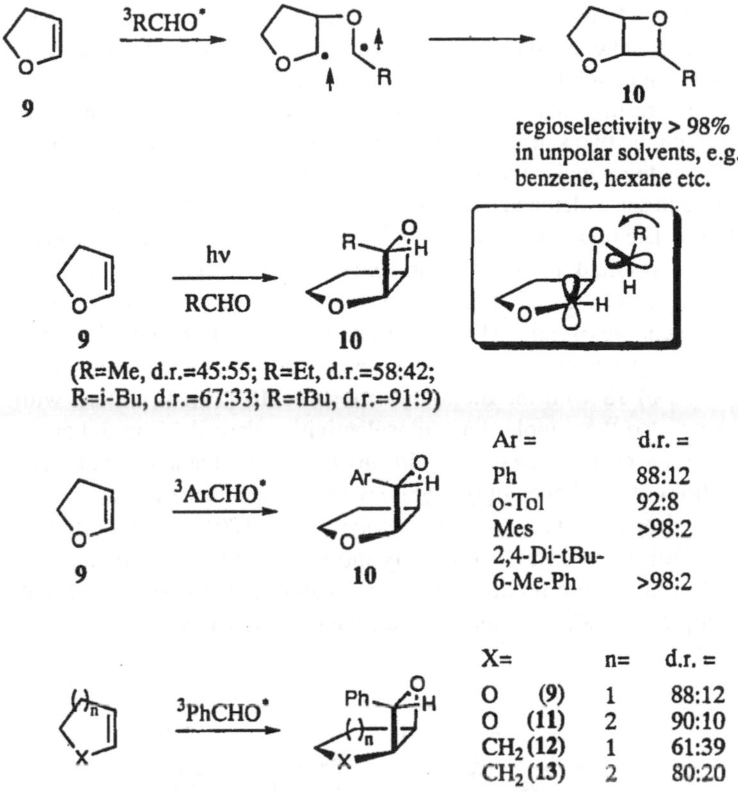

10

regioselectivity > 98%
in unpolar solvents, e.g.:
benzene, hexane etc.

9   10
(R=Me, d.r.=45:55; R=Et, d.r.=58:42;
R=i-Bu, d.r.=67:33; R=tBu, d.r.=91:9)

9   10

Ar =	d.r. =
Ph	88:12
o-Tol	92:8
Mes	>98:2
2,4-Di-tBu-6-Me-Ph	>98:2

X=		n=	d.r. =
O	(9)	1	88:12
O	(11)	2	90:10
CH_2	(12)	1	61:39
CH_2	(13)	2	80:20

Scheme 5   Reaction of monoalkenes with benzaldehydes.

reaction but solvent effects account for several different regioselectivity-determining factors [25]. In unpolar solvents the photocycloadditions proceeded with good chemical yields and surprising diastereoselectivities. The DHF addition to acetaldehyde resulted in a 45:55 mixture of endo and exo diastereoisomers **10**. With increasing size of the α-carbonyl substituent (Me-Et-iPr-Ph-tBu), the simple diastereoselectivity increased with preferential formation of the endo stereoisomer. The benzaldehyde addition which was most intensively investigated gave a 88:12 mixture of endo and exo diastereoisomers [26]. Thus, the thermodynamically less stable stereoisomers (>1.5 kcal/mol, ab initio calculations) were formed preferentially. To further enlarge the phenyl substituent, we used *ortho*-tolyl- and mesitylaldehyde as well as 2,4-di-*tert*-butyl-6-methylbenzaldehyde and actually the diastereoselectivity did further increase [27]. No trace (>99.5:0.5 by NMR) of the exo diastereoisomer was found in the Paternò–Büchi reaction of the latter carbonyl reagent. This contrathermodynamic trend was also found with other cycloalkenes, such as 2,3-dihydropyran (**11**), cyclohexene (**12**), cyclopentene (**13**), and cyclobutene (**14**) [28] (Scheme 6).

Especially interesting alkenes were 1,3-dioxolenes because additional differentiation between the two sites of the molecule could be achieved by variation of the substituents at the acetal center C-2. In agreement with the results described above, an increase in steric demand of one of the α substituents led to an increase in endo-selectivity. The photocycloaddition of benzaldehyde to 2,2-bisisopropyl-1,3-dioxolene (**15**) resulted in 65:35 endo/exo mixture of oxetanes **16** [24]. Again, an increase in steric demand of the α substituent at the carbonyl component improved the endo/exo ratio. Methyl trimethylpyruvate (**17**) with the sterically more demanding *tert*-butyl group gave solely the endo-diastereoisomeric oxetane (d.s. > 98%) **18** with the dioxolene substrate **15** [29]. In accord with these results were also the simple diastereoselectivities described by Scharf et al. for the phenylglyoxylate photocycloadditions (i.e., *endo*-phenyl selectivity). Photocycloadditions of ethyl phenylglyoxylate with cyclopentene (**12**), cyclohexene (**13**), 2,3-dihydropyran (**11**), and 1,3-cyclohexadiene (**20**) were reported by Hu and Neckers [30]. In all cases, exclusively the *endo*-phenyl diastereoisomers **21** were formed. Again, these products were less stable (0.5–1.0 kcal/mol) than the corresponding exo-diastereoisomers which were not detected.

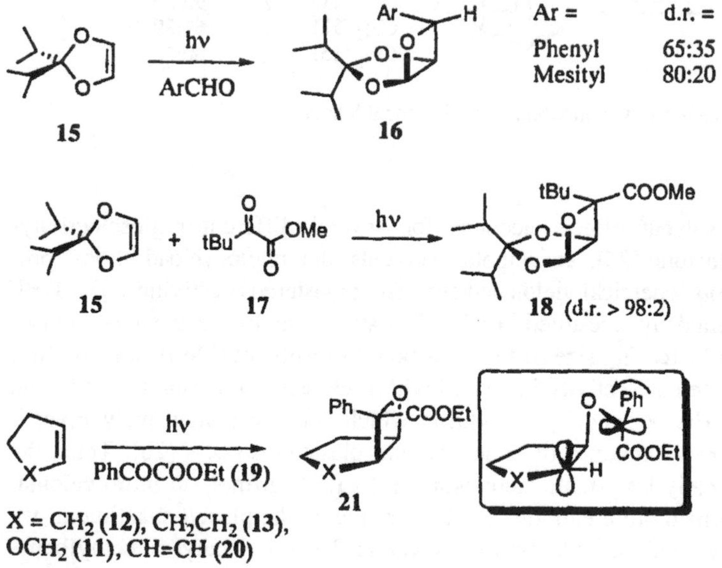
hν
ArCHO
Ar
H
Ar =
d.r. =
Phenyl
65:35
Mesityl
80:20
15
16
+
tBu
OMe
hν
tBu
COOMe
15
17
18 (d.r. > 98:2)
hν
PhCOCOOEt (19)
Ph
COOEt
21
Ph
COOEt
H
X = CH2 (12), CH2CH2 (13),
OCH2 (11), CH=CH (20)

Scheme 6 The endo-selectivity effect.

1. ISC-Reactive Conformations Determining Product Stereochemistry

Most of the results described above were striking in the sense that classical text-book analysis of triplet photocycloaddition reactions leads to a completely different expectation [31]. The lifetime of many triplet biradical intermediates (trimethylenes, tetramethylenes, or 2-oxatetramethylenes) is high enough to enable bond rotations, and the formation of the thermodynamically favored product is expected because the radical–radical combination step should no longer be influenced by the approach geometry, i.e., memory effects should be erased due to the relatively long lifetimes. On the other hand, converting triplet biradical to closed-shell products means intersystem crossing, which is a spin-forbidden process and requires special geometrical and energetic conditions. These criteria were described for the first time in the legendary publication by Salem and Rowland emphasizing the role of spin-orbit coupling (SOC) as the decisive interactive mechanism for triplet-to-singlet intersystem crossing in 1, n biradicals [32]. In contrast to other interactions such as electron-nuclear hyperfine coupling (HFC) and spin-lattice relaxation (SLR), SOC strongly depends on the geometry of the triplet biradical. The rules postulated in the paper by Salem and Rowland were as follows: 1) SOC decreases with increasing distance between the two spin-bearing atoms. Because of additional through-bond interactions in the 1,n-biradical, not only is the actual distance between the two radical centers important but so also is the number of bonds (n − 1); 2) conservation of the total angular momentum demands that the axes of the p orbitals at the radical centers be oriented orthogonal to each other and these in turn mutually orthogonal to the axes around which the orbital angular momentum is changed; 3) SOC is proportional to the ionic character of the corresponding singlet biradical state. Summarizing these three rules, a pronounced conformational and structural dependence should result for the lifetime of triplet 1, n biradicals. A numerical equation for SOC was reported by Carlacci and Doubleday et al.: $SOC = B(R) |S| \sin \phi$. $B(R)$ is a function of the distance R between the radical centers, ϕ is the angle between the localized p orbitals at these positions, and $|S|$ is the overlap integral of these orbitals [33]. Spin-orbit coupling controls the rate of ISC for tetramethylene, 2-oxatetramethylene, or trimethylene triplet biradicals, i.e., strong SOC enhances the ISC rate and lowers the lifetime of the triplet biradicals. In recent publications by Michl [34] and Adam [35] and co-workers, these rules, which help in estimating the magnitude of SOC, were modified due to new experimental and theoretical results. The spatial orientation of the two singly occupied orbitals has been determined to be highly important for the biradical lifetimes, whereas the through-space distance between the radical centers plays a subordinate role and the degree of ionic contribution in the corresponding singlet state often seems to be overestimated. This clearly indicates that for flexible, 1,4-triplet biradicals *not only one conformational arrangement* is responsible for facilitating ISC, but rather many are. After transition from the triplet to the singlet potential energy surface, immediate product formation is expected. *Thus, the ISC is expected to proceed in concert with the formation of a new bond or the cleavage of the primarily formed single bond.* This picture resembles the concept by Turro of "tight" and "loose" geometries of biradicals (originally used for singlets and triplets) which are precursors to cycloaddition and cleavage products, respectively [36]. As long as we are interested in high product stereoselectivity and not in the quantum yields of product formation, only the tight-biradicals geometry must be considered as relevant for biradical coupling. The reactive conformers do not have to represent energy minima but rather are representations of the conformational situations most probable for triplet–singlet interconversion (Scheme 7).

The conformational minima structures preceding these transition points are in rapid equilibria with each other with a rate of rotation k_{rot} higher than the rate of intersystem crossing k_{isc}. Consequently, the lifetime of the triplet biradical (about 1–7 ns for 2-oxatetramethylenes) [37] is given by the reciprocal sum of the two intersystem crossing rates for the two "tight" conformers and the ISC rate for anticonformers (which have, however, substantially lower SOC values [38]). The transition can directly produce a vibrationally hot ground state of the product or a longer lived singlet 1,4 biradical with low bond order between the radical centers. The latter possibility might exist for highly stabilized (allylic or benzylic) 1,n biradicals which have already been reported in the literature [39]. Be it as it may, the singlet biradicals should be too short-lived to enable rotation about the endocyclic C–O or C–C bonds; therefore, conformational memory effects on the stereochemistry of the products are expected. This situation is depicted for the Paternò-Büchi reaction of a triplet-excited RCHO with a five-membered cycloalkene. Minimization of steric interaction between the substituents at the terminal centers favors structure syn-^{3}A over syn'-^{3}A which collapses after ISC to give the endo-diastereoisomer in competition to cleavage reaction. During the formation of the second C–C bond, steric interaction increases between the ring skeleton atoms and the substituent R, which might also favor the turnoff into the cleavage channel. Thus, the stereoselectivity of the Paternò–Büchi reaction as well as of any other reaction involving triplet 1,n biradicals (with $n > 3$) is the result of a combination of several rate constants for cyclization vs. cleavage reaction. In a triplet 1,3 biradical, steric interactions between substituents at C–1 and C–3 are continuously increasing during bond formation because the maximal value for sin ϕ is obtained for an in-plane conformation of the p orbitals and therefore no contrathermodynamic effect is expected.

This model leads to several predictions which should be discussed in the following part, e.g., that alkylation of the unsubstituted site of the double bond in cycloalkenes or double steric crowding of one at the biradical termini lowers

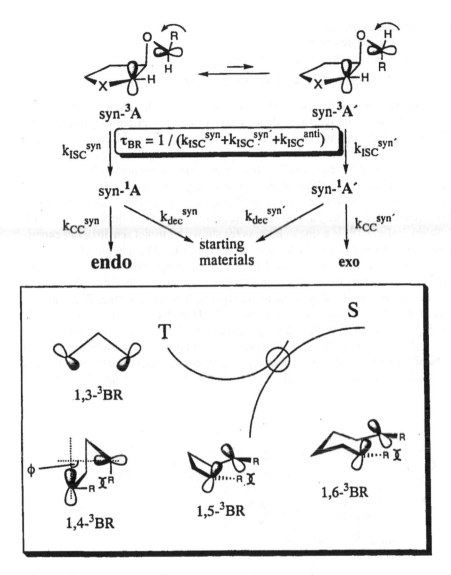

Scheme 7 SOC-controlled intersystem crossing geometries.

the stereoselectivity and might even invert the endo/exo ratio (vide infra). The principle, as indicated in Scheme 7, might be valid also for 1,5 and even 1,6 biradicals, albeit much less conformationally defined species.

2. Further Examples, Singlet Reactions, and PET Reactions

Whatever the mechanism is for the corresponding singlet photoreactions, ISC geometries cannot be relevant. Thus, if singlet and triplet photocycloadditions with identical chemo- and regioselectivity are compared, the differences in diastereoselectivities allow a conclusion about the role of ISC geometries. Naphthaldehydes can be used in Paternò–Büchi reactions as singlet excited carbonyl components [40]. The reaction efficiencies and chemical yields are normally much lower than with other aromatic ketones or aldehydes, but chemo- and regioselectivities are often identical. The Paternò–Büchi reaction of 2-naphthaldehyde (**22**) with 2,3-dihydrofuran (**9**) is a singlet process (as shown by triplet sensitization experiments) and gives exclusively the exo-diastereoisomer **23** (Scheme 8).

Similar results were obtained for the 2,2-bisisopropyl-1,3-dioxolene (**15**): the endo/exo ratio is completely inverted when going from the triplet excited mesitylaldehyde to the singlet excited naphthaldehyde as substrate. The comparison of these experiments with the benzaldehyde cycloadditions reveals the role of ISC geometries in triplet photocycloadditions. Two structural features made the conformational analysis of spin-inversion geometries straightforward: the two sites of the alkene part were well differentiated concerning degree of substitution and steric hindrance. The alkenes investigated were always cyclic, thus reducing

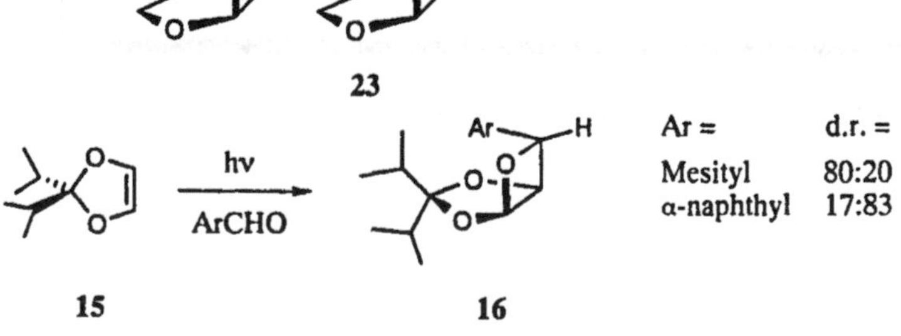

Ar =	d.r. =
Mesityl	>98:2
α-naphthyl	<2:98
β-naphthyl	<2:98

Ar =	d.r. =
Mesityl	80:20
α-naphthyl	17:83

Scheme 8 Paternò–Büchi reactions with naphthaldehydes.

the conformational flexibility at the biradical stage. The situation became more complex when one or both of these structural features is modified. Methyl-substituted cycloalkenes have two ISC-reactive sites and thus the endo/exo ratios drop significantly [28] (Scheme 9). An extreme example is 1,2-dimethylcyclobutene (24): the Paternò–Büchi reaction of this substrate with benzaldehyde resulted solely in the exo-diastereoisomer 25. This is exactly predicted by the ISC geometry model, because the bismethylated site of the cycloalkene is now sterically more demanding (compared to unsubstituted starting materials) and the biradical combination trajectory involves the approach from the less shielded cyclobutene plane.

Steric hindrance can also reach a critical value during bond formation and might favor the formation of the thermodynamically stable product. In a report by Park and co-workers the photocycloaddition of benzaldehyde to 2,2-diethoxy-3,4-dihydro-2H-pyran (26) is described and this process led preferentially to the exo-phenyl product (d.e. = 92%) 27 [41].

Another feature which might oppose the ISC geometry model is primary photoinduced electron transfer (PET). If this process is energetically feasible, the geometrical restrictions might be circumvented, i.e., intersystem crossing can occur at the stage of the radical ion pair and a singlet 1,4 biradical or a 1,4 zwitterion can be formed, depending on the reaction conditions. In polar solvents, the assumption of a 1,4 zwitterion as decisive intermediate is reasonable. This situation then resembles the sequence observed for ET-induced thermal [2 + 2] cycloaddition reactions [42]. Both regio- and diastereoselectivity are influenced by this mechanistic scenario. The regioselectivity is now a consequence of maximum charge stabilization and no longer a consequence of the primary interaction be-

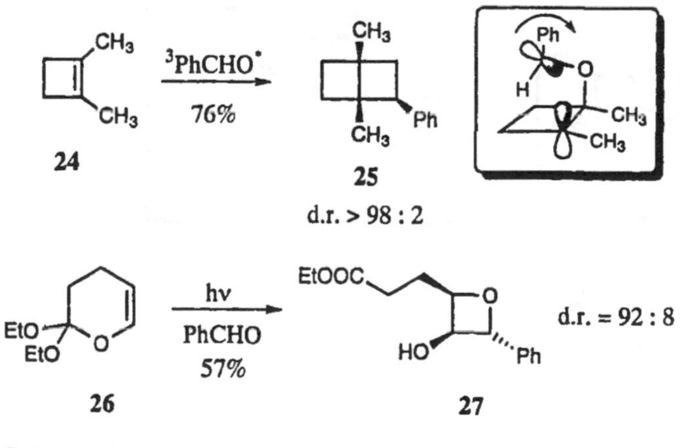

Scheme 9 Further examples of the endo-selectivity effect.

tween excited carbonyl compound and alkene. Whereas 3-alkoxyoxetanes are preferentially formed from triplet excited aldehydes and enolethers, 2-alkoxyoxetanes result from the reaction of triplet excited ketones or aldehydes and highly electron-rich ketene silylacetals **28** [43] (Scheme 10).

In the second case, PET which gives the carbonyl radical anion and the ketene acetal radical cation is energetically feasible [44]. PET might be followed by ISC and formation of a highly stabilized 1,4-zwitterion intermediate (aldol intermediate). Eventually, C–O bond formation leads to the oxetanes **29** with correct (with respect to the experimental results) regiochemistry. Substituent tuning is not the only possibility to influence the regioselectivity of the Paternò–Büchi reaction. By processing the photocycloaddition in a highly polar solvent which reduces the coulombic term in the Rehm–Weller equation [45], PET becomes compatible with radical pathways. This effect was observed with 2,3-dihydrofuran (**9**) as electron-rich substrate which selectively gave the 3-alkoxyoxetane **10** when reacted with triplet excited aliphatic aldehydes in unpolar

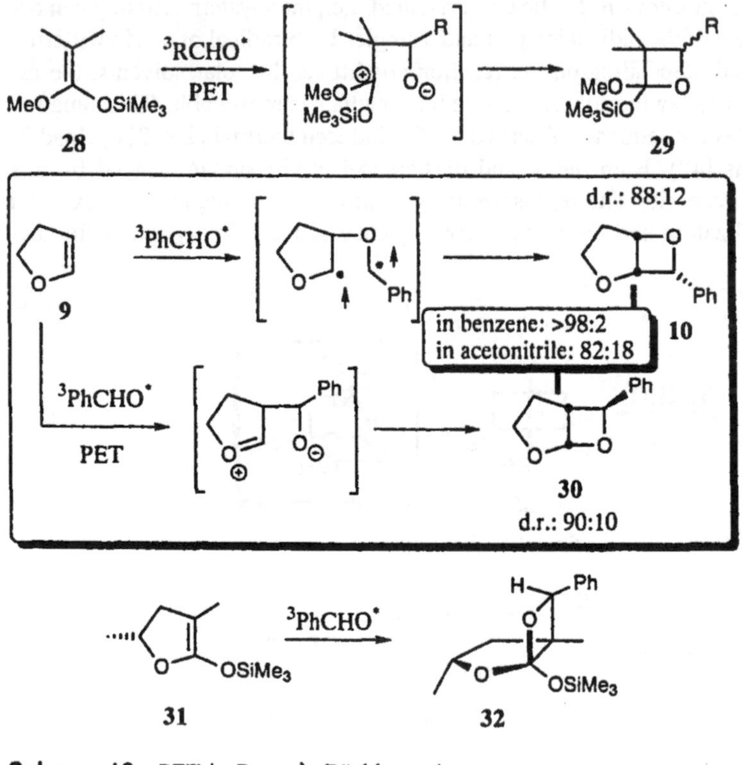

Scheme 10   PET in Paternò–Büchi reactions.

solvents. In acetonitrile, however, the corresponding 2-alkoxyoxetane 30 was also detected. The relative amount of this product correlated with solvent polarity parameters, thus indicating PET as the responsible mechanism. If a 1,4 zwitterion is formed during the reaction, the stereochemistry is no longer controlled by ISC geometries but by the orientation of the two prostereogenic carbon centers of the radical ions when forming the primary C–C bond. The major diastereoisomer obtained *from the PET cycloaddition* of benzaldehyde with 2,3-dihydrofuran was the *exo*-phenyl isomer (d.r. = 85:15). [25] Thus, a switch from 1,4 biradical to 1,4 zwitterion path leads to an inversion of regio- *and* diastereoselectivity. A mechanism which involves a sequence of PET, formation of a contact ion pair and charge recombination to give a triplet 1,4 biradical [46], also explains the change in regioselectivity (due to the different orientation of radical anion and radical cation in the contact ion pair) but is in contradication to the observed diastereoselectivity (*15:85 endo/exo ratio*) which should also be *endo*-phenyl from the alternative 1,4-triplet biradical [similar to the biradical collapse from the cyclopentene/benzaldehyde reaction (*61:39 endo/exo ratio*)].

Abe and co-workers have also observed this stereochemical effect in the Paternò–Büchi reaction of aromatic aldehydes with cyclic ketene acetals [47]. The addition of benzaldehyde to the 5-silyloxy-substituted 2,3-dihydrofuran 31 resulted in the *exo*-phenyl product 32, albeit in low yields. Higher yields and higher selectivities were obtained with naphthaldehydes and acceptor-substituted benzaldehydes as carbonyl addends.

Interesting substrates in this context were also the 2,3-dialkylated ascorbic acid acetonides 33 investigated by Kulkarni and co-workers [48]. Photocycloaddition of benzaldehyde with these substrates resulted in the formation of two regioisomeric products 34 and 35 (Scheme 11). Both oxetanes were formed exclusively with the *exo*-phenyl configuration. The observed regio- as well as diastereoselectivity are in accord with the assumption of a PET process involving the oxidation of the ascorbic acid derivatives and the formation of the carbonyl radical anions. In these special cases, 1,4-biradical and 1,4-zwitterion stabilization result in the same product regiochemistry. The relative configuration of the products favors the assumption of a PET process.

Alkenes with moderate oxidation potentials were investigated by Torsten Bach and his co-workers in the last decade. They have intensively studied the complex stereoselectivity of Paternò–Büchi reactions of acyclic trialkylsilylenol ethers 36 with aromatic aldehydes and developed an impressive set of synthetic applications [49]. Both structural restrictions described above are absent in these substrates: the substrates are acyclic, the two sites of the alkenes are not well differentiated, and all investigated groups are sterically demanding. The synthetic approach to these compounds allowed broad variation of the substituent at C-1 and at the silyl group. A series of photocycloadditions of benzaldehyde to trimethylsilyl (TMS) enol ethers showed a stereoselectivity trend which at first sight

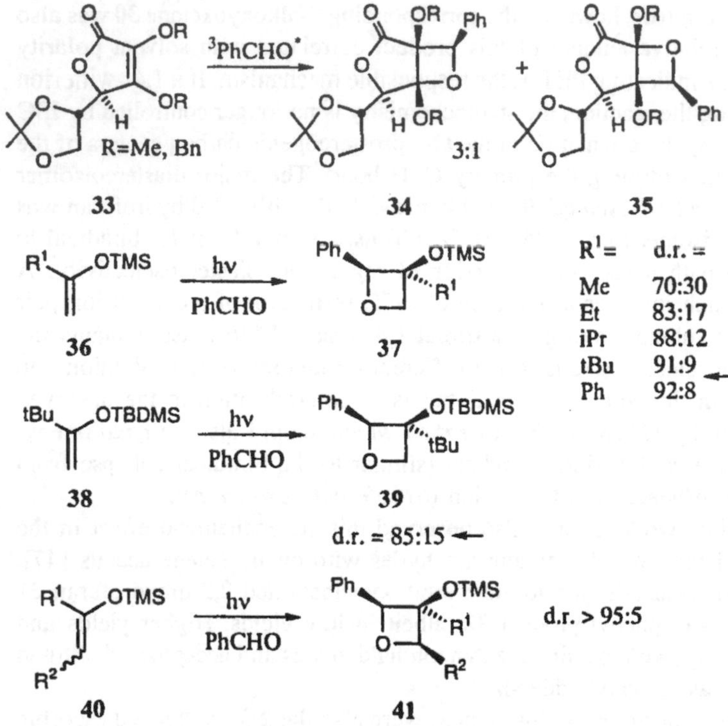

Scheme 11 Enolethers as substrates in Paternò–Büchi reactions.

was in contradiction to the rules described above, i.e., the thermodynamically more stable trans-stereoisomers **37** (with respect to the C substituents at C-2 and C-3) were formed preferentially and the trans/cis ratio increases with increasing size of the C-3 substituent [50]. That indeed the difference in steric demand of the respective groups at C-1 is correlated with the diastereoselectivity is shown for substrate **38**: an increase in group size of the silyloxy substituent (from TMS to *tert*-butyldimethylsilyl in **38**) reduces the d.r. for the *tert*-butylated substrate from 91:9 to 85:15 (oxetane **39**).

This peculiar stereoselectivity might be attributed to a memory effect from the approach geometry between the triplet excited benzaldehyde and the alkene. That this cannot be the case was demonstrated by a series of experiments with β-substituted TMS enol ethers **40**. Independent of their starting configuration (E/Z with respect to the enol ether double bond) the alkyl substituents align trans in the oxetane products **41** [51]. The simple diastereoselectivity found for β,β′-unsubstituted alkenes was also observed here. The lifetime of the intermediary

triplet 1,4 biradical must therefore be long enough to enable complete bond rotation about the C-1/C-2 bond and consequently also about the C3–O bond. By
combination of these stereodirecting effects with facial selectivity induced by
a stereogenic center already present in the starting molecule, the synthesis of
enantiomerically highly enriched products became possible. A prominent example is the photocycloaddition of benzaldehyde with the glycerinaldehyde-derived
substrate **42**, which gave the oxetane **43** in 90% d.s. and 70% yield [52]
(Scheme 12).

One can speculate about why the ISC geometry model does not apply here.
The explanation given by Bach et al. is that the simple stereoselectivity is dominated by the retrocleavage steps following the ISC process. This possibility is
already indicated in the mechanistic Scheme 7: the endo/exo ratio is influenced
not only by the rates of ISC from the two reactive conformers but also by the
conformational equilibrium *and* the relative rates of retrocleavage. The latter reaction becomes apparently important when steric interactions increase early in
the terminal bond-forming step and lead to an increase in cleavage rate.

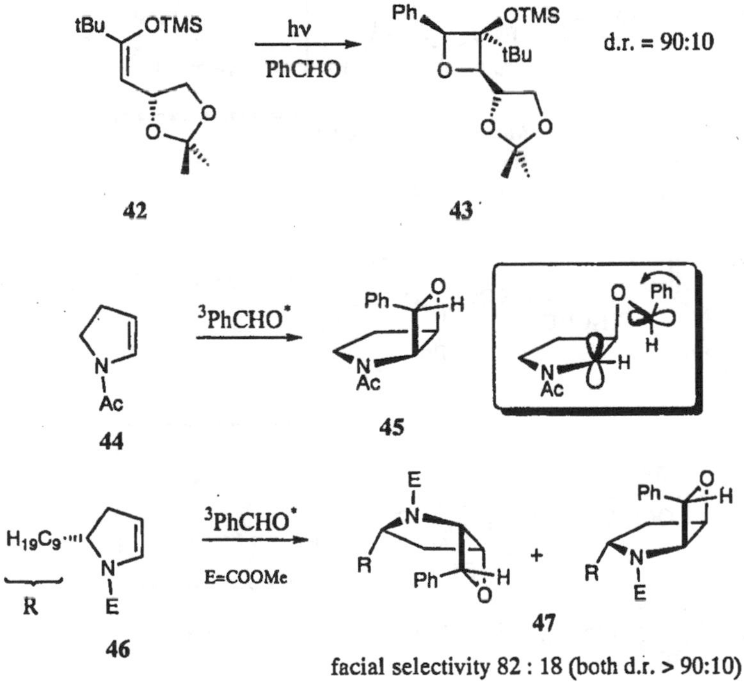

Scheme 12 Enamides as substrates in Paternò–Büchi reactions.

Recent investigations from the Bach group were focused on enamides as substrates. The resulting 3-aminooxetanes were used for the synthesis of chiral 1,2-aminoalcohols [53]. The cyclic substrate *N*-acetylpyrrolidine **44** (a hetero analog to 2,3-dihydrofuran) resulted exclusively in the formation of the *endo*-phenyl product **45**, i.e., the simple diastereoselectivity is even higher than in the case of 2,3-dihydrofuran. A striking effect was found when the chiral substrate **46** bearing a nonyl substituent at C-5 was applied as alkene addend: the facial selectivity was 82:18 in favor of the cis product **47** [54]. This result might be attributed to hydrophobic interactions stabilizing the syn orientation of the alkyl and the phenyl group. Both major Paternò–Büchi adducts were formed with *endo*-phenyl configuration, thus preserving the ISC geometries for final bond formation (Scheme 13).

Unsubstituted acyclic enamides **48** also behave accordingly. The Paternò–Büchi reaction of benzaldehyde with *N*-acylenamines proceeded with good to excellent diastereoselectivities, and even the photoaddition of acetaldehyde with the *N*-benzyl-*N*-BOC-substituted enamine **50** resulted in a 65:35 endo/exo mix-

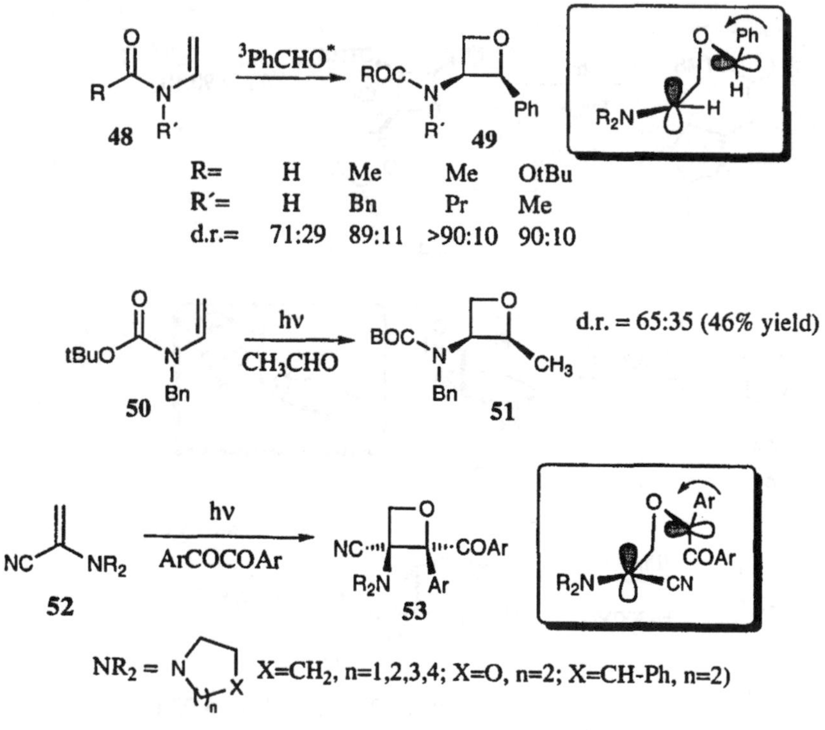

48
³PhCHO*
ROC
Ph
49
R= H Me Me OtBu
R′= H Bn Pr Me
d.r.= 71:29 89:11 >90:10 90:10
R₂N
tBuO
Bn
50
hν
CH₃CHO
BOC
CH₃
51
d.r. = 65:35 (46% yield)
NC
NR₂
52
hν
ArCOCOAr
COAr
Ar
53
CN
NR₂ = X=CH₂, n=1,2,3,4; X=O, n=2; X=CH-Ph, n=2)

Scheme 13 Enamides and captodative alkenes.

ture of oxetanes **51** [55]. Thus, the simple diastereoselectivities for these group of substrates nicely corroborate the ISC geometry model due to the site differentiation at the alkene double bond.

Captodative-substituted alkenes should prefer the reaction path via triplet 1,4 biradicals with one carbon center stabilized by the captodative effect. When 2-aminopropenenitriles **52** were irradiated with benzil, oxetanes **53** were formed regio- and diastereoselectively [56]. The regiochemistry matches the expectation of a captodative stabilized 2-oxatetramethylene biradical, and the diastereoselectivity can be explained by the ISC geometry model. The role of the cyano substituent in orthogonal interactions is somewhat problematic because the active space is larger than in traditional models used for determination of steric demands. On the other hand, the dipolar groups cyano and benzoyl can align parallel to each other and minimize steric interactions in the reactive 1,4 biradical.

At first sight, a remarkable exception from the stereodirecting rules discussed above exists for an exceedingly important class of alkene substrates for Paternò–Büchi reactions: furan [57] and furan derivatives [58], as well as other heteroaromatic substrates such as thiophenes [59], pyrroles [60], imidazoles [61], and thiazoles [62], and even carbocyclic dienes such as 1,3-cyclopentadiene [63] or 1,3-cyclohexadiene [64]. The photocycloaddition of these 1,3-dienes **54** with aromatic and aliphatic aldehydes proceeds with unusual high exo-diastereoselectivity to give the bicyclic oxetanes **55** in good yield (Scheme 14).

For most examples, no exact d.r. values are reported in the literature; however, the exo-diastereoselectivity definitely exceeds 90% in all cases. We have explicitly investigated the Paternò–Büchi reaction of furan (**56**) with acetaldehyde [27] propionic aldehyde [65] and with benzaldehyde [25]. The diastereoselectivities (exo/endo ratios) were 19:1, 82:1, and 212:1. Thus, the extraordinarily high stereoselectivity of the benzaldehyde/furan photocycloaddition is not a result of π interactions with the phenyl substituent. It was therefore of interest to study the influence of α substituents in benzoyl compounds on the stereoselectivity of the Paternò–Büchi reaction with furan. Recently, it was reported that methylbenzoate (**57**) reacts with furan with a high degree of exoselectivity [66]. When repeating this reaction, we found two diastereoisomeric products **58** in a ratio of 95:5. The spectral data and definitely the x-ray analysis of the major diastereoisomer showed that the *endo*-phenyl isomer predominated. Surprisingly, the exchange of the hydrogen in benzaldehyde by a methoxy group completely inverts the diastereoselectivity in the photocycloaddition with furan.

Further modification of the α substituent in the benzoyl substrates uncovered a distinct dependence of the exo/endo ratio (**59/60**) on the size of this substituent. The acetophenone reaction with **56** gave only one product, whereas a 77:23 mixture of diastereoisomers resulted from the addition of benzoyl cyanide. Increasing the size of the aryl group from phenyl to mesityl in aroyl cyanides led to an increase in exo-diastereoselectivity from 3.9:1 up to 16:1 [67]. Thus,

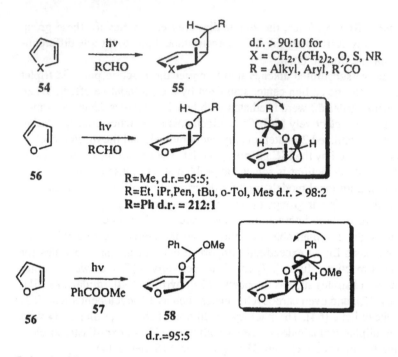

Scheme 14 Cyclodienes as substrates in Paternò–Büchi reactions.

by changing the α substituent from H or Me to CN, COOMe, OMe, and eventually COOR, the direction (*exo*-phenyl vs. *endo*-phenyl) of the diastereoselectivity was completely switched. Again, this cannot be a result of π interactions with the phenyl substituent because an identical sequence was obtained when using the corresponding *tert*-butyl compounds, i.e., pivaldehyde gave solely the exo [2 + 2] cycloadduct, whereas methyl trimethylpyruvate exclusively resulted in the formation of the endo [2 + 2] cycloadduct (exo, endo with respect to the *tert*-butyl group). What might be the reason for this peculiar behavior? (See Scheme 15.)

As already described for cycloalkene, ISC from **A** and **C** are expected to lead to endo- and exo-diastereoisomers, respectively. Using density functional theory (DFT), we determined if structure **C(C′)** with an orthogonal arrangement of the two spin-bearing orbitals corresponds to an energy minimum. As model reaction we used the 1,4 biradical from furan and triplet acetaldehyde. Structure **A** represents the global energy minimum (fully optimized); the restricted conformation **C** is 9.9 kJ/mol higher in energy, separated by an activation barrier of 28 kJ/Mol. This activation energy is obviously too high due to the conformational

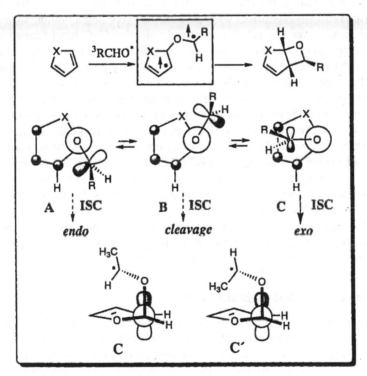

$R^2 =$	H	Me	CN	COOMe	OMe	COOR
R^1 = Ph	212	>49	3.7	0.11	0.05	<0.02
R^1 = Mes			16			
R^1 = tBu	>250			49		

59/60-ratios

Scheme 15 Furan cycloaddition reactions.

restriction for rotation about the exocyclic O–C(H, Me) bond. The alternative arrangement **C′** with the methyl group pointing toward the ring plane is much higher in energy and should be less significant in this reaction. If structure **C** (and not **A**) represents the main contribution to product formation, a high exoselectivity is expected due to the nearly exclusive formation of the "methyl-up" conformation **C**. The torque induced in conformer **C** rotates the large substituent (methyl or any other substituent) in opposite direction with respect to the ring plane and results in the formation of the major (exo-) diastereoisomer.

What happens if the hydrogen in structure **C** is exchanged by a substituent larger than methyl, i.e., ketones are used as triplet carbonyl substrates? Following the arguments from above, both structures **C** and **C′** are no longer relevant and only structure **A** remains as the SOC-controlling arrangement for the ISC process. Thus, all triplet carbonyl substrates with substituents exceeding a limiting degree of steric interaction were expected to give endoselective C–C bond formation. This model offers an explanation for the dichotomy that some carbonyl substrates (e.g., phenylglyoxylates) give idential diastereoselectivity with cyclo-1,3-dienes *and* with the corresponding cycloalkenes, whereas other carbonyls (e.g., aldehydes or acetophenone) show a completely different selectivity pattern. An alternative explanation for the high exoselectivity in furan–aldehyde photocycloadditions, which we suggested earlier, was the extended lifetime of the single 1,4 biradical which is formed after ISC. However, this concept predicts thermodynamic control for the formation of *all* cycloaddition products, whether they are formed from triplet excited aldehydes or ketones, esters, etc. Obviously, this is not the case (see Scheme 15). Thus, an interaction between the allylic and the exocyclic radical in the 1,4 triplet biradical (as depicted in structures **C**) must be crucial for the dominance of this biradical geometry for rapid ISC. This effect can be described as a secondary orbital interaction which facilitates intersystem crossing by means of an increase in spin-orbit coupling.

B. Enone–Ene Photocycloadditions

The first example of an enone–ene photocycloaddition also came from Italy: Ciamician and Silber reported in 1908 the intramolecular photocycloaddition of carvone [68]. This reaction type has been used in numerous application for natural product synthesis as well as for the synthesis of complex strained molecules and is summarized in excellent reviews [69].

Intermolecular reactions can give rise to two regioisomeric and eight diastereoisomeric cyclobutanes. Thus, the complexity of this reaction is higher than that of the Paternò–Büchi reaction. In analogy to the carbonyl–ene reaction, triplet 1,4 biradicals can be postulated as the decisive intermediates; however, in contrast to the former reaction, all possible four structures are plausible and none can be excluded from the beginning (Scheme 16).

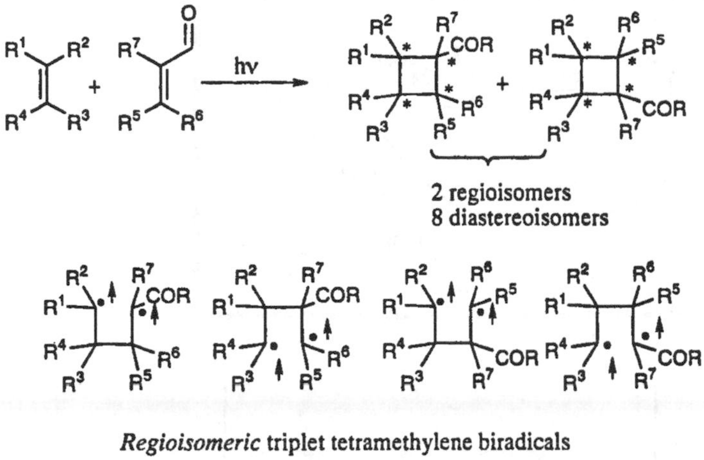

Regioisomeric triplet tetramethylene biradicals

Scheme 16 The enone–ene photocycloaddition.

The enone-ene cycloadditions seems to proceed via a triplet exciplex as shown by Caldwell et al. for the 4,4-dimethylcyclohexenone/1,1-diphenylethene system [70]. This intermediate has a remarkably long lifetime of about 50 ns and might be formed in competition with or prior to the triplet 1,4 biradical. These tetramethylene biradicals have comparable lifetimes of 50–100 ns [71] and can be intercepted with radical traps. If the 1,4 biradicals are the only precursors to the cyclobutanes, efficient trapping of these transient species allows the determination of the product formation/retrocleavage ratios. This concept has been intensively processed by A. Weedon and co-workers. An excellent reagent was hydrogen selenide, a highly reactive hydrogen-donating radical trap. The "rule of five" was nicely examined by use of this reagent [72] (Scheme 17).

The dienone **61** upon irradiation gives a single diastereoisomer **62** [73]. This result might mean that the regiochemistry of the product is controlled by the formation of the more stable five-membered biradical structure and the stereochemistry of the product is controlled by the approach geometry, i.e., already existing interactions between the α-carbonyl center and the terminal unsaturated carbon atom of the alkenyl chain. The trapping experiments gave the two spiro products **63** and **64** in a 37:63 ratio, which clearly showed that the regiochemistry is indeed determined by the rule of five; however, the product stereochemistry is completely determined at the stage of the intermediate 1,4 biradicals. Because only the syn-1,4 biradical can undergo a second C–C bond–forming step, at least 37% of the biradicals formed must fragment to the starting material. The hydrogen selenide trapping reaction unambiguously showed that a selective preorienta-

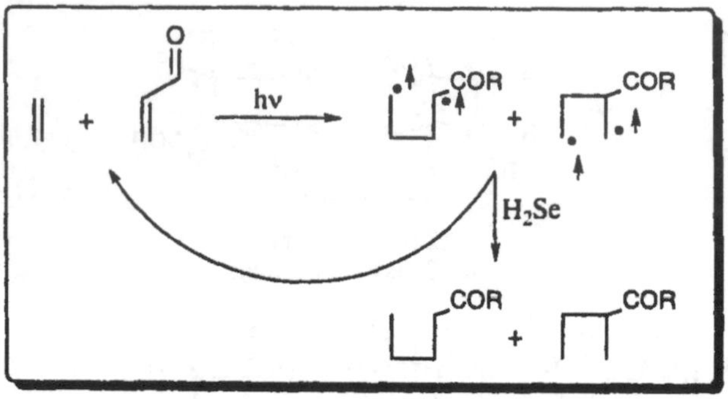

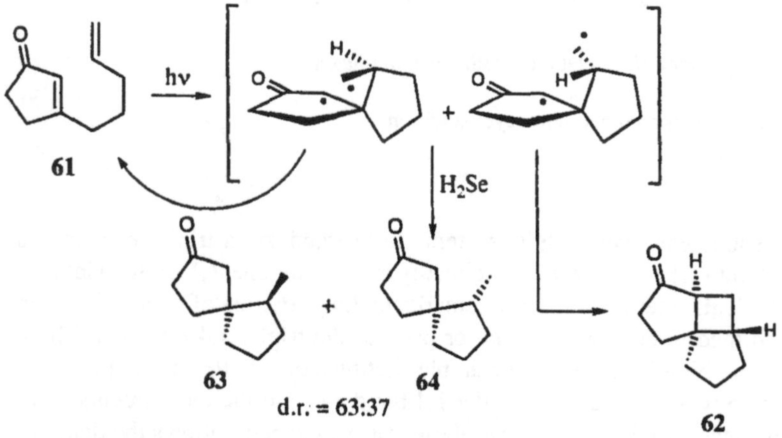

Scheme 17 Mechanism of biradical trapping reactions.

tion between the electronically exited enone and the ene component of the substrate does not necessarily exist. The Weedon group has intensively investigated intermolecular enone–ene photocycloadditions [74] and concluded that the originally proposed exciplex preorientation model is not necessary to explain the product regiochemistry [75]. The results are described here because they resemble the situation which has already been described for the carbonyl–ene photocycloaddition and stereochemical features were also expected to run parallel (Scheme 18).

The authors investigated the photocycloaddition of isobutylene with 2-cyclopentenone (**65**) and the monomethyl derivatives **68** and **71**. In all cases, the major biradical intermediates are formed by bonding of the less substituted terminus of the isobutylene preferentially to the 3-position of the enone. The major

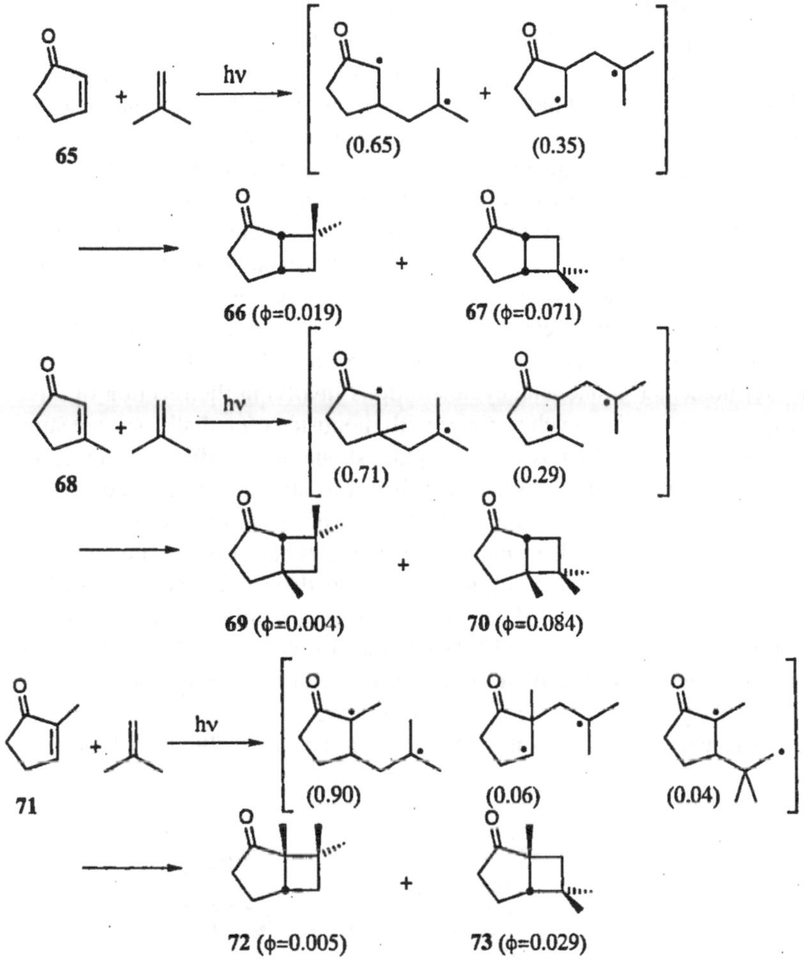

Scheme 18 Trapping reactions and regioselectivity.

cyclobutane regioisomers, however, derive from the intermediates which are composed from bonding to the 2-position of the enone, i.e., fragmentation to give the starting materials or hydrogen migration products preferentially occurred from the major biradical intermediates. Andrew and Weedon rationalized this phenomenon by assuming different radical geometries with the odd electron density at position 2 of the cyclopentenone residue in a p-type orbital and at position 3 in a sp^3-type orbital. This might influence the rate of intersystem crossing as well as the respective ISC geometries as already postulated by Shaik and Epiotis

[76]. A detailed FMO analysis was performed by Somekawa and co-workers using two-center HSOMO-LUMO and LSOMO-HOMO interactions [77]. On the other hand, the product-determining factors should also influence the stereoselectivity of the enone–ene reactions due to the differences in ISC rates and geometries. The photocycloaddition of ethylvinyl ether (74a) with cyclopentenone (65) is a simple test reaction for the analysis of regio- and stereoselectivity [78]. The head-to-head cycloadduct 75a was the minor product, also from isopropyl- and the phenylvinyl ether 74b and 74c, respectively [79] (Scheme 19). The major products were the head-to-tail cyclobutanes 76. All cyclobutanes were formed with moderate diastereoselectivity, the head-to-tail cyclobutanes showed (contrathermodynamic) 3:2 endo/exo ratios. The preceding triplet 1,4 biradicals are formed in a 1:1 ratio as determined by trapping reaction [80]. Many more examples of intermolecular enone–ene cycloadditions involving the reaction of three prostereogenic centers, i.e., cyclic enones with monosubstituted alkenes, are summarized in the review by Mattay et al. [69] and should not be discussed in detail. Apparently, the simple diastereoselectivity is not as strongly influenced by conformational orientation phenomena as in the Paternò–Büchi reaction.

Dan Becker and N. Haddad and their co-workers have investigated intramolecular enone–ene photocycloaddition in order to determine the first bonding event and the stereochemistry of the second bond formation by using radical-clock-substituted starting materials [81] (Scheme 20). Substrate 77, which is analogous to the already discussed 61, gave only cyclopropyl ring-opening products 78 indicating that the phenyl-activated rearrangement reaction is faster than 10^{11} s^{-1}. The corresponding 2-alkylated 2-cyclohexenone 79 gave beside re-

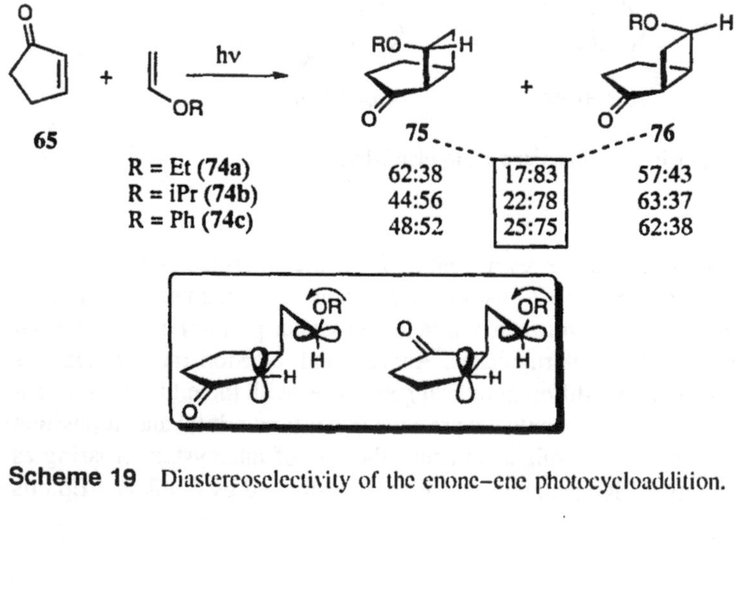

	75		76
R = Et (74a)	62:38	17:83	57:43
R = iPr (74b)	44:56	22:78	63:37
R = Ph (74c)	48:52	25:75	62:38

Scheme 19 Diastereoselectivity of the enone–ene photocycloaddition.

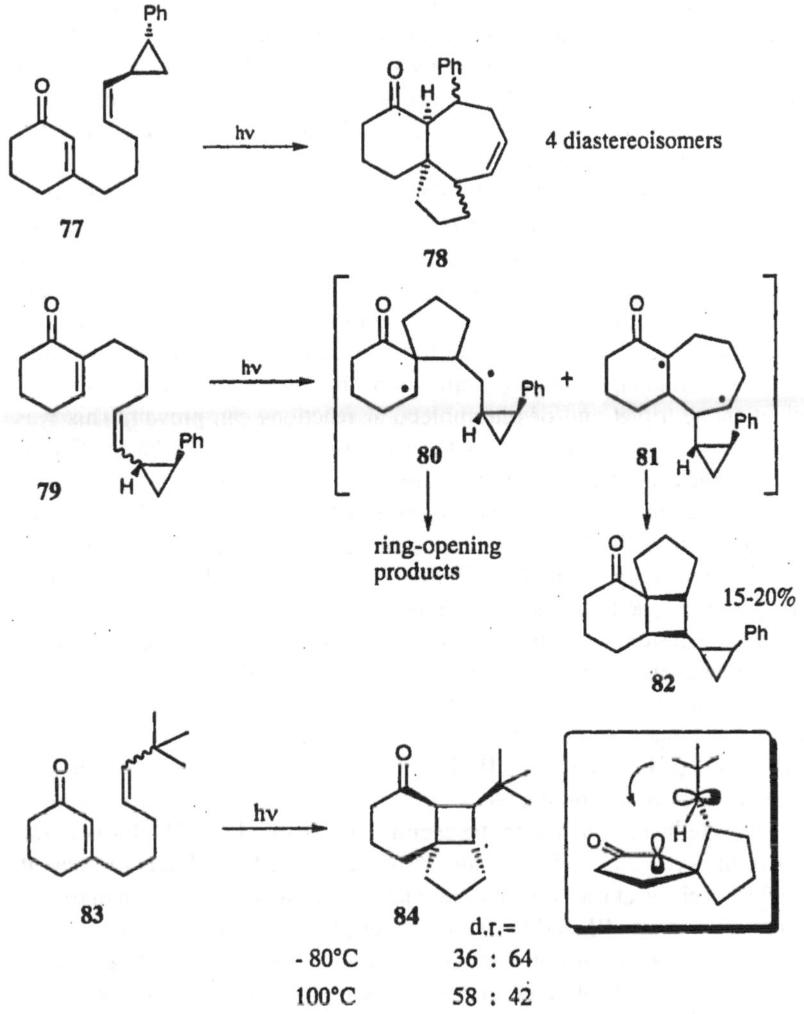

Scheme 20 Intramolecular cyclohexenone–ene photocycloadditions.

arrangement products also 15–20% of the annulated cyclobutanes **82**. Following the kinetic analysis for **77**, the 1,4 biradical following the rule of five cannot be the source for the cyclobutanes **82**. Thus, an appreciable amount of the 1,4-cycloheptadiyl **81** must be formed in competition to **80**. A stereochemical analysis was published using the *tert*-butyl functionalized 3-alkylated cyclohexenone **83** [82]. At low temperatures, the exo-diastereomer (exo with respect to the *tert*-

butyl group) **84** dominated, at higher temperatures the diastereoisomeric ratio was shifted in direction of the endo-diastereomer. From the Eyring correlation it was deduced that the activation enthalpy for exo ring closure is smaller than for the endo process. In contrast, the activation entropy for endo closure is more positive than for exo closure. This temperature dependence fits well in the ISC geometry model described above.

Numerous applications of intramolecular enone–ene photocycloadditions have been reported in the actual literature, especially in the field of diastereo- and enantioselective synthesis. Facial selectivity can be successfully induced by stereogenic centers already present in the substrate [83]. The influence of the enone–ene connecting linker and its conformational flexibility on the efficiency and selectivity of the [2 + 2] cycloaddition is crucial [84]. If conformationally less flexible substrate units became elements of the linker chain, deactivation of the excited enone triplet and/or intermolecular reactions can prevail. This was nicely demonstrated by Piva et al. for the ester-linked substrates **85** to **87** [85] (Scheme 21). The combination of a short and conformationally unfavorable (preferentially s-cis) ester chain in **85** leads to complete deactivation, whereas *gem*-dimethyl substitution (in **86**) as well as chain elongation (in **87**) restored the photochemical reactivity. Cyclodimerization in the latter case became favored at lower temperatures due to the decrease in conformational mobility necessary for the intramolecular step. Cyclobutane and cyclohexaneone rings were cis-fused in all cyclodimers and intramolecular cycloadducts **88** and **89**.

Two elegant concepts for controlling facial diastereoselectivity have been successfully developed for the 1,3-dioxen-4-one skeleton: 1) the concept of self-regeneration of stereogenic centers [86] using (*R*)- or (*S*)-β-hydroxybutyric acid as enantiomerically pure building block, and 2) the use of chiral ketones as auxiliaries which were incorporated into the acetal functionality [87]. The former concept allows the synthesis of both 2-*tert*-butyl substituted 1,3-dioxen-4-ones **90** and *ent*-**90** from cheap chiral substrates, the latter leads to the generation of diastereoisomeric dioxenones **91** and **92** which, after photoaddition to the enone part and hydrolysis of the acetal function, give both enantiomers of the target molecules from only one enantiomer of the inductor molecule. When comparing ground-state with excited-state reactivity of these substrates, a switch in facial selectivity can be observed. Ground-state reactions such as cuprate additions preferentially occur from the (apparently) sterically more hindered side, whereas intermolecular photocycloadditions prefer the approach from the higher substituted side of the dioxenone [88]. The photochemical addition of cyclohexene to the 2-methyl-2-*tert*-butyl-substituted dioxenone **93** proceeded exclusively from the *tert*-butyl substituted side of the enone double bond [89] (Scheme 22).

A plausible mechanistic explanation for this reversion in facial selectivity is the excited-state conformation of dioxenones and more general electronically excited conjugated cyclohexenones. It is assumed that the internal ring double

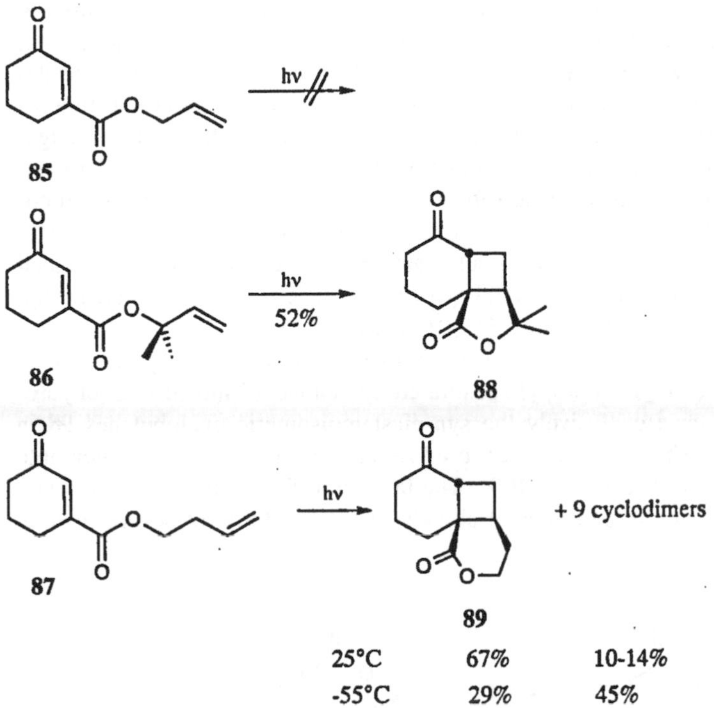

Scheme 21 Ester spacer in intramolecular enone–ene photocycloadditions.

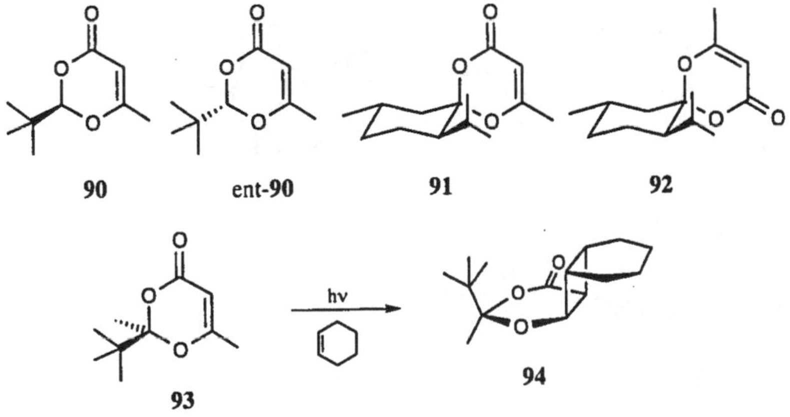

Scheme 22 Enantiomerically pure cyclohexenones.

bond is strongly twisted in the triplet state and the β carbon already strongly pyramidalized [88]. This changes the conformation of the cyclohexanone from a twist boat to a sofa conformation and by this pushes the sterically more demanding groups in pseudoequatorial positions. Thus, the pseudoaxial substituents (methyl in **93**) shield one side of the enone chromophore. A thorough investigation of the intramolecular dioxenone–ene photocycloaddition with emphasis on the stereodirecting effect of acetal substituents was reported by Haddad and coworkers [90]. They connected the unsaturated carbon tether with C_β of the cycloenone. The chain-unfunctionalized substrate **95** cyclized with moderate facial stereoselectivity showing preferential syn (with respect to the *tert*-butyl group) attack to give the tricyclic product **96** [91] (Scheme 23).

Even higher facial selectivity was obtained with dihydro-4-pyrone–linked substrates. The methyl, isopropyl-substituted dioxenone **97** linked to a dihydropyrane moiety gave exclusively the syn-diastereoisomeric cyclobutanes **98** in high yields [92]. These products were used for acid-catalyzed rearrangement into complex spiro ketals (e.g., **99→100**). The latter transformation has been developed by Winkler et al. and applied already in the synthesis of several natural

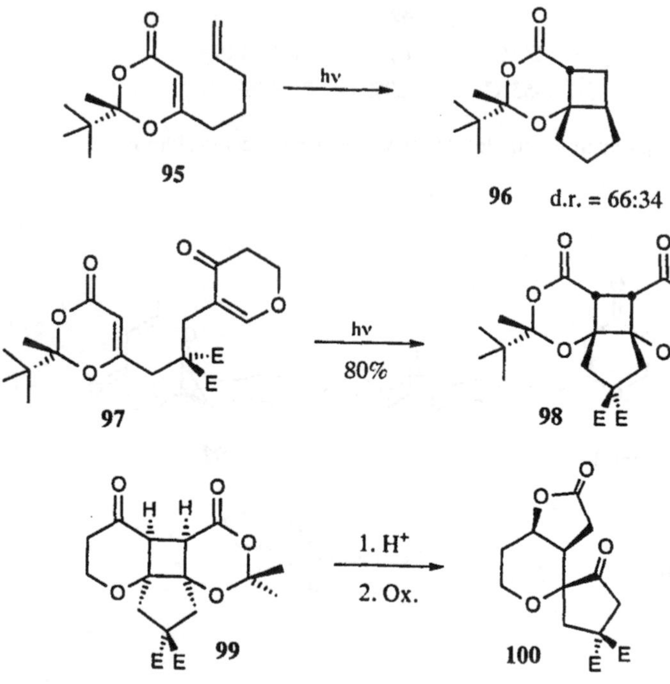

Scheme 23 Dioxenone-based enone–ene photocycloadditions.

products [93]. In these syntheses, the dioxenone part serves as a temporary chromophore which incorporates a C_2 unit into a cyclopentene and results in the formation of a 2,7-bridged cycloheptanone as present in the diterpene ingenol **102**, which was synthesized from **101** in a multistep procedure. A similar synthetic protocol was published for the synthesis of the sandin skeleton [94] (Scheme 24).

The anti, trans cycloadduct **104** was formed with a diastereoselectivity of 2.5:1 in this application from the C_β-connected dioxenone **103**. Acid-catalyzed ring opening of the annulated cyclobutane led to the β-branched cyclohexanone **105** in high yields.

Concerning the regioselectivity of the intramolecular enone–ene photocycloaddition, head-to-tail adducts were formed with β-linked dioxenones, whereas head-to-head adducts resulted from α-linked substrates. A third possibility for linking unsaturated carbon tethers was studied by Muller et al. [95]. They used substrates **106** in which a C_4 up to a C_6 chain was linked to the acetal center of 1,3-dioxen-4-ones and observed a distinct dependence of the regioselectivity on the lengths of the tether chain: two-carbon tethers gave head-to-head products **107** while substrates with three- or four-carbon tethers gave predominantly head-to-tail products. The facial stereoselectivity is for these substrates already completely determined by the relative configuration at the acetal carbon atom (Scheme 25).

Dritz and Carreira have reported a useful procedure for the synthesis of functionalized tetrahydrofurans (e.g. **111**) and tetrahydropyrans by using β-tethered dioxinones **109** with an additional ether functionality in the linker chain.

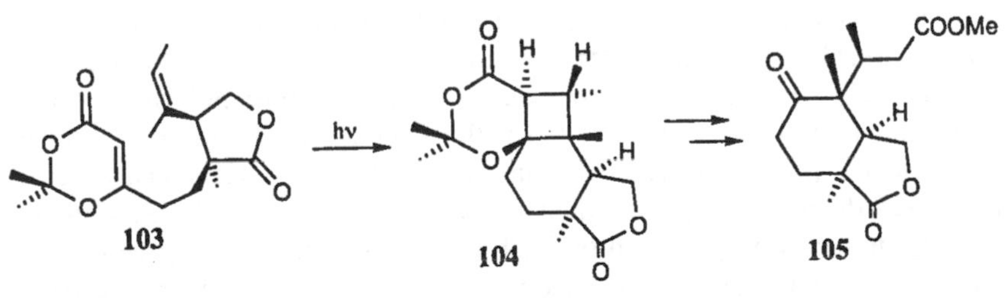

Scheme 24 Applications in natural product synthesis.

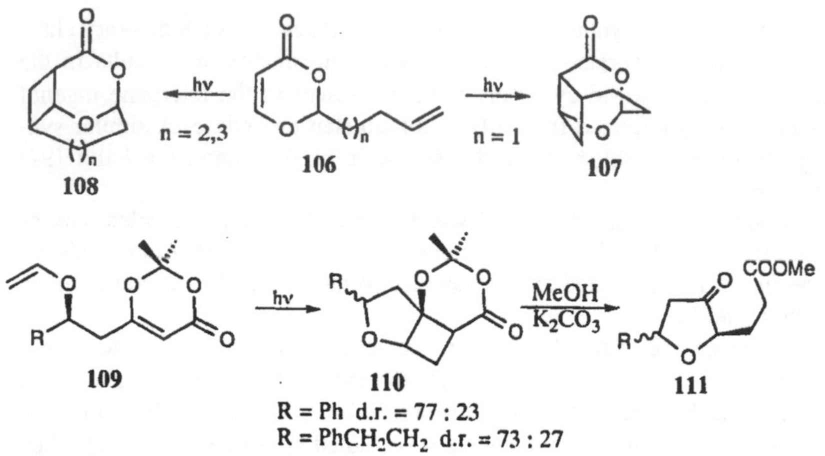

Scheme 25 Acetal-linked dioxenone photocycloadditions.

The diastereoselectivity in the resulting tricyclic adducts **110** is in these cases already determined at the first C–C bonding step and asymmetrical induction from an additional stereogenic center is only moderate [96].

An attractive concept for achieving high asymmetrical induction in enone–ene photocycloadditions using chiral spacer groups which separate the cyclo-enone and the acyclic ene part was developed by Piva and Pete [97]. As already mentioned (vide supra), a short ester tether restricts the conformational flexibility to such an extent that no intramolecular photocycloaddition can be observed. In contrast to that, the "di-malic ester" substrate **112** gave rise to two photocycload-ducts: the head-to-head product **113** and the head-to-tail product **114** in 27% and 49% yield, respectively. In both cases, the facial diastereoselectivity was very high, leading to enantiomerically pure products **115** and **116** after saponification of the tethers and further functionalization. This new route is highly useful be-cause the inexpensive chiral auxiliary is used as temporary linker and can easily be removed after photocycloaddition (Scheme 26).

Small changes in the constitution of the tether chain or ring can sometimes lead to remarkable effects on the stereoselectivity of the photocycloaddition. This has been nicely demonstrated by Winkler et al. for the intramolecular photo-cycloaddition of a vinylogous amide **117** with a cyclohexene in the course of the synthesis of the polycyclic alkaloid manzamine [98]. The unsaturated azocine linker in **117** distorted the preferred chair-boat conformation of the eight-mem-bered ring in such a way that an optimal alignment of the reacting enone/ene parts of the substrate resulted which led to the desired product **118** preferentially in the configuration shown.

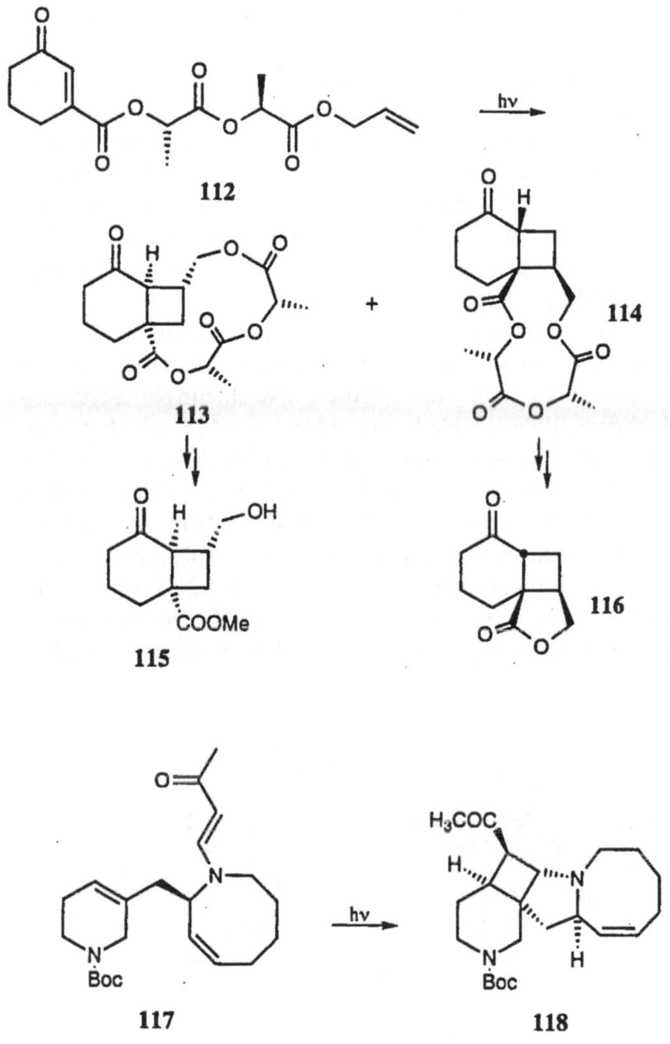

Scheme 26   Stereogenic elements in the tether chain.

The synthesis of eight-membered carbocycles via de Mayo fragmentation [99] of bicyclo[4.2.0]octanes was in the focus of recent work published by Booker-Milburn et al. [100]. An exceedingly easy approach to the synthetic precursors additionally functionalized at three or even at all four cyclobutane carbon atoms was the intermolecular photocycloaddition of alken-or alkyn-3-ols with

3,4,5,6-tetrahydrophthalic acid anhydride (**119**) [101] (Scheme 27). The exo/ endo diastereoselectivities were as high as 10:1 (for the addition of allylic alcohol to tetrahydrophthalimide) and the yields were remarkably high (74–97%). As reason for this high stereocontrol, electrostatic repulsion between the allylic hydroxy (and as well alkoxy) and the carbonyl groups at the stage of the 1,4 biradical was proposed, which is expected to lead to high exo selectivity (i.e., preferred formation of **120**) after the second bonding event. The intramolecular version was also feasible starting from substrate **122** with a conformationally flexible ether linkage and led to the tetracyclic adduct **123**, which was subsequently ring-opened to give the bis-annulated cyclooctanone **124**.

C$_2$-symmetrical bis-α,β-butenolides have been investigated as the respective enantiopure butenolide dimers which were applied, depending on the substituent pattern in the tether chain, for highly diastereoselective alkene photocycloadditions. D-Mannitol is the readily available building block for **125**, which proved to be the most powerful substrate in this investigation [102] (Scheme 28). The photocycloaddition of ethylene gave the dimeric (also C$_2$-symmetrical) adduct **126** which was subsequently cleaved to give **127** with more than 98% enantiomeric excess. Thus, this procedure using more complex enone substrates is superior to the ethylene addition to monomeric homochiral 5-alkyl-2[5H]-furanones which proceeds with diastereoselectivities below 79% [103]. The monomeric cycloadducts are direct precursors to grandisol, a sex pheromone and attractive

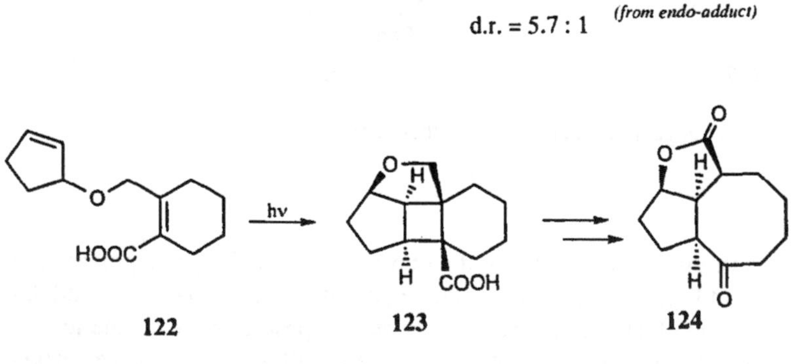

d.r. = 5.7 : 1   *(from endo-adduct)*

Scheme 27   Tetrahydrophthalic acid-based photocycloadditions.

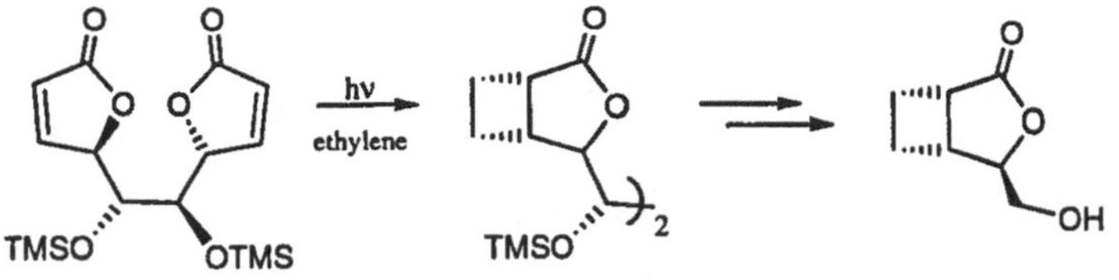

Scheme 28 Butenolides in enone–ene photocycloadditions.

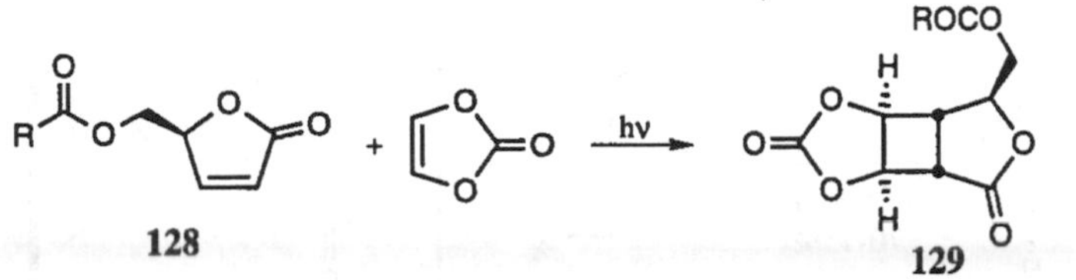

target molecule for numerous photochemical investigations [104]. An interesting facet of intermolecular enone–ene reactions was also discovered by the Font group [105]: enantiomerically pure γ-alkyl-α,β-butenolides **128** add vinylene carbonate exclusively to give the exo adducts with different degrees of facial diastereoselectivities. The introduction of a phenyl group in the substituent at C_γ led to an remarkable increase in selectivity. This phenomena was interpreted as originating from a π-stacking interaction between phenyl and enone π systems, strongly improving the shielding character of the respective substituent. The products could be selectively transformed into γ-butyrolactones of carbohydrates [106].

In some cases, unbranched aliphatic substituents are enough to completely shield one diastereotopic face, as shown by Ohfune and co-workers for the [2 + 2] photocycloaddition of ethylene to the γ-lactam **130** [107]. The resulting diastereoisomers **131** were converted to cyclobutylglycines which serve as conformationally restricted glutamate analogs. A comparable degree in facial diastereoselectivity was also reported from Meyers and Fleming for ethylene addition to enantiomerically pure oxazolinones [108] (Scheme 29).

Intramolecular enone–ene photocycloadditions have been intensively investigated for natural product synthesis by several research groups. Stereogenic centers in the linking chain can be introduced by group transformations using substrates from the chiral pool or by asymmetrical syntheses using prostereogenic elements in the starting materials. Numerous applications were reported by the Crimmins group, e.g., the photochemistry of 2-acyloxy-2,3-hexenoylcyclohexenones **132** to give cis-fused cycloadducts **133** with high facial and relative (sim-

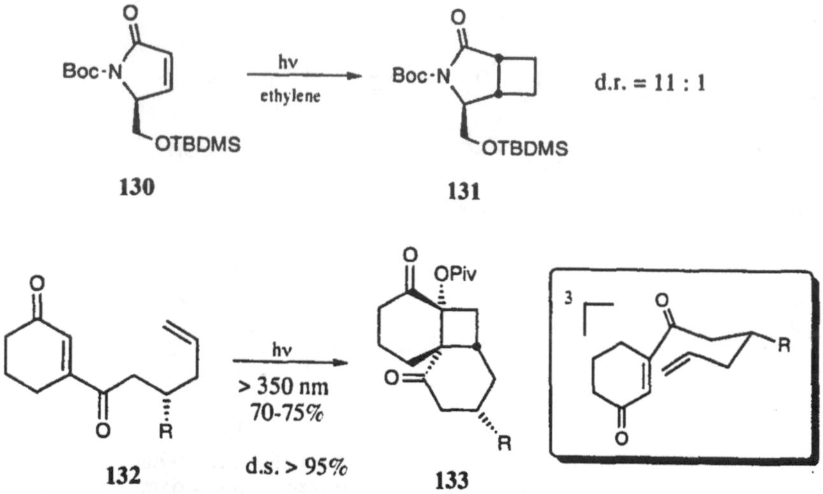

Scheme 29 Butyrolactame- and enedione-based photocycloadditions.

ple) diastereoselectivity [109]. As reason for the high facial stereocontrol a chair-like approach geometry is assumed, which leads to the formation of a 1,4-triplet biradical with defined relative conformation with respect to the facial approach. The same trend came out of the calculation of the primary 1,8-ring closure, i.e., for both assumptions pseudoequatorial chain substituents direct the relative configuration. An impressive example of the potential of this reaction is the synthesis of the phytoalexin ($\pm$)-lubiminol **136** [110] (Scheme 30). The key reaction is the photocycloaddition of **134**, which proceeded with high (double) facial diastereoselectivity and high simple stereoselectivity to give the tetracyclic adduct **135**, which was converted in the target molecule **136** in several steps.

An *immolative* photocycloaddition reaction, i.e., a reaction in which the formation of a new stereogenic element is coupled with the loss of the asymmetry-inducing stereogenic element, was reported by Shepard and Carreira [111]. Most elegantly, they applied enantiomerically pure allenylsilanes **137** and converted them to tricyclic photoadducts **138**. With respect to the resulting vinylsilane, diastereoisomeric E,Z mixtures (2–1.5:1) were formed. Both isomers were produced in high enantiomeric excess and could be easily converted to one product by protodesilylation to give the methylenecyclobutane **139**.

C. Ene–Ene Photocycloadditions

Intramolecular [2+2] photocycloadditions of 1,*n*-dienes which are not activated via conjugation to carbonyl or carbonyl–analog groups can be achieved by the use of Lewis acid catalysis. This strategy has been reported by Evers and Mackor

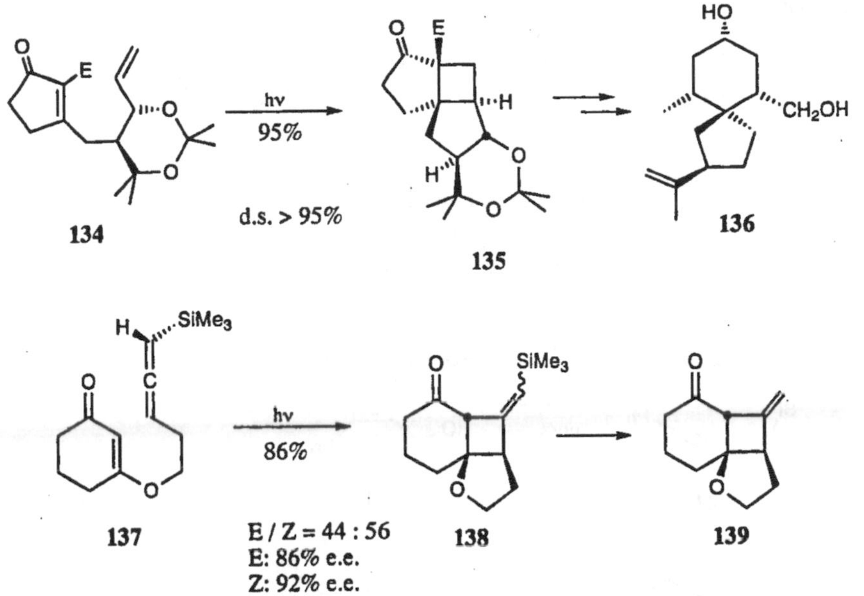

Scheme 30 Stereogenic centers and axes in enone–ene photocycloadditions.

[112] and subsequently examined by the groups of Salamon [113] and Rosini [114]. Recently, Ghosh and co-workers have widely explored the Cu(I)-calayzed photocycloaddition as synthetic route to carbo- and heterocyclic product structures. In order to set the right atom pattern for the synthesis of vicinally substituted cyclopentanones, bisallyl ether was applied with one of the alkene parts activated as enol ether [115]. Irradiation of **140** in the presence of cuprous triflate gives the bicyclo[3.2.0]heptanes **141** in good yield with good to excellent simple diastereoselectivity. A chairlike approach geometry again rationalizes this result by assuming the preference of bulky substituent to be equatorial and thus the ethoxy group adopting the syn position. Acid-catalyzed cyclobutylcarbinyl rearrangement subsequently leads to gem-disubstituted cyclopentanones **142** (Scheme 31). By using dihydrofuran or dihydropyran as the enol ether part, tricyclic photoadducts were produced which can be ring-opened in an analogous way to give 1,1,2,4-tetrasubstituted cyclopentanones [116]. A stereocontrolled approach to highly substituted cyclopentanones starts with 6-oxa-1,3,8-octatrienes as substrates which were oxidatively degraded after photocycloaddition and can be transformed into linear fused triquinanes [117].

Oxygen as the central element of the tether chain can be exchanged by silicon [118]. This element improves the chain flexibility and even eneynes **143**

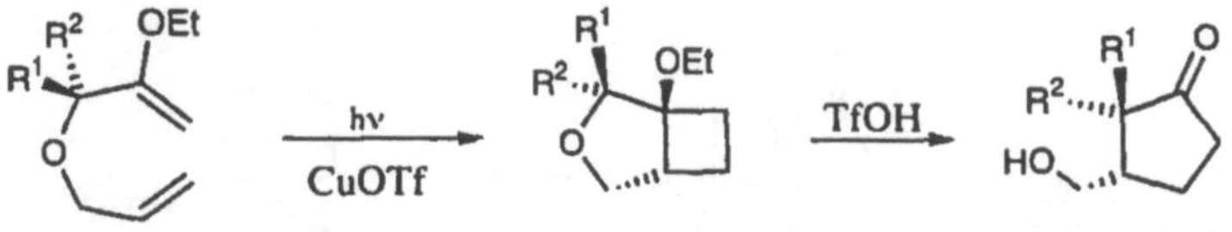

140   **141**   **142**

R¹=Me, R²=Et   d.r. = 3.8:1
R¹=Me, R²=(CH₂)₂Ph   d.r. = 5:1
R¹=Me, R²=Ch₂Ph   d.r. = 19:1

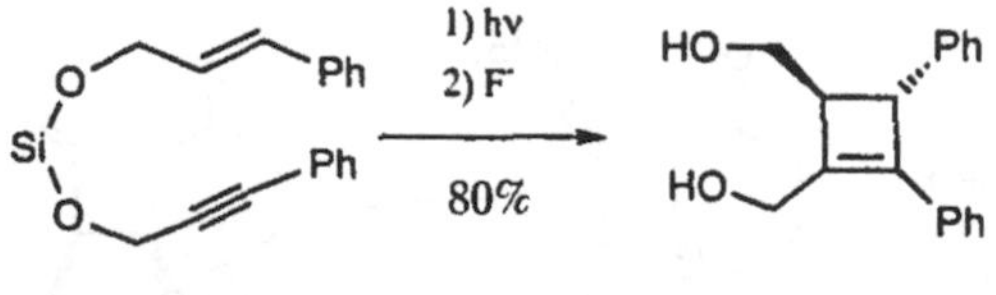

143   **144**

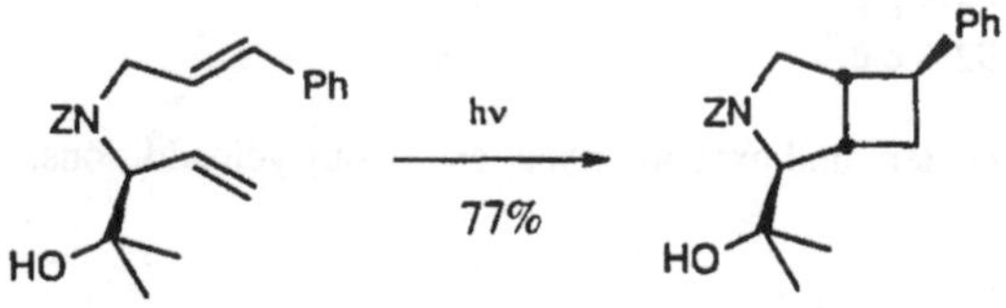

145   **146**

Scheme 31 Intramolecular ene–ene photocycloadditions.

can be photocyclized in high yields to give cyclobutenes **144** with stereospecific formation of the two stereogenic cyclobutene centers. Amides and carbamates can also serve as linking groups such as the Z-protected secondary amine **145** derived from vinylglycine which gave upon photolysis with more than 95% facial and simple diastereoselectivity solely exo product **146** [119].

Sieburth [120] has collected impressive data on the application of [4+4] photocycloadditions with emphasis on the stereoselective synthesis and further transformations of the primary formed cycloocta-1,5-dienes. Intermolecular homodimerization of pyridones led preferentially to head-to-tail adducts **147**, a regiochemical preference which can also be enforced by use of 3,3′-coupled substrates **148** [120] (Scheme 32). The thereby formed bicyclo[6.3.0]undecane skeleton **149** can be reductively opened to give 11-membered carbocycles [121]. Heterodimerization of pyridones was recently reported for the intermolecular [122] and the intramolecular case [123]. The latter reaction serves as an efficient

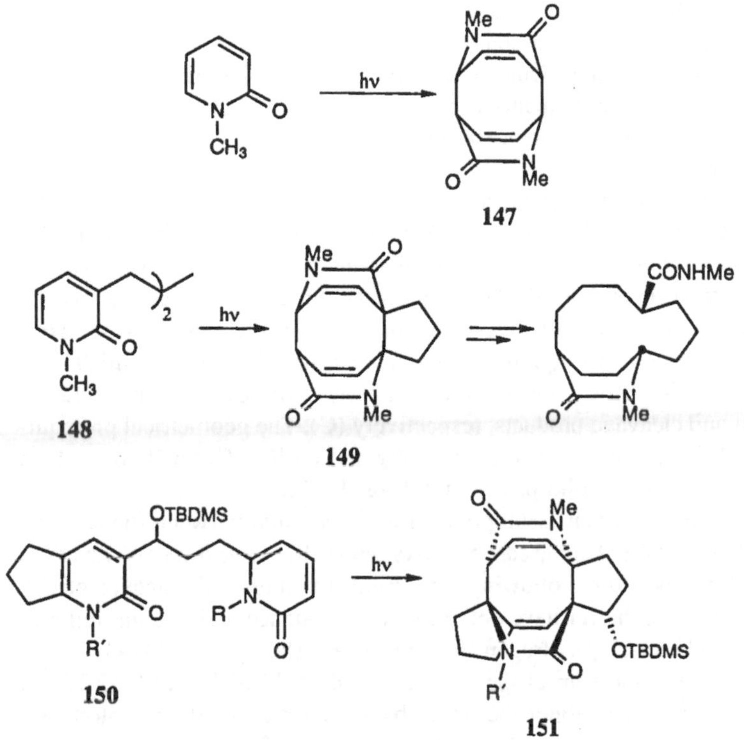

Scheme 32   Intra- and intermolecular pyridone photocycloadditions.

means for generation of the fusicoccin ring skeleton **151** (dicyclopenta[*a*,*d*]-cyclooctane) via [4 + 4] photocycloaddition of the C_3-fused pyridone hetero-dimer **150**.

IV. PHOTOCYCLIZATION REACTIONS

A. Yang Cyclization

The Yang reaction is defined as the photochemical formation of cyclobutanols from acyclic or cyclic carbonyl compounds initiated by an intramolecular γ-hydrogen abstraction (Norrish type II reaction) [124]. The mechanism and the synthetic potential of intramolecular hydrogen abstractions have been intensively investigated and are summarized in excellent reviews [125]. Acyclic carbonyl substrates often tend to cleave when electronically excited: beside the α cleavage (Norrish I), β cleavage often follows the formation of a 1,4 biradical of the

1-hydroxytetramethylene type. Thus, the Yang reaction is limited to those cases where the formation of a new carbon–carbon bond can efficiently compete with the cleavage of the β,γ bond. Due to the experimentally established facts that the first step of this reaction sequence (i.e., transfer of a hydrogen atom) is reversible but the cleavage process is not, special prerequisites are necessary for high-yielding Yang reactions.

The selectivity-determining factors of the Yang cyclization reaction can be described using the three selection stages A–C, which are depicted in Scheme 33. After electronic excitation and intersystem crossing of the carbonyl compound, a chairlike conformation initiates the Norrish type II hydrogen transfer process (A) [126]. The primarily formed structure is that of a skew 1,4-triplet biradical. Biradical dynamics subsequently equilibrate syn- and anti-conformers (B) which have different intrinsic ISC rates (vide supra) and subsequently result in the formation of cyclization and cleavage products, respectively (C). The geometrical prerequisites for step A have been extensively investigated by Scheffer and co-workers for liquid- and especially solid-phase photolyses [127].

Following the classical assumption of an interaction between the reactive CH bond and the p_y orbital at the carbonyl oxygen [128], α substituents can adopt a pseudoequatorial position. Following the crystal structure–solid-state reactivity method, Ihmels and Scheffer have determined the distance and angular requirements for photochemical γ-hydrogen transfer [129]. The optimal $C{=}O \cdots H_\gamma$ distance d is close to the sum of the van der Waals radii of H and O (2.72 Å), the optimal ω angle (describing the angle by which the γ-hydrogen atom lies outside the mean plane of the carbonyl group) is $52 \pm 5°$, the optimal Δ-angle

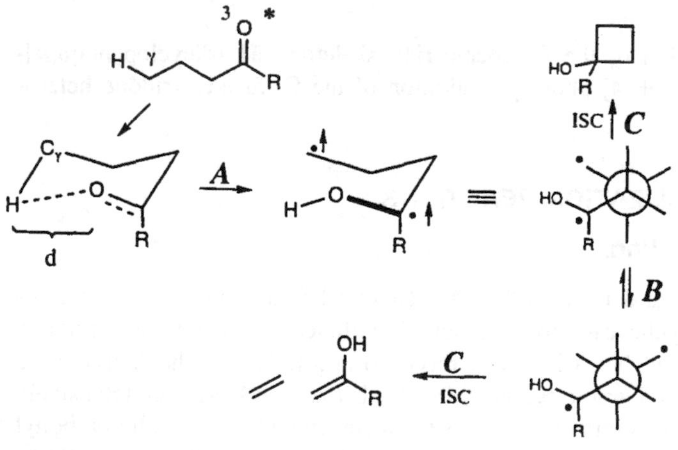

Scheme 33 Mechanism of the Yang reaction.

(describing the C=O · · · · H angle) is 83 ± 4°, and the optimal θ angle (describing the C_γ-H · · · O angle) is 115 ± 2°. These values have been extracted from a series of carbonyl substrates which have been analyzed in solid state and in fluid phase. The introduction of another sp^2 center into the chairlike transition state can, however, alter these values. This was the case of N-alkylated phthalimides where the optimal θ angle was calculated to be 100 ± 5° [130]. Be it as it may, the electronically excited carbonyl substrate has to pass through a chairlike transition state in order to achieve efficient γ-hydrogen transfer. Norrish type I reactions always accompany this reaction and because of less restricted geometrical prerequisites often dominate. Concerning the nature of the activated CH bond, a 1:30:200 per bond primary secondary tertiary selectivity ratio is characteristic [131].

Our interest in this process originates from current work on photochemical amino acid transformations [132]. In order to generate C-activated derivatives, α-amino acids were converted to N-acylated aromatic ketones. The photochemistry of these substrates 152 allowed the detailed analysis of processes B and C with respect to the stereoselectivity of the Yang cyclization [133]. In all reactions, the 2-acetylamino-p-methylacetophenone 154, resulting from the Norrish type II cleavage reaction, was formed to different extents (beside approximately 10–20% of α-cleavage products) (Scheme 34). Both the tert-leucine-derived substrate 152a and the leucine-derived substrate 152b gave solely the cis-diastereoisomeric cyclobutanes 153a and 153b. The cyclization to cleavage ratios (CCR) were also similar: 63:37 and 58:42, respectively. The stereochemical situation became more complex for the valine-derived substrate 152c: not only was the cis/trans selectivity high, but also the diastereoselectivity concerning the formation of the third stereogenic center was surprisingly high (>20:1). Additionally, a high CCR of 90:10 was detected, which made this Yang reaction the most effective and selective one studied in this series. An additional increase in complexity occurred with the isoleucine-derived substrates 152d. These starting materials already exhibit two stereogenic centers and two chemically different γ-CH positions (in contrast to 152a–152c where only homotopic [for 152a] or diastereotopic [for 152c] methyl groups delivered the γ hydrogens).

From a mixture of diastereoisomers (2RS,3S)-152d a CCR of 58:42 and high diastereoselectivity (d.r. = 9:1) for the cyclobutane 153d was observed. Because this stereoselectivity was in contradiction to the fact that two diastereoisomers were applied in the photoreaction and no photoepimerization should be expected, the enantiomerically pure (2S,3S)-152d was investigated. When irradiated under the standard conditions, this diastereoisomer gave only the ketone 154, i.e., a CCR of <0.1. Consequently, the cyclization product 153d originated from the 2R,3S diastereoisomer. Thus, the isoleucine substrate (2R,3S)-152d gives rise to a rare example of an efficient regioselective and stereoselective Yang

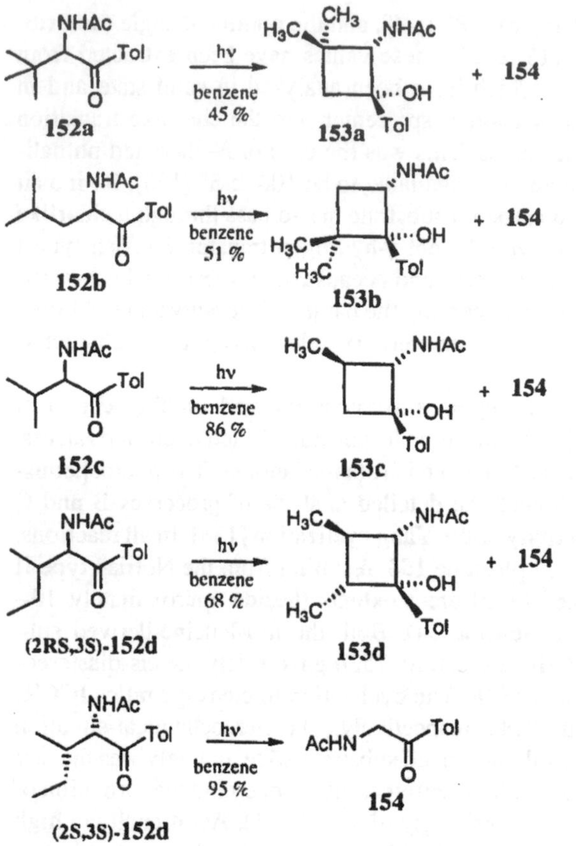

Scheme 34 Yang cyclizations of α-amidoketones.

reaction with *three new stereogenic centers* which are generated during the course of this reaction in a highly selective manner.

The product stereochemistry requires a high 1,2-asymmetrical induction, which is convincingly explained by assuming a hydrogen bond formed already at the triplet biradical stage. The lifetimes of 1-hydroxytetramethylenes reported in the literature [71] are in the 50-ns region and thus long enough to guarantee conformational flexibility at room temperature. Hydrogen bonds stabilizing Norrish II biradicals have been discussed several times due to experimental data [134] and theoretical calculations [135]. Calculations also resulted in the biradical ³**BR′** as global minimum. This syn biradical is additionally in equilibrium with the anti conformer by rotation about the C-2/C-3 bond. The equilibrium constant together with the hydrogen back-transfer efficiency (from the syn structure) and the orbital orientation (p,p relative to the central C–C single bond) control the cyclization/cleavage ratio. The assumption is widely accepted that after intersystem crossing the singlet biradicals maintain conformational memory of their triplet precursors [136], i.e., the anti–1,4 biradical conformer gives exclusively cleavage products whereas the syn–1,4 biradicals can cleave and cyclize. For ^{3}**BR′** this means that process C is responsible for the CCR as well as for the stereoselectivity of the formation of additional new stereogenic centers (vide infra). The latter aspect is not relevant for substrates **152a** and **152b** where the 1,2-asymmetrical induction is dictated by the hydrogen bond and no further stereogenic centers are created. A second stereogenic center is created during the photolysis of the valine-derived substrate **152c**. The calculation (PM3) for the syn biradical ^{3}**BR′$_c$** resulted in a global minimum conformation with a strong hydrogen bond (C=O $\cdots$ HO: 1.80 Å C(sp^2)-C(sp^2): 2.91 Å) (Scheme 35).

In ^{3}**BR′$_c$** two gauche interactions exist whereas in the alternative structure ^{3}**BR″$_c$** three gauche interactions destabilize the conformer. The latter intermediate is expected to isomerize into its anti isomer, which subsequently undergoes cleavage. Another mechanistic alternative is the assumption of a highly stereoselective hydrogen abstraction from only one of the two diastereotopic methyl groups. At this stage of our research we cannot distinguish between these two alternatives.

With the isoleucine-derived substrates **152d** our speculations made above were confirmed. The γ hydrogen is preferentially abstracted from the methylene rather than from the methyl group and the cyclobutane formation occurs stereospecifically only with the 2R,3S diastereoisomer. After isomerization of the pri-

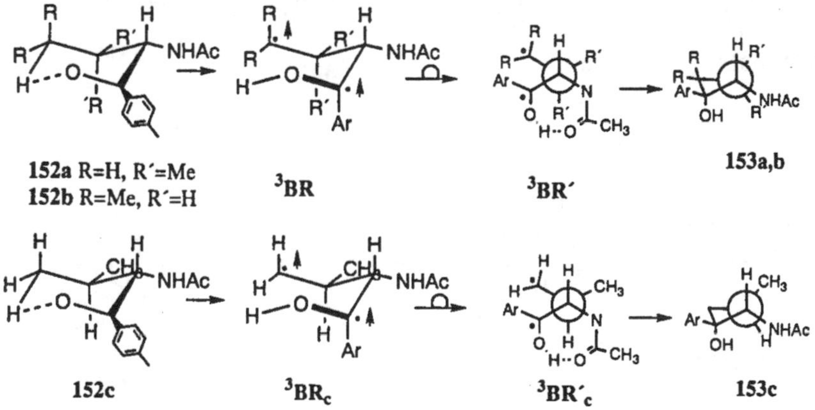

152a R=H, R′=Me
152b R=Me, R′=H

^{3}**BR**

^{3}**BR′**

153a,b

152c

^{3}**BR$_c$**

^{3}**BR′$_c$**

153c

Scheme 35 Mechanism of the Leu, tLeu, and Val photocyclizations.

marily formed triplet biradical ^{3}**BR**$_1$ the hydrogen-bonded structure ^{3}**BR'$_1$** is formed (biradical dynamics = stage B). Again, this structure exhibits only two gauche interactions and thus is preferred over other rotamers (Scheme 36). Eventually, C–C bond formation leads to the cyclobutane **153d**. Prior to this terminating step, the triplet biradical has to undergo spin inversion to cross into the singlet energy surface. In singlet photochemical reactions, stereoselectivity is often controlled by the optimal geometries for radical–radical combinations, whereas in triplet photoreactions the geometries most favorable for intersystem crossing (ISC) are considered to be of similar relevance. These geometries can be quite different from the former ones due to differences in spin–orbit coupling (SOC) values. This phenomenon was described for [2 + 2] photocycloadditions (vide supra). The situation is very much the same for the terminal step of the Yang reaction. Calculations resulted in large SOC values for syn geometries and lower values for anti geometries [38]. Assuming that in the syn conformer the hydroxy group localized at the benzylic radical is sterically more demanding, the SOC principle (Scheme 7) predicts the correct relative product stereochemistry for C-3/C-4 (cis configuration with respect to methyl and hydroxy). On the other hand, the diastereoisomeric substrate (2S,3S)-**152d** gave only cleavage products. The explanation for this is obvious from the analysis of the biradical dynamics: the hydrogen-bonded triplet syn biradical which was primarily formed suffers from three gauche interactions and isomerizes to the more stable trans biradical, which subsequently cleaves to give **154**.

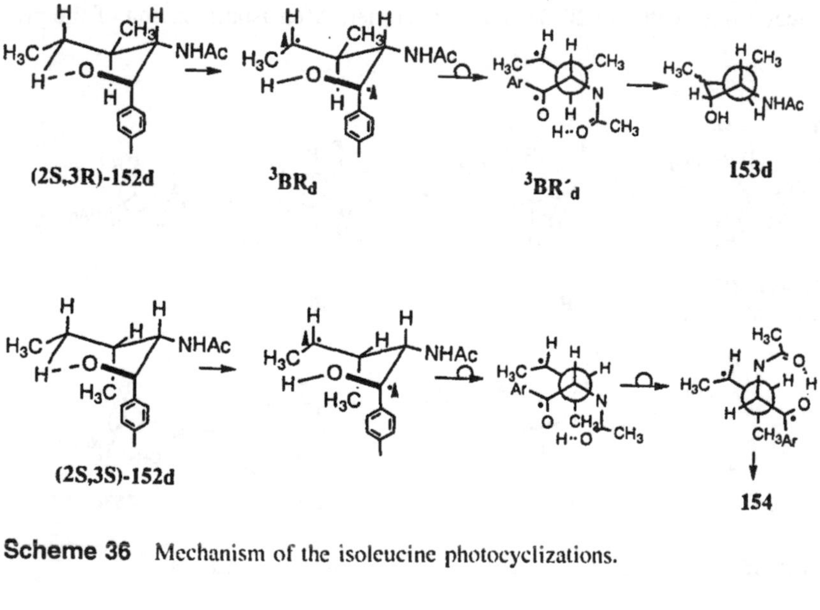

Scheme 36 Mechanism of the isoleucine photocyclizations.

Summarizing the experimental results and the proposed reaction model, a three-stage selection protocol results which enables the prediction of the correct chemo- and stereoselectivity. The hydrogen abstraction (step A) differentiates between diasterotopic hydrogens due to the strict geometrical prerequisities, biradical dynamics (step B, i.e., rotation about C-1/C-2 and C-2/C-3 for substrates 152) control the 1,2-asymmetrical induction *as well as* the cyclization/cleavage ratio. Eventually, SOC-controlled ISC (step C) directly leads to the formation of the final C–C bond and thus determine the stereochemistry of the conversion of the two radical centers into new stereogenic centers.

The Yang cyclization has recently been applied for the synthesis of target molecules using substrates from the "chiral pool." A route to both (2R)- and (2S)-azetidine 2-carboxylic acid derivatives 156 from enantiomerically pure aminoalcohols has been developed [137] (Scheme 37).

Analogous to the Yang reactions described above, the cyclization of 155 leads to *cis*-product 156. In order to circumvent charge transfer quenching which is known from aromatic α-aminoketones, the amino function is carbamate-protected (and thus redox-deactivated). Structure–reactivity relationships for Yang reactions of a thioketone [138] and differences between solid-state and solution chemistry of macrocyclic 1,2-diketones [139] were reported from the Scheffer group. Another report on solid-state photochemistry involved the use of prochiral organic acids which formed salts with chiral enantiopure amines to give crystals in chiral space groups [140]. In many cases these substrates can be photolysed with high product enantiomeric excesses. Application of this principle to the Yang cyclization proved to be very successful [141]. The achiral 14-membered aminoketone 157 gave in solution the two Yang products 158 and 159 (total $\Phi = 0.15$), whereas solid-state photolysis of the malic acid salt of 157 resulted in an approximately 1:1 ratio of 158 and the Norrish type II cleavage product. The cyclobutanol 158 was formed after 20% conversion with 95.3% e.e. A covalently linked chiral group was used by Ito et al. in the solid-state Yang reaction of 2,4,6-triisopropylbenzophenone derivatives 160 [142]. These compounds readily cyclized to give hydroxybenzocyclobutenes 161 with high diastereomeric excess even after prolonged irradiation.

B. 1,3-, 1,5-, and 1,6-Photocyclizations

A highly enantioselective 1,5-photocyclization was reported by Irngartinger et al. for 1-benzoyl-8-benzylnaphthalene 162 [143]. The substrate crystallizes in a chiral space group and, depending on what enantiomorphic crystal was used, the corresponding enantiomer of the photoproduct acenaphthene 163 was formed. This is one of the rare examples of crystal-to-crystal absolute asymmetrical synthesis (AAS) in solid-state photochemistry. Scheffer and co-workers have also successfully contributed to this approach to enantiomerically enriched compounds [144] (Scheme 38).

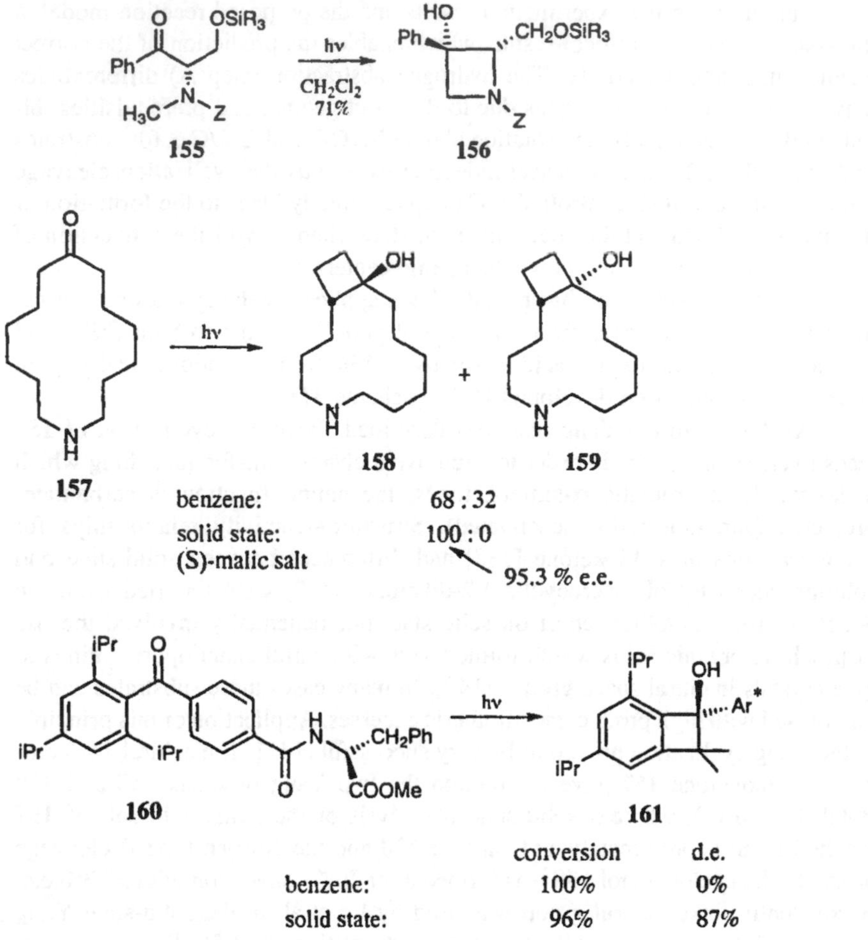

Scheme 37   Solid-state Yang cyclizations.

In general, carbonyl photocyclizations initiated by hydrogen atom transfer from a position other than a γ-CH group can compete with the Norrish type II reaction under two circumstances: 1) the γ-CH group does not exist at all, i.e., this position is blocked by a heteroatom or any other group lacking γ hydrogens, or, 2) a photo-induced electron transfer process oxidizes one part of the molecule much faster than γ-CH transfer and subsequent mesolytic transformations active nearer or more remote positions in the substrate [145]. Both possibilities exist for the 1,3 cyclization of β-aminoketones 164 as reported by Weigel and Wag-

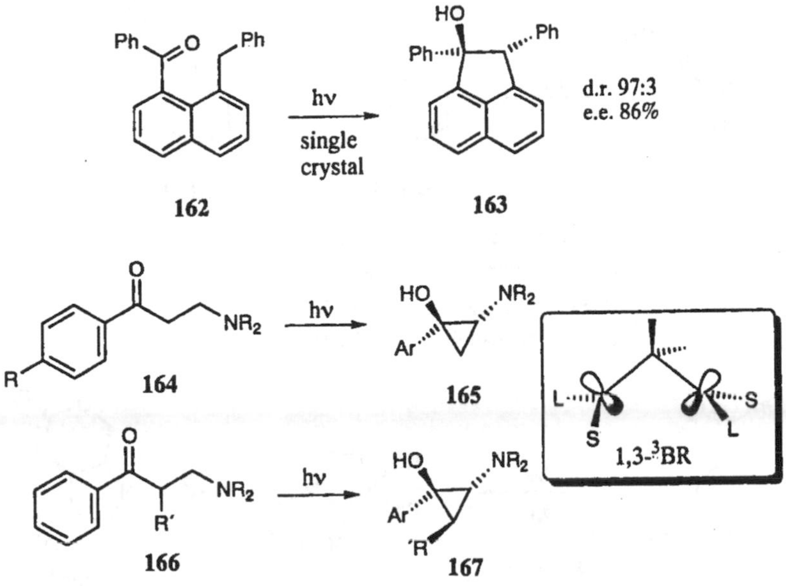

Scheme 38 The 1,3 and 1,5 photocyclizations.

ner [146]. The amino group excludes γ-CH transfer and likewise can be easily oxidized to give the radical cation which rapidly undergoes proton transfer. Both processes lead to 1,3-triplet biradicals which subsequently cyclize diastereoselectively to give the (E)-2-aminocyclopropanols 165 in good yields. Chiral substrates 166 also reacted with high simple and additionally high induced diastereoselectivity to give the aminocyclopropanols 167. Both properties can be rationalized with the structure of the 1,3 biradical approaching maximum spin-orbit coupling geometries, which brings the eclipsed substituents at C-1 and C-3 in an eclipsed conformation.

The formation of indanols via 1,5 cyclizations starting from α-(o-alkyl) acetophenones 168 has been intensively and thoroughly investigated by Wagner and co-workers [147]. High simple diastereoselectivities were obtained with the monoethyl substrate 168a as well as with the triethyl substrate 168b with preferential formation of the (Z)-indanols 169 (Scheme 39). The complex conformational situation at the 1,5-triplet biradical level together with the assumption of preferred orthogonal ISC geometries has been investigated using temperature and solvent effects. An essential question raised by P. Wagner was whether ISC is static and leads to singlet biradicals which have additional (small) barriers to bonding vs. cleavage or ISC occurs dynamically during cyclization rather than

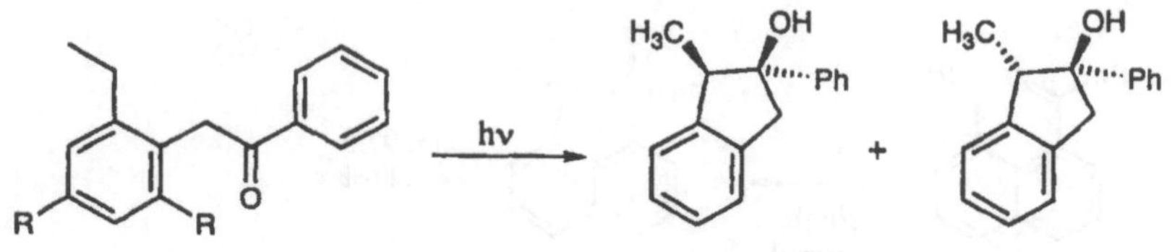

168 **E-169** **Z-169**

R = H / benzene	95 : 5
R = H / methanol	67 : 33
R = Et / benzene	97 : 3
R = Et / methanol	81 : 19

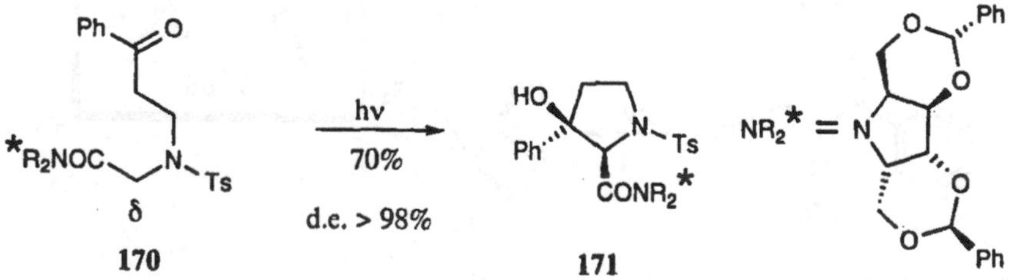

170 **171**

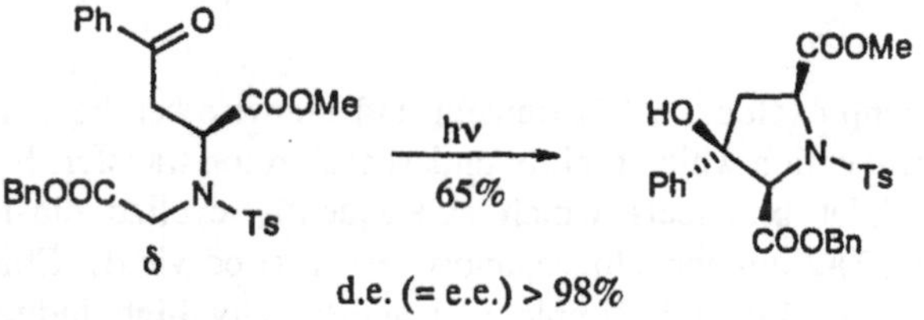

172 **173**

Scheme 39 Diastereoselective 1,5 photocyclizations.

statically beforehand. From the discussion of Paternò-Büchi biradicals in Section III.A.1 (see Scheme 7), the second alternative is more attractive.

There has been considerable interest in using 1,5 cyclizations for the synthesis of unnatural proline derivatives using substrates from the chiral pool of acyclic amino acids. If chiral auxiliaries were incorporated in the N-connected side chain of the β-aminopropiophenone derivative **170**, high asymmetrical induction was observed in the 3-hydroxyproline derivative **171** [148]. The chiral inductor can be easily synthesized from mannitol. A second possibility for inducing highly stereoselective 1,5-cyclizations is the use of stereogenic elements in the chain connecting carbonyl and amino group. Substrate **172**, available from aspartic acid, photocyclizes with high yields and with excellent diastereoselectivity to give the enantiopure product **173** [149]. These experiments set the stage for photocyclization experiments involving dipeptide substrates [150] (Scheme 40).

A series of dipeptides Gly-X **174** were investigated which were activated at the central amide nitrogen by means of a benzoylethyl group. Photocyclization of these compounds proceeded with high regioselectivity (glycine-selective ε-CH abstraction) and high simple diastereoselectivity (cis with respect to hydroxy and amido groups at the two new stereogenic centers formed during the C–C coupling step) with formation of the δ-lactames **175** [151]. Additionally, the induced diastereoselectivity was moderate to good depending on the second amino acid residue. Reducing the chain length in the chromophor linker increased the induced diastereoselectivity in the 1,5 photocyclization of substrates **176** to γ-lactams **177**. Having now a useful and high-yielding route to highly functionalized chiral lactams in their hands, the Giese group went one step further and investigated CH abstraction reactions from stereogenic centers [152]. As substrates the diastereoisomeric Ala-Val dipeptides **178** and **181** were used. However, from both substrates the diastereoisomeric γ-lactams **179** and **180** were formed in completely different ratios. A long-lived 1,5-triplet biradical which can undergo all possible bond rotations is not compatible with this result and a

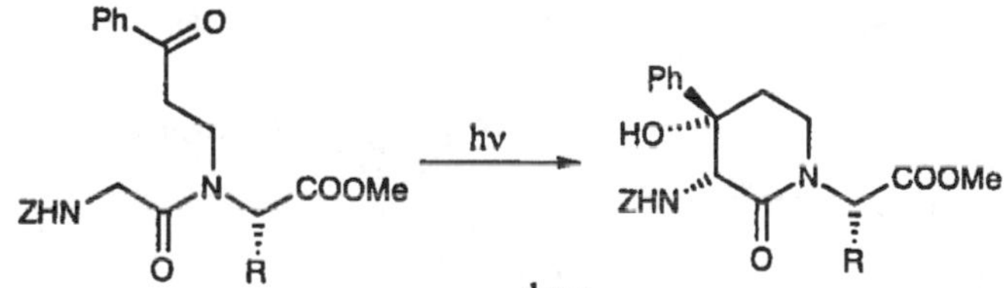

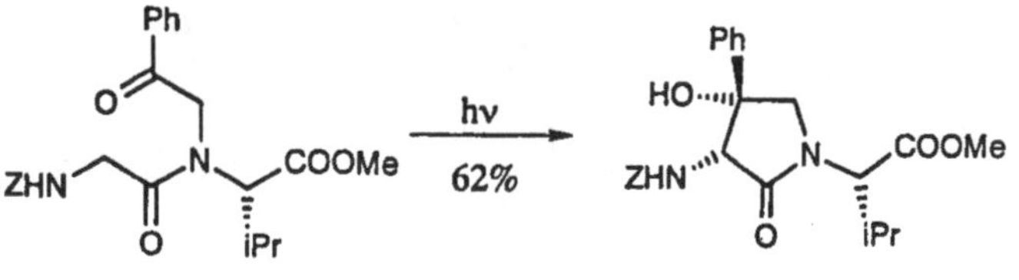

Scheme 40   Diastereoselective 1,5 and 1,6 photocyclizations.

memory effect of chirality was postulated which might be a highly valuable concept for further 1,*n*-biradical cyclization reactions (Scheme 41).

Substrate controlled asymmetrical syntheses of δ-lactams were also described from aspartic acid–derived amides. Substrates **182** gave the bicyclic lactams **183** in moderate diastereoselectivities [153]. The N,N-unsymmetrically substituted amides **184** gave the δ-lactams **185** and **186** with excellent simple (for **185** and induced diastereoselectivities with moderate regioselectivities [154]. The SOC-controlled ISC conformation model again served as a powerful tool to rationalize these results.

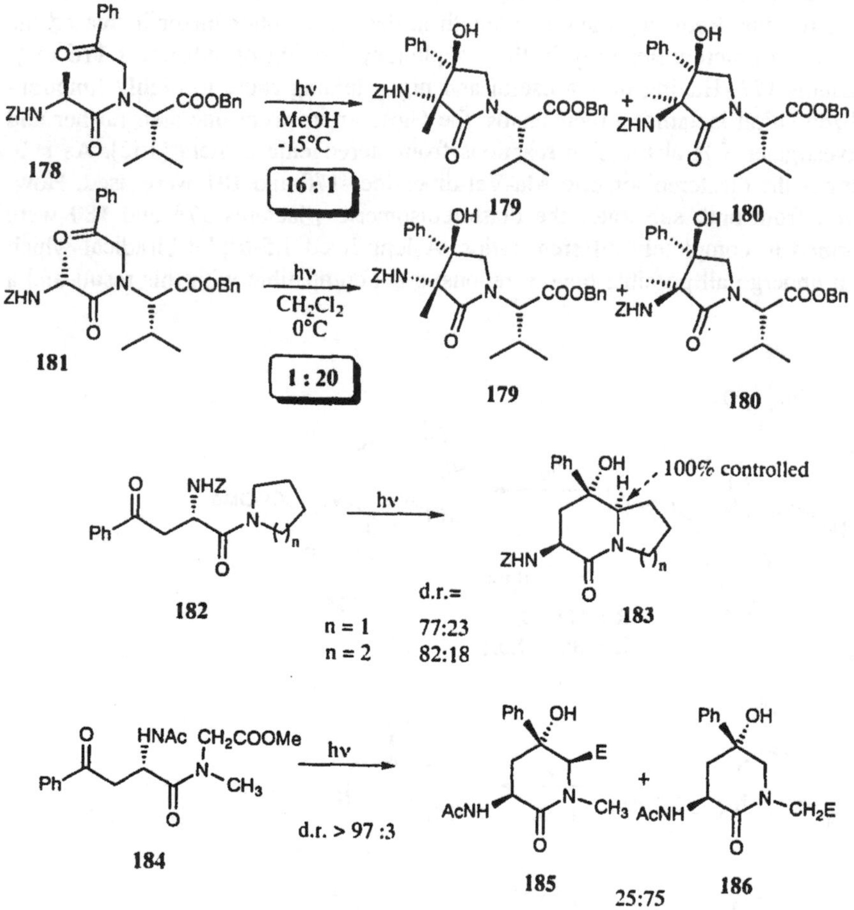

Scheme 41 Memory of chirality and δ-lactam synthesis.

C.   Remote Photocyclizations

The dominant CH activation path for excited carbonyl groups substituted with flexible hydrocarbon chains is the γ-hydrogen abstraction following the preferred six-membered transition state for Norrish II reactions. By exchanging the critical methylene group by a quaternary carbon, an aryl group, or by heteroatoms, it is possible to influence this reactivity pattern. Alternatively, conformational restricted substrates with especially reactive CH groups at remote positions can give higher (1,n) biradicals by homolytic cleavage or mesolytic cleavage.

Kraus and Wu have developed a photochemical strategy for radical cyclizations via 1,10-hydrogen atom abstraction [155]. The photolysis of the α-keto ester **187** afforded only the eight-membered ring lactone **188** in 51% yield. The 1,8 diradical preceding the lactone **188** could be formed directly or by γ-hydrogen atom abstraction leading to a 1,4-biradical and subsequent 1,5-hydrogen transfer. That the latter mechanism is not operating was shown by irradiation of the deuterated analog: the corresponding bis-deuterated lactone **188** was isolated in 62% yield. Conformational rigidity is required for this remote hydrogen abstraction: only the cis stereoisomer was photochemically active. In **189** the benzylic hydrogen atoms are more easily activated than the γ-methin CH group. The benzannulated oxepanone **190** was isolated in 74% yield (Scheme 42). Less effective was the photocyclization of the ketal derivative **191**: the tricyclic product **192** was isolated in 22% yield. This example shows that α-heteroatom-substituted methylene groups are also sufficiently activated for homolytic CH abstraction from remote positions. Unexpected was the result for the unsaturated keto ester **193**. Only the eight-membered lactones **194** were formed in moderate yields. From a synthetic point of view the eight-membered oxacyclic ring systems are valuable targets because they constitute common subunits in marine natural products.

During the last two decades photo-induced electron transfer chemistry has developed as a highly active field in organic photochemistry. It was shown that many photochemical transformations, earlier thought to proceed via homolytic steps, are initiated by electron transfer (vide supra for Paternò–Büchi reactions). Several excellent reviews covering the synthetic applications of these reactions are available [156]. As starting materials for remote photocyclizations donor–acceptor pairs are used which are linked by a flexible hydrocarbon chain. In most cases, mesolytic cleavage [157] of a CH bond proximate to the radical cation (i.e., oxidized heteroatom, alkene, or arene group) proceeds the primary electron transfer. The resulting (1,n) biradicals combine to give medium- to large ring systems. A very useful carbonyl chromophore which has been intensively investigated in the last two decades is the phthalimido group [158]. The reduction potential (about −1.4 V) is remarkably low in comparison with aromatic carbonyl compounds. In the presence of electron donor groups (thioether, amines, alkenes, arenes) an exergonic electron transfer can occur after electronic excitation [45].

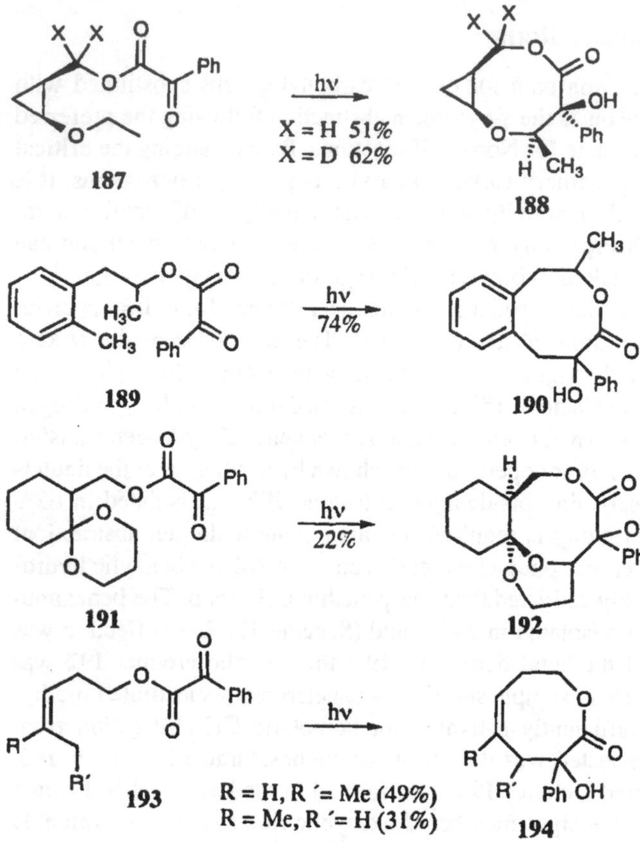

Scheme 42 Remote photocyclizations of phenylglyoxylates.

The course of the intramolecular photoreaction of carbonyl compounds with electron-rich alkenyl or aryl substituents in the side chain is dictated essentially by the thermodynamics of the electron transfer step. For phthalimides, this relationship was intensively studied. For these cases where G_{ET}^0 is positive, $[\pi^2 + \sigma^2]$-cycloaddition reactions were observed with alkenyl substitutents and classical Norrish II chemistry for aryl-substituted substrates. For negative values G_{ET}^0, electron transfer products dominate. This transition was described for enantiomerically pure γ-aryl-substituted phthalimides [159]. Macrocyclization reactions of phthalimides with remote styryl substituents **195** were studied by Kubo and co-workers [160] (Scheme 43).

When these photolyses were performed in methanol as solvent, the styrene radical cations were trapped by methanol in an anti-Markovnikov fashion and

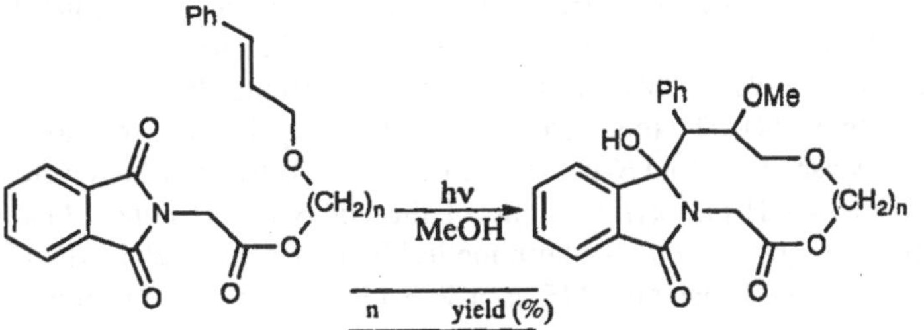

195 196

n	yield (%)
2	68
4	55
5	65
6	35

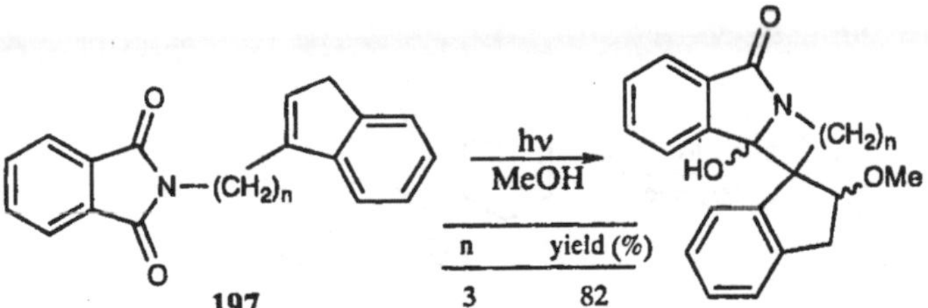

197 198

n	yield (%)
3	82
5	76
10	86

Scheme 43 Remote PET photocyclizations of styryl substrates.

the resulting biradicals combined to give macrocyclic lactones **196** in moderate to good yields. Spiroannulated products were formed by irradiation of the corresponding indenyl-substituted starting materials **197** in methanol [161]. The yields for the heterocyclic products **198** were excellent even for the 13-membered representative (86%). More ambiguous were the results for the alkenyl-activated substrates [162].

Intramolecular PET cyclizations of aminoalkyl-substituted phthalimides were investigated by the groups of Kanaoka and Coyle [163]. Due to the relative low oxidation potentials of tertiary amines, these electron transfer reactions are highly exergonic [164]. The regioselectivity of the CH activation step is remarkably high for the *N*-methyl, *N*-phenyl-substituted substrates **199**.

This effect has been rationalized by the higher kinetic acidity of the *N*-methyl vs. an *N*-methylene group in amine radical cations. The chemical yields for the hydroxyisoindolinones **200**, however, were only poor (10–15%) for all cases (ring sizes between 5 and 16). The yields of these cyclization reactions could be improved by use of aromatic carbonyl groups as electron acceptors instead of phthalimides. In this context, Hasegawa and co-workers used aromatic β-oxoesters as substrates. Medium-sized azalactones **202** and lactones **203** were available via photolysis of *N,N*-dialkylaminoalkylbenzoylacetates **201** in moderate yields [165] (Scheme 44). The regioselectivity of these cyclization reactions could be improved by use of twofold *N*-benzyl-substituted substrates, which were exclusively activated by CH mesolysis of one of the benzylic CH bonds. One example is the photolysis of *N,N*-dibenzylaminoethyl benzoylacetate **204**, which led to the eight-membered azalactone **205** as a 2:3 mixture of cis- and trans-diastereoisomers [166].

In recent publications, Neckers [167] and Hasegawa [168] were able to show that phenylglyoxylates are also capable of macrocyclization reactions initiated via photo-induced electron transfer. The N,N-disubstituted aminoethylben-

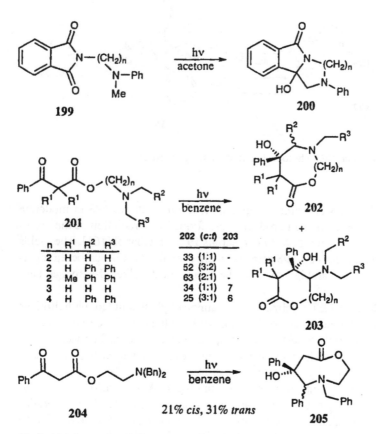

n	R^1	R^2	R^3	202 (c:t)	203
2	H	H	H	33 (1:1)	-
2	H	Ph	Ph	52 (3:2)	-
2	Me	Ph	Ph	63 (2:1)	-
3	H	H	H	34 (1:1)	7
4	H	Ph	Ph	25 (3:1)	6

21% cis, 31% trans

Scheme 44   Remote PET photocyclizations of amino substrates.

zoylformates **206** efficiently underwent remote photocyclization to give the seven-membered azalactones **207** with low diastereoselectivity.

The widely investigated cyclohexenone chromophore can also serve as the electron-accepting group in the presence of a dialkylamino group, e.g., in substrate **208** [169] (Scheme 45). The regioselectivity of the mesolytic CH cleavage was again controlled by use of the *N,N*-dimethylamino group. In contrast to the reactions described before, conjugate addition was observed in this case. As major product, the cis-fused 3-azabicyclo-[5.2.2.]undecan-8-one **209** was isolated in 50% yield.

Electronically excited carbonyl compounds with an alkylthio group in the side chain undergo facile photocyclization to give a variety of azathiacyclols. With relatively high simple diastereoselectivity, which even increased with increasing length of the hydrocarbon tether, the sulfide-containing phenylglyoxylates **210** were efficiently cyclized to give the thialactones **211** with sizes up to 15-ring atoms [170] (Scheme 46). This type of PET macrocyclization is, in contrast to analogous transformations described above, remarkably efficient and tolerates many different functional groups in the hydrocarbon spacer. Scope and limitation of this important method were intensively investigated by Kanaoka and co-workers for phthalimides as the electron-accepting carbonyl species [171]. The limitations concerning the maximum ring size of the macrocycles available have not yet been exactly determined. The chemical yields of unbranched azathiacyclols **213** decreased with increasing spacer lengths in the substrates **212**.

The ratio for methyl vs. methylene (leading to the branched products **214**) CH activation varies between 15 and 20. This process allowed the synthesis of

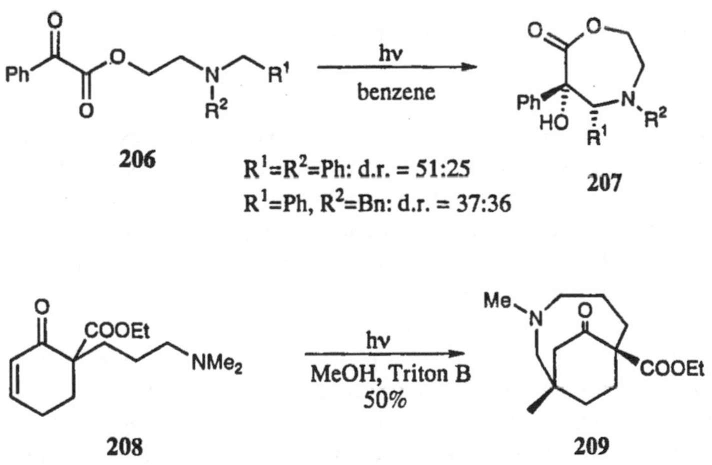

$R^1 = R^2 = Ph$: d.r. = 51:25
$R^1 = Ph, R^2 = Bn$: d.r. = 37:36

Scheme 45   Diastereoselective PET photocyclizations of amino substrates.

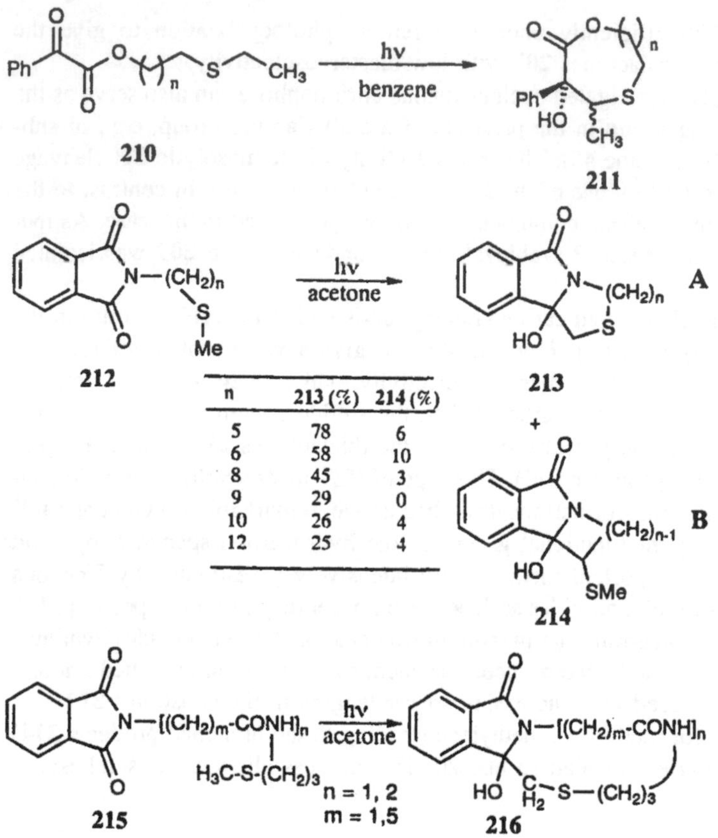

n	213 (%)	214 (%)
5	78	6
6	58	10
8	45	3
9	29	0
10	26	4
12	25	4

Scheme 46 Remote PET photocyclizations of thioether substrates.

cyclopeptide and macrolide [172] model compounds with ring sizes up to 37. The acyclic starting materials **215** were conveniently cyclized to give **216** in yields between 25% and 50%. One major problem connected with this thioether method is the incorporation of the electron-donating group into the macrocyclic ring systems. A practical way to overcome this disadvantage was developed by Kanaoka et al. using the 1,3-dithiolanyl group as electron donor substituent which can be removed to give the sulfur-free compounds.

The molecular systems for the remote photocyclizations, as described in the previous chapters, consist of an electron-accepting and an electron-donating group connected by a flexible hydrocarbon chain. Both donor and acceptor groups are inserted into the newly formed macrocyclic ring system and have to be eliminated in a second reaction step. This detour makes these reactions less attractive

for C–C coupling steps, e.g., for macrolide synthesis. An improved route used
ω-aminocarboxylic acids as substrates [173]. Electron transfer activation from
remote positions became energetically feasible when the corresponding carboxylates were used. An illustrative example is the transformation of the ester-linked
N-phthaloyl-ω-aminocarboxylic acid **217** [174]. This compound, easily available
by reaction of glutaric anhydride with *N*-hydroxyethylphthalimide, was converted
(67% yield) to the macrocyclic lactone **218** when the corresponding potassium
salt was irradiated in an acetone–water mixture. Likewise, the crown ether **219**
was formed with excellent yield (73%) from the acyclic percursor **220**. That the
electron transfer step occurred predominately in an *intramolecular* fashion
is probably due to complexation effects in the ground state of the substrates
(Scheme 47).

Using starting materials from the chiral pool, benzopyrrolizidines **222** were
synthesized from the glutamic acid derivative **221** with moderate diastereoselec-

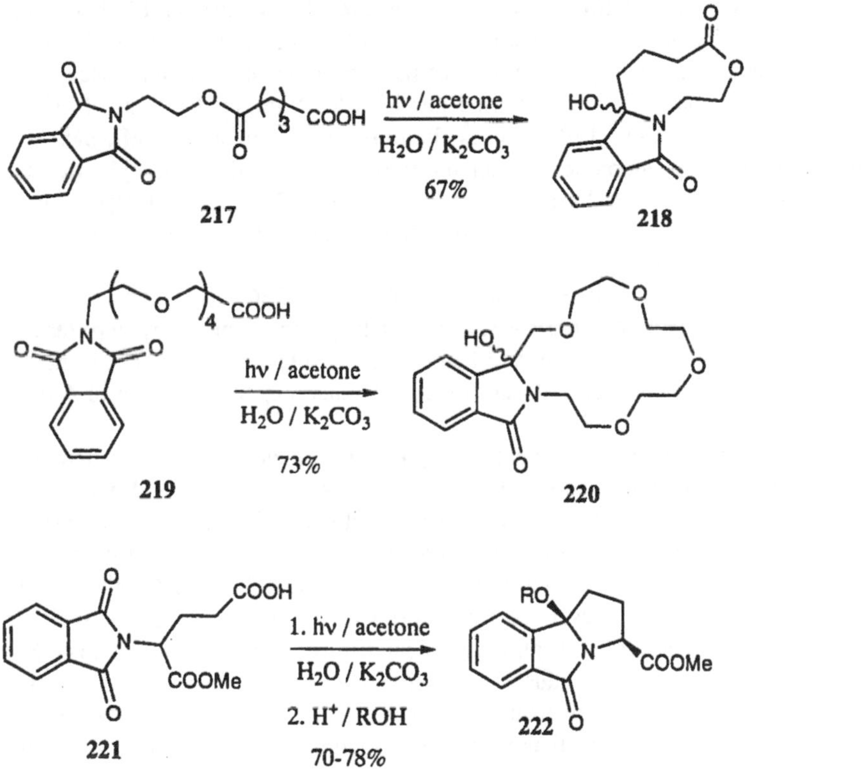

Scheme 47   Remote PET photocyclizations of carboxylates.

tivity, which could be improved by subsequent epimerization of the newly formed stereogenic centers [175]. Further intra- and intermolecular applications of the photodecarboxylative addition reactions were recently summarized [176].

V. RECENT DEVELOPMENTS

Several reports on stereoselective photocyclizations induced by hydrogen transfer steps have recently appeared. The photocyclization of substituted o-amino phenylketones **223** yielded dihydroindanols **224** in high yields as well as with solvent- and substituent-dependent diastereoselectivity [177]. We are interpreting these results by applying the SOC-ISC model previously described. Protic solvents increase the bulk of the OH group by solvation, and thus the simple (*trans*) diastereoselectivity is also increasing. For R' = COOR, intramolecular hydrogen bonding stabilizes the triplet biradical and leads to higher *cis*-selectivity. The ISC geometry model was also used by Hu and Neckers to rationalize the results of arylglyoxylate cycloaddition reactions with cyclo-1,3-dienes [178]. Directing stereoselectivity of photocycloaddition and photocyclization reactions via intra- and intermolecular hydrogen bonding seems to become an important concept for future applications. We have reported this idea for Yang-cyclizations of amino acid–derived substrates [179]. Paternò–Büchi reactions can also be influenced by hydrogen bonding interactions either intramolecularly in the carbonyl substrate (which differentiates the diastereotopic faces in a chiral aldehyde [180]) or intermolecularly between carbonyl compound and the alkene (e.g., allylic alcohol and benzophenone cycloaddition [181]).

A highly functionalized "chimeric" amino acid **226** was synthesized using a hydrogen abstraction/cyclization reaction as the photochemical key step starting from the 4-hydroxyproline–derived ketone **225** as substrate [182]. The same research group also reported a diastereoselective pinacolization reaction of 4-oxo-4-phenylbutanamides [183].

The synthesis of enantiometically pure molecules *via* photochemical key steps has been reported following an electron-transfer–induced addition of tertiary amines with the chiral pool–derived substrate 5-menthyloxy-2[5H]-furanone [184]. This route allowed the efficient synthesis of the enantiomerically pure pyrrolizidine alkaloids laburnine and isoretronecanol [185]. Another highly promising concept was described by Scheffer and co-workers: the use of ionic chiral sensitizers in solid-state photochemical di-π-methane rearrangements [186] as well as the asymmetric induction in solid state photochemistry by using ionic chiral auxiliaries in Yang cyclization [187], 1,5-hydrogen abstraction/cyclization [188], and benzocyclohexadienone ring contractions [189]. The third and most spectacular approach to asymmetric synthesis was reported by the Giese group [190].

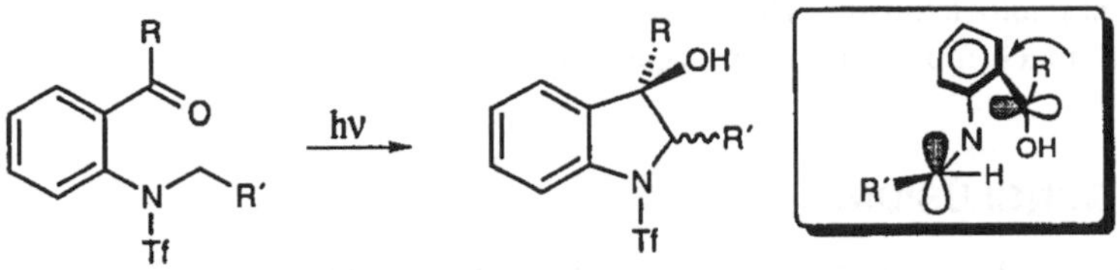

223   d.r.=cis/trans   224
E = COOCH₃

d.r. (hexane) d.r. (CH₃CN)

R=Ph, R´=E   92 : 8   52 : 48
R=E, R´=Ph   6 : 94   4 : 96

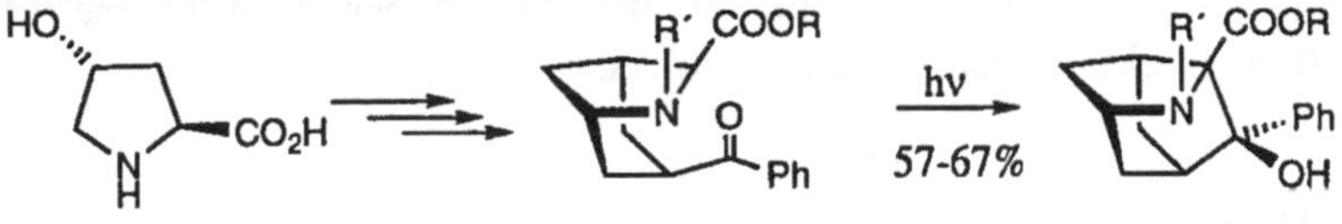

225   226

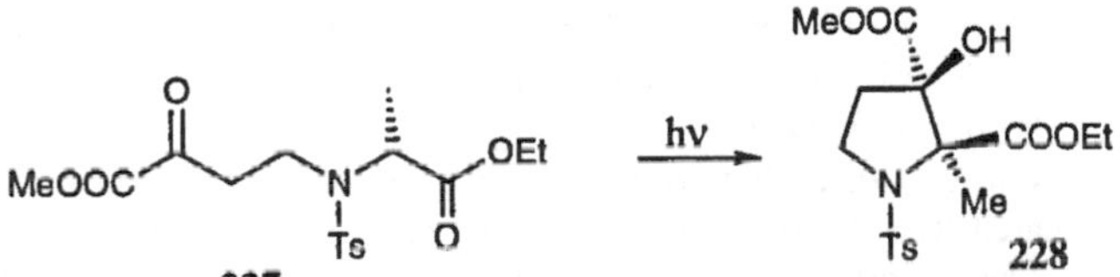

227   228

	e.e.	cis/trans
PhH/1M naphthalene	24 : 1	5.7 : 1
PhH/1M benzophenone	1.4 : 1	0.8 : 1

Scheme 48 Diastereoselective photocyclizations: actual examples.

As substrate, the alanine-derived ketoester **227** was used which under photoexcitation gave the diastereoisomeric *cis/trans*-pyrrolidines **228** following a
1,5-hydrogen transfer/cyclization sequence [190]. In the presence of 1M of
naphthalene as triplet quencher, exceedingly high enantiomeric excesses (and
also high simple diastereoselectivities) were observed, whereas in the presence
of 1M of benzophenone as triplet sensitizer nearly zero e.e. and low d.e. were
detected. This example of a *chiral memory effect* is not without precedent in

organic photochemistry. The inherent efficiency and elegance of this effect will initiate much more activity in organic synthesis.

VI. CONCLUSIONS

The study of stereochemical aspects of photochemical reactions leads to an improved mechanistic understanding of these processes and consequently to an enormous increase in possible applications for synthetic organic chemistry. In the last decade, detailed analyses of substituent, temperature, and spin effects have shown that the conformational behavior of biradicals, the specific energetic and geometrical prerequisites for intersystem crossing from the triplet to the singlet potential energy surface, as well as the competition between product formation and deactivation to restore ground-state molecules from reactive intermediates result in a convincing picture of the three-dimensional course of transformations of electronically excited molecules.

REFERENCES

1. Fleming, S. A.; Bradford, C. L.; Gao, J. J. in Molecular and Supramolecular Photochemistry, Ramamurthy, V.; Schanze, K. S., eds., Marcel Dekker: New York, Basel, Hong Kong, 1997, Vol. 1, 187. Everitt, S. R.; Inoue, Y in Molecular and Supramolecular Photochemistry, Ramamurthy, V.; Schanze, K. S., eds., Marcel Dekker: New York, Basel, Hong Kong, 1999, Vol. 3, 71.
2. Wagner, P. J. *The Spectrum* **1998**, *11(4)*, 9.
3. Paternò, E.; Chieffi, G. *Gazz. Chim. Ital.* **1909**, *39*, 431.
4. Porco, J. A.; Schreiber, S. L. in *Comprehensive Organic Synthesis*, Trost, B., ed., Pergamon Press: New York, 1991, *5*, 151. Mattay, J.; Conrads, R.; Hoffmann, R. in *Methoden der Organischen Chemie* (Houben-Weyl) Helmchen, G.; Hoffmann, R. W.; Mulzer, J.; Schaumann, E., eds., Thieme Verlag: Stuttgart, 1995, E21c, 3133. Bach, T. *Synthesis* **1998**, 683. Carless, H. A. J. in *Synthetic Organic Photochemistry*, Horspool, W. M., ed., Plenum Press: New York, **1984**, 425.
5. Griesbeck, A. G. in *Handbook of Organic Photochemistry and Photobiology*, Eds.: Horspool, W. M.; Song, P.-S., eds., CRC Press: Boca Raton, 1995, 550. Griesbeck, A, G. in *Handbook of Organic Photochemistry and Photobiology*, Horspool, W. M.; Song, P.-S., eds., CRC Press: Boca Raton, 1995, 755.
6. Freilich, S. C.; Peters, K. S. *J. Am. Chem. Soc.* **1981**, *103*, 6255. Freilich, S. C.; Peters, K. S. *J. Am. Chem. Soc.* **1985**, *107*, 3819.
7. Sengupta, D.; Chandra, A. K.; Nguyen, M. T. *J. Org. Chem.* **1997**, *62*, 6404.
8. Adam, W.; Kliem, U.; Mosandl, T.; Peters, E.-M.; Peters, K.; von Schnering, H. G. *J. Org. Chem.* **1988**, *53*, 4986. Wilson, R. M.; Wunderly, S. W.; Walsh, T. F.; Musser, A. K.; Outcalt, R.; Geiser, F.; Gee, S. K.; Brabender, W.; Yerino, L.; Conrad, T.; Thorp, G. A. *J. Am. Chem. Soc.* **1982**, *104*, 4429.
9. Wagner, P. J.; Cheng, K. L. *J. Am. Chem. Soc.* **1994**, *116*, 7945.

10. Neckers, D. C.; Hu, S. *J. Org. Chem.* **1997**, *62*, 755.
11. Gan, C. Y.; Lambert, J. N. *J. Chem. Soc., Perkin Trans 1* **1998**, 2363.
12. Hammond, G. S.; Liu, R. S. *J. Am. Chem. Soc.* **1967**, *89*, 4930. Carlough, K. H.; Srinivasan, R. *J. Am. Chem. Soc.* **1967**, *89*, 4930.
13. Sauers, R. R.; Rousseau, A. D.; Byrne, B. *J. Am. Chem. Soc.* **1975**, *97*, 4947.
14. Dvorak, C. A.; Dufour, C.; Iwasa, S.; Rawal, V. H. *J. Org. Chem.* **1998**, *63*, 5302. Dvorak, C. A.; Rawal, V. H. *Chem. Commun.* **1997**, 2381. Rawal, V. H.; Eschbach, A.; Dufour, C.; Iwasa, S. *Pure Appl. Chem.* **1996**, *68*, 675.
15. Palmer, I. J.; Ragazos, I. N.; Bernardi, F.; Olivucci, M.; Robb, M. A. *J. Am. Chem. Soc.* **1994**, *116*, 2121. Minaev, B. F.; Agren, H. *J. Mol. Struct.: Theochem.* **1998**, *434*, 193.
16. Jones II, G. *Org. Photochem.*, Padwa, A., ed., **1981**, *5*, 1.
17. Koch, H.; Scharf, H.-D.; Runsink, J.; Leismann, H. *Chem. Ber.* **1985**, *118*, 1485. Weuthen, M.; Scharf, H.-D.; Runsink, J.; Vass Ben, R. *Chem. Ber.* **1988**, *121*, 971.
18. Pelzer, R.; Jütten, P.; Scharf, H.-D. *Chem. Ber.* **1989**, *122*, 487.
19. Pelzer, R.; Scharf, H.-D.; Buschmann, H.; Runsink, J. *Chem. Ber.* **1989**, *122*, 1187.
20. Buschmann, H.; Scharf, H.-D.; Hoffmann, N.; Plath, M.; Runsink, J. *J. Am. Chem. Soc.* **1989**, *111*, 5367.
21. Giese, B. *Acc. Chem. Res.* **1984**, *17*, 438.
22. Buschmann, H.; Scharf, H.-D.; Hoffmann, N.; Esser, P. *Angew. Chem. Int. Ed. Engl.* **1991**, *30*, 477.
23. Gypser, A.; Norrby, P.-O. *J. Chem. Soc., Perkin Trans. 2*, **1997**, 939.
24. Griesbeck, A. G.; Mauder, H.; Stadtmüller, S. *Acc. Chem. Res.* **1994**, *27*, 70.
25. Griesbeck, A. G.; Buhr, S.; Fiege, M.; Schmickler, H.; Lex, J. *J. Org. Chem.* **1998**, *63*, 3847.
26. Griesbeck, A. G.; Stadtmüller, S. *J. Am. Chem. Soc.* **1990**, *112*, 1281.
27. Griesbeck, A. G.; Stadtmüller, S. *Chem. Ber.* **1990**, *123*, 357.
28. Griesbeck, A. G.; Stadtmüller, S. *J. Am. Chem. Soc.* **1991**, *113*, 6923.
29. Buhr, S.; Griesbeck, A. G.; Lex, J.; Mattay, J.; Schröer, J. *Tetrahedron Lett.* **1996**, *37*, 1195.
30. Hu, S.; Neckers, D. C. *J. Org. Chem.* **1997**, *62*, 564.
31. Gilbert, A.; Baggott, J. Essentials of Molecular Photochemistry, Blackwell Scientific Publications: London, 1991, 340.
32. Salem, L.; Rowland, C. *Angew. Chem. Int. Ed. Engl.* **1972**, *11*, 92.
33. Carlacci, L.; Doubleday, C., Jr.; Furlani, T. R.; King, H. F.; McIver, J. W. *J. Am. Chem. Soc.* **1987**, *109*, 5323. Doubleday, C., Jr.; Turro, N. J.; Wang, J.-F. *Acc. Chem. Res.* **1989**, *22*, 199.
34. Michl, J. *J. Am. Chem. Soc.* **1996**, *118*, 3568.
35. Kita, F.; Nau, W. M.; Adam, W.; Wirz, J. *J. Am. Chem. Soc.* **1995**, *117*, 8670.
36. Turro, N. J. Modern Molecular Photochemistry, Benjamin/Cummings Publishing Co.; Menlo Park 1978, 434 pp.
37. Johnston, L. J.; Scaiano, J. C., *Chem. Rev.* **1989**, *89*, 521.
38. Klessinger, M. *Pure Appl. Chem.* **1997**, *69*, 773.
39. Reynolds, J. H.; Berson, J. A.; Kumashiro, K. K.; Duchamp, J. C.; Zilm, K. W.;

Scaiano, J. C.; Berinstain, A. B.; Rubello, A.; Vogel, P. *J. Am. Chem. Soc.* **1993**, *115*, 8073.

40. Griesbeck, A. G.; Mauder, H.; Peters, K.; Peters, E.-M.; von Schnering, H. G. *Chem. Ber.* **1991**, *124*, 407.

41. Park, S.-K.; Lee, S.-J.; Baek, K.; Yu, C.-M. *Bull. Korean Chem. Soc.* **1998**, *19*, 35.

42. Kim, T.; Sarker, H.; Bauld, N. L. *J. Chem. Soc. Perkin Trans. 2* **1995**, 577.

43. Abe, M.; Shirodai, Y.; Nojima, M. *J. Chem. Soc., Perkin Trans. 1* **1998**, 3253.

44. Sun, D. L.; Hubig, S. M.; Kochi, J. K. *J. Org. Chem.* **1999**, *64*, 2250. Hubig, S. M.; Sun, D. L.; Kochi, J. K. *J. Chem. Soc., Perkin Trans. 2* **1999**, 781. Bosch, E.; Hubig, S. M.; Kochi, J. K. *J. Am. Chem. Soc.* **1998**, *120*, 386.

45. Weller, A. *Z. Phys. Chem. N. F.* **1982**, *130*, 129. Weller, A. *Z. Phys. Chem. N. F.* **1982**, *133*, 93. Rehm, D.; Weller, A. *Isr. J. Chem.* **1970**, *8*, 259.

46. Eckert, G.; Goez, M. *J. Am. Chem. Soc.* **1994**, *116*, 11999. Goez, M. Eckert, G. *J. Am. Chem. Soc.* **1996**, *118*, 140.

47. Abe, M.; Masayuki, I.; Nojima, M. *J. Chem. Soc., Perkin Trans. 1* **1998**, 3261.

48. Thopate, S. R.; Kulkarni, M. G.; Puranik, V. G. *Angew. Chem. Int. Ed. Engl.* **1998**, *37*, 1110.

49. Bach, T. *Liebigs Ann.* **1997**, 1627.

50. Bach, T. *Tetrahedron Lett.* **1991**, *32*, 7037. Bach, T.; Jödicke, K. *Chem. Ber.* **1993**, *126*, 2457.

51. Bach, T. *Liebigs Ann.* **1995**, 855.

52. Bach, T., Jödicke, K.; Kather, K.; Fröhlich, R. *J. Am. Chem. Soc.* **1997**, *119*, 2437.

53. Bach, T. *Angew. Chem. Int. Ed. Engl.* **1996**, *35*, 884. Bach, T.; Schröder, J.; Brandl, T.; Hecht, J.; Harms, K. *Tetrahedron* **1998**, *54*, 4507.

54. Bach, T.; Brummerhop, H. *Angew. Chem. Int. Ed. Engl.* **1998**, *37*, 3400.

55. Bach, T.; Schröder, J. *J. Org. Chem.* **1999**, *64*, 1265.

56. Döpp, D.; Fischer, M.-A. *Recl. Trav. Chim. Pays-Bas* **1995**, *114*, 498.

57. Ogata, M.; Watanabe, H.; Kano, H. *Tetrahedron Lett.* **1967**, 533. Toki, S.; Sakurai, H. *Tetrahedron Lett.* **1967**, 4119. Zamojski, A.; Kozluk, T. *J. Org. Chem.* **1977**, *42*, 1089. Evanega, G. R.; Whipple, E. B. *Tetrahedron Lett.* **1967**, 2163. Heatchcock, C. H.; Pirrung, M. C.; Montgomery, S. H.; Lampe, J. *Tetrahedron* **1981**, *37*, 4087. Schreiber, S. L.; Satake, K. *J. Am. Chem. Soc.* **1983**, *105*, 6723. Schreiber, S. L., Satake, K. *J. Am. Chem. Soc.* **1984**, *106*, 4186. Schreiber, S. L.; Hoveyda, A. H.; Wu, H.-J., *J. Am. Chem. Soc.* **1984**, *106*, 1148. Schreiber, S. L.; Hoveyda, A. H. *J. Am. Chem. Soc.* **1984**, *106*, 7200.

58. Carless, H. A. J.; Halfhide, A. F. E. *J. Chem. Soc. Perkin Trans. 1* **1992**, 1081. Schreiber, S. L.; Desmaele, D.; Porco, Jr., J. A. *Tetrahedron Lett.* **1988**, *29*, 6689. Schreiber, S. L.; Porco, Jr., J. A. *J. Org. Chem.* **1989**, *54*, 4721. Sekretár, S.; Rudá, J.; Stibrányi, L. *Coll. Czech. Chem. Commun.* **1984**, *49*, 71.

59. Rivas, C.; Pacheco, D.; Vargas, F.; Ascanio, J. *J. Heterocyclic Chem.* **1981**, *18*, 1065.

60. Rivas, C.; Bolivar, R. A. *J. Heterocyclic Chem.* **1976**, *13*, 1037. Jones II, G.; Gilow, H. M.; Low, J. *J. Org. Chem.* **1979**, *44*, 2949.

61. Matsuura, T.; Banba, A.; Ogura, K., *Tetrahedron* **1971**, *27*, 1211.

62. Nakano, T.; Rodriquez, W.; de Roche, S. Z.; Larrauri, J. M.; Rivas, C.; Pérez, P. *J. Heterocyclic Chem.* **1980**, *17*, 1777.

63. Hoye, T. R.; Richardson, W. S. *J. Org. Chem.* **1989**, *54*, 688.

64. Shima, K.; Kubota, T.; Sakurai, H. *Bull. Chem. Soc. Jpn.* **1976**, *44*, 2567.

65. Griesbeck, A. G.; Fiege, M. unpublished results.

66. Cantrell, T. S., Allen, A. C., Ziffer, H. *J. Org. Chem.* **1989**, *54*, 140.

67. Zagar, C., Scharf, H.-D. *Chem. Ber.* **1991**, *124*, 967.

68. Ciamician, G.; Silber, P. *Chem. Ber.* **1908**, *41*, 1928.

69. Mattay, J.; Conrads, R.; Hoffmann, R. in *Methoden der Organischen Chemie* (Houben-Weyl) Helmchen, G.; Hoffmann, R. W.; Mulzer, J.; Schaumann, E., eds., Thieme Verlag: Stuttgart, 1995, E21c, 3133. Baldwin, S. W. in *Org. Photochem.*, Padwa, A. ed., Marcel Dekker: New York, 1981, *5*, 123. Demuth, M., Gamal, M. *Synthesis* **1989**, 145. Schuster, D. I. in *The Chemistry of Enones*, Vol. 2, Patai, S., Rappoport, Z., eds., **1989**, 623. Crimmins, M. T. *Chem. Rev.* **1988**, *88*, 1453. De Keukeleire, D.; He, S.-L. *Chem. Rev.* **1993**, *93*, 359.

70. Caldwell, R. A.; Hrncir, D. C.; Munoz, Jr., T.; Unett, D. J. *J. Am. Chem. Soc.* **1996**, *118*, 8741.

71. Johnston, L. J.; Scaiano, J. C., *Chem. Rev.* **1989**, *89*, 521. Rudolph, A.; Weedon, A. C. *Can. J. Chem.* **1990**, *68*, 1590. Kaprinidis, N. A.; Lem. G.; Courtney, S. H.; Schuster, D. I. *J. Am. Chem. Soc.* **1993**, *115*, 3324. Schuster, D. I.; Lem, G.; Kaprinidis, N. A. *Chem. Rev.* **1993**, *93*, 3.

72. Maradyn, D. J.; Weedon, A. C. *J. Am. Chem. Soc.* **1995**, *117*, 5359.

73. Agosta, W. C.; George, C. F.; Matlin, A. R.; Wolff, S. *J. Am. Chem. Soc.* **1986**, *108*, 3385.

74. Andrew, D.; Weedon, A. C. *J. Am. Chem. Soc.* **1995**, *117*, 5647.

75. Corey, E. J.; Bass, J. D.; LeMahieu, R.; Mitra, R. B. *J. Am. Chem. Soc.* **1964**, *84*, 5570. Loutfy, R. O.; de Mayo, P. *J. Am. Chem. Soc.* **1977**, *99*, 3559.

76. Shaik, S. S.; Epiotis, N. D. *J. Am. Chem. Soc.* **1978**, *100*, 18. Shaik, S. S. *J. Am. Chem. Soc.* **1979**, *101*, 2736. Shaik, S. S. *J. Am. Chem. Soc.* **1979**, *101*, 3184.

77. Suishu, T.; Shimo, T.; Somekawa, K. *Tetrahedron* **1997**, *53*, 3545.

78. Hill, E. A.; Theissen, R. J.; Cannon, C. E.; Miller, R.; Guthrie, R. B.; Chen, A. T. *J. Org. Chem.* **1976**, *41*, 1191. Termont, D.; De Keukeleire, D.; Vandewalle, M. *J. Chem. Soc., Perkin Trans. 1* **1977**, 2349.

79. Griesbeck, A. G.; Stadtmüller, S.; Bringmann, G.; Busse, H.; Buddrus, J. *Chem. Ber.* **1992**, *125*, 933.

80. Hastings, D. J.; Weedon, A. C. *J. Am. Chem. Soc.* **1991**, *113*, 8525. Andrew, D.; Hastings, D. J.; Oldroyd, D. L.; Rudolph, A.; Weedon, A. C.; Wong, D. F.; Zhang, B. *Pure Appl. Chem.* **1992**, *64*, 1327.

81. Becker, D.; Galili, N.; Haddad, N. *Tetrahedron Lett.* **1996**, *37*, 8941. Haddad, N.; Galili, N. *Tetrahedron Lett.* **1997**, *38*, 6083.

82. Becker, D.; Cohen-Arazi, Y. *J. Am. Chem. Soc.* **1996**, *118*, 8278.

83. Pete, J.-P. *Advances in Photochemistry*, Neckers, D.C.; Volman, D. H.; von Bünau, G., eds. Wiley: New York, 1996, *21*, 135.

84. Pirrung, M. C.; Thomson, S. A. *Tetrahedron Lett.* **1986**, *27*, 2703.

85. Piva-Le Blanc, S.; Pete, J.-P.; Piva, O. *Chem. Commun.* **1998**, 235.

86. Seebach, D.; Sting, A. R.; Hoffmann, M. *Angew. Chem. Int. Ed. Engl.* **1996**, *35*, 2708.
87. Demuth, M.; Palomer, A.; Sluma, H. D.; Dey, A. K.; Krüger, C.; Tsay, Y. H. *Agnew. Chem. Int. Ed. Engl.* **1986**, *25*, 1117.
88. Seebach, D.; Zimmermann, J.; Gysel, U.; Ziegler, R.; Ha, T. K. *J. Am. Chem. Soc.* **1988**, *110*, 4763.
89. Organ, M. G.; Froese, R. D. J.; Goddard, J. D.; Taylor, N. J.; Lange, G. L. *J. Am. Chem. Soc.* **1994**, *116*, 3312.
90. Haddad, N.; Abramovich, Z.; Ruhman, I. *Tetrahedron Lett.* **1996**, *37*, 3521.
91. Haddad, N.; Abramovich, Z. *J. Org. Chem.* **1995**, *60*, 6883.
92. Haddad, N.; Rukhman, I.; Abramovich, Z. *J. Org. Chem.* **1997**, *62*, 7629.
93. Winkler, J. D.; Henegar, K.; Williard, P. *J. Am. Chem. Soc.* **1987**, *109*, 2850. Winkler, J. D.; Henegar, K.; Hong, B.; Williard, P. *J. Am. Chem. Soc.* **1994**, *116*, 4183. Winkler, J. D.; Hong, B.-C.; Bahador, A.; Kazanietz, M. G.; Blumberg, P. M. *J. Org. Chem.* **1995**, *60*, 1381.
94. Winkler, J. D.; Doherty, E. M. *Tetrahedron Lett.* **1998**, *39*, 2253.
95. Muller, C. L.; Bever, J. R.; Dordel, M. S.; Kitabwalla, M. M.; Reineke, T. M.; Sausker, J. B.; Seehafer, T. R.; Li, Y.; Jasinski, J. P. *Tetrahedron Lett.* **1997**, *38*, 8663.
96. Dritz, J. H.; Carreira, E. M. *Tetrahedron Lett.* **1997**, *38*, 5579.
97. Faure, S.; Piva-Le Blanc, S.; Piva, O.; Pete, J.-P. *Tetrahedron Lett.* **1997**, *38*, 1045.
98. Winkler, J. D.; Axten, J.; Hammach, A. H.; Kwak, Y.-S.; Lengweiler, U.; Lucero, M. J.; Houk, K. N. *Tetrahedron* **1998**, *54*, 7045.
99. de Mayo, P. A. *Acc. Chem. Res.* **1971**, *4*, 41. Petasis, N. A.; Patane, M. A. *Tetrahedron* **1992**, *48*, 5757.
100. Booker-Milburn, K. I.; Cowell, J. K.; Harris, L. J. *Tetrahedron* **1997**, *53*, 12319.
101. Booker-Milburn, K. I.; Cowell, J. K.; Jiménez, F. D.; Sharpe, A.; White, A. J. *Tetrahedron* **1999**, *55*, 5875.
102. de March, P.; Figuerdo, M.; Font, J.; Raya, J. *Tetrahedron Lett.* **1999**, *40*, 2205.
103. Alibés, R.; Bourlande, J. L.; Font, J.; Gregori, A.; Parella, T. *Tetrahedron* **1996**, *52*, 1267.
104. Alibés, R.; Bourlande, J. L.; Font, J.; Parella, T. *Tetrahedron* **1996**, *52*, 1279.
105. Gregori, A.; Alibés, R.; Bourlande, J. L.; Font, J. *Tetrahedron Lett.* **1998**, *39*, 6961.
106. Gregori, A.; Alibés, R.; Bourlande, J. L.; Font, J. *Tetrahedron Lett.* **1998**, *39*, 6963.
107. Tsujishima, H.; Nakatani, K.; Shimamoto, K.; Shigeri, Y.; Yumoto, N.; Ohfune, Y. *Tetrahedron Lett.* **1998**, *39*, 1193.
108. Meyers, A. I.; Fleming, S. A. *J. Am. Chem. Soc.* **1986**, *108*, 306.
109. Crimmins, M. T.; King, B. W.; Watson, P. S.; Guise, L. E. *Tetrahedron* **1997**, *53*, 8963.
110. Crimmins, M. T.; Wang, Z.; McKerlie, L. A. *J. Am. Chem. Soc.* **1998**, *120*, 1747.
111. Shepard, M. S.; Carreira, E. M. *J. Am. Chem. Soc.* **1997**, *119*, 2597.
112. Evers, J. T.; Mackor, A. *Tetrahedron Lett.* **1978**, 821.
113. Raychaudhuri, S. R.; Ghosh, S.; Salomon, R. G. *J. Am. Chem. Soc.* **1982**, *104*, 6841.

114. Rosini, G.; Geier, M.; Marotta, E.; Petrini, M.; Ballini, R. *Tetrahedron* **1986**, *42*, 6027.
115. Patra, D.; Ghosh, S. *J. Org. Chem.* **1995**, *60*, 2526. Ghosh, S.; Patra, D.; Samajdar, S. *Tetrahedron Lett.* **1996**, *37*, 2073.
116. Haque, A.; Ghatak, A.; Ghosh, S.; Ghoshal, N. *J. Org. Chem.* **1997**, *62*, 5211.
117. Samajdar, S.; Patra, D.; Ghosh, S. *Tetrahedron* **1998**, *54*, 1789.
118. Bradford, C. L.; Fleming, S. A.; Ward, S. C. *Tetrahedron Lett.* **1995**, *36*, 4189.
119. Bach, T.; Pelkmann, C.; Harms, K. *Tetrahedron Lett.* **1999**, *40*, 2103.
120. Sieburth, S. McN.; Siegel, B. *Chem. Commun.* **1996**, 2249.
121. Sieburth, S. McN.; Al-Tel, T. H., Rucando, D. *Tetrahedron Lett.* **1997**, *38*, 8433.
122. Sieburth, S. McN.; Lin, C.-H. *Tetrahedron Lett.* **1996**, *37*, 1141. Sieburth, S. McN.; Rucando, D.; Lin, C.-H. *J. Org. Chem.* **1999**, *64*, 954.
123. Sieburth, S. McN.; MeGee Jr., K. F.; Al-Tel, T. H. *J. Am. Chem. Soc.* **1998**, *120*, 587.
124. Yang, N. C.; Yang, D.-H. *J. Am. Chem. Soc.* **1958**, *80*, 2913.
125. Wagner, P. J.; Park, B.-S. *Org. Photochem.*, Padwa, A., ed., **1991**, *11*, 227. Wagner, P. J. in *Handbook of Photochemistry and Photobiology*, Horspool, W. M., Song, P.-S., eds., CRC Press: Boca Raton, 1995, 449. Wagner, P. J. in *Molecular Rearrangements in Ground and Excited States*, de Mayo, P., ed., Wiley-Interscience: New York, 1980, Chapter 20.
126. Dorigo, A. E.; McCarrick, M. A.; Loncharich, R. J.; Houk, K. N. *J. Am. Chem. Soc.* **1990**, *112*, 7508. Sauers, R. R.; Edberg, L. A. *J. Org. Chem.* **1994**, *59*, 7061.
127. Scheffer, J. R.; Pokkuluri, in *Photochemistry in Organized and Constrained Media*, Ramamurthy, V., ed., VCH: New York, 1991, 185. Scheffer, J. R.; Garcia-Garibay, M.; Nalmasu, O. *Organic Photochemistry*, Padwa, A., ed., Marcel Dekker: New York, 1987, 8. Scheffer, J. R. in *Organic Solid State Chemistry*, G. R. Desiraju, ed., Elsevier: Amsterdam, 1987, 1.
128. Zimmerman, H. E. *Science* **1966**, *153*, 837.
129. Ihmels, H.; Scheffer, J. R. *Tetrahedron* **1999**, *55*, 885.
130. Griesbeck, A. G.; Henz, A.; Kramer, W.; Wamser, P.; Peters, K.; Peters, E.-M. *Tetrahedron Lett.* **1998**, *39*, 1549–1552.
131. Wagner, P. J. in *Handbook of Photochemistry and Photobiology*, Horspool, W. M., Song, P.-S., eds., CRC Press: Boca Raton, 1995, 459.
132. Griesbeck, A. G. *Chimia* **1998**, *63*, 3847. Griesbeck, A. G. *EPA Newletters* **1998**, *27*, 70. Griesbeck, A. G. *Liebigs Ann. Chem.* **1996**, 1951.
133. Griesbeck, A. G.; Heckroth, H.; Lex, J. *Chem. Commun.* **1999**, 1109.
134. Yoshida, K.; Fueno, T. *Bull. Soc. Chim. Jpn.* **1987**, *55*, 229.
135. Leroy, C.; Peeters, D.; Wilante, C. *Theochemistry* **1982**, *5*, 217.
136. Scaiano, J. C. *Tetrahedron* **1982**, *38*, 819. Wagner, P. J.; Zand, A.; Park, B.-S. *J. Am. Chem. Soc.* **1996**, *118*, 12856.
137. Wessig, P.; Schwarz, J. *Helv. Chim. Acta* **1998**, *81*, 1803.
138. Fu, T. Y.; Scheffer, J. R.; Trotter, J. *Tetrahedron Lett.* **1996**, *37*, 2125.
139. Olovsson, G.; Scheffer, J. R.; Trotter, J.; Wu, C.-H. *Tetrahedron Lett.* **1997**, *38*, 6549.
140. Gamlin, J. N.; Jones, R.; Leibovitch, M.; Patrick, B.; Scheffer, J. R.; Trotter, J. *Acc. Chem. Res.* **1996**, *29*, 203.

141. Cheung, E.; Netherton, M. R.; Scheffer, J. R.; Trotter, J. *J. Am. Chem. Soc.* **1999**, *121*, 2919.

142. Ito, Y.; Kano, G.; Nakamura, N. *J. Org. Chem.* **1998**, *63*, 5643.

143. Irngartinger, H.; Fettel, P. W.; Siemund, V. *Eur. J. Org. Chem.* **1998**, 2079.

144. Leibovitch, M.; Olovsson, G.; Sundarababu, G.; Ramamurthy, V.; Scheffer, J. R.; Trotter, T. *J. Am. Chem. Soc.* **1996**, *118*, 1219; Leibovitch, M.; Olovsson, G.; Scheffer, J. R.; Trotter, T. *J. Am. Chem. Soc.* **1997**, *119*, 1462.

145. Henning, H.-G. in *Handbook of Photochemistry and Photobiology*, Horspool, W. M., Song, P.-S., eds., CRC Press: Boca Raton, 1995, 484.

146. Weigel, W.; Wagner, P. J. *J. Am. Chem. Soc.* **1996**, *118*, 12858. Weigel, W.; Schiller, S.; Henning, H.-G. *Tetrahedron* **1997**, *53*, 7855.

147. Wagner. P. J.; Zand, A.; Park, B.-S. *J. Am. Chem. Soc.* **1996**, *118*, 12856. Zand, A.; Park, B.-S.; Wagner, P. J. *J. Org. Chem.* **1997**, *62*, 2326.

148. Wessig, P.; Wettstein, P.; Giese, B.; Neuburger, M.; Zehnder, M. *Helv. Chim. Acta* **1994**, *77*, 829.

149. Steiner, A.; Wessig, P. *J. Inf. Recording* **1996**, *23*, 17.

150. Wyss, C.; Batra, R.; Lehmann, C.; Sauer, S.; Giese, B. *Angew. Chem. Int. Ed. Engl.* **1996**, *35*, 2529.

151. Sauer, S.; Staehelin, C.; Wyss, C.; Giese, B. *Chimia* **1997**, *51*, 23.

152. Sauer, S.; Schumacher, A.; Barbosa, F.; Giese, B. *Tetrahedron Lett.* **1998**, *39*, 3685.

153. Lindemann, U.; Wulff-Molder, D.; Wessig, P. *Tetrahedron: Asymmetry* **1998**, *9*, 4459.

154. Griesbeck, A. G.; Heckroth, H.; Schmickler, H. *Tetrahedron Lett.* **1999**, *40*, 3137.

155. Kraus, G. A.; Wu, Y. *J. Am. Chem. Soc.* **1992**, *114*, 8705.

156. Pandey, G., *Top. Curr. Chem.* **1993**, *168*, 175–221. Mariano, P. S., Stavinoha, J. L. in *Synthetic Organic Photochemistry* (Horspool, W. A., ed.), 1984, Plenum Press, New York, London. Kavarnos, G. J. *Fundamentals of Photoinduced Electron Transfer*, VCH: Weinheim, 1993; Chapter 3. Mattay, J., Vondenhof, M. *Topics in Current Chemistry*, Mattay, J., Ed.; Springer: Berlin, 1991, *159*, 219.

157. Maslak, P.; Narvacz, J. N. *Angew. Chem. Int. Ed. Engl.* **1990**, *29*, 302.

158. Kanaoka, Y. *Acc. Chem. Res.* **1978**, *11*, 407. Griesbeck, A. G.; Mauder, H. in *CRC Handbook of Organic Photochemistry and Photobiology*, Horspool, W. M.; Song, P. S., eds, CRC Press: New York, 1995, 513.

159. Griesbeck, A. G.; Henz, A.; Hirt, J.; Ptatschek, V.; Engel, T.; Löffler, D.; Schneider, F. W. *Tetrahedron* **1994**, *50*, 701–714.

160. Maruyama, K.; Kubo, Y. *J. Am. Chem. Soc.* **1978**, *100*, 7772. Maruyama, K.; Kubo, Y. *Chem. Lett.* **1978**, 851. Maruyama, K.; Kubo, Y.; Machida, M.; Oda, K.; Kanaoka, Y.; Furuyama, K. *J. Org. Chem.* **1978**, *43*, 2303.

161. Machida, M.; Oda, K.; Kanaoka, Y. *Tetrahedron* **1985**, *41*, 4995.

162. Maruyama, K.; Kubo, Y. *J. Am. Chem. Soc.* **1978**, *100*, 7772.

163. Machida, M.; Takechi, H.; Kanaoka, Y. *Heterocycles* **1980**, *14*, 1255. Machida, M.; Takechi, H.; Kanaoka, Y. *Heterocycles* **1977**, *7*, 273. Coyle, J. D.; Newport, G. L. *Tetrahedron Lett.* **1977**, 899. Coyle, J. D.; Newport, G. L. *J. Chem. Soc. Perkin Trans. 1* **1980**, 93.

164. Machida, M.; Minamikawa, S.; Wilson, P. *J. Org. Chem.* **1979**, *14*, 1186. Machida, M.; Takechi, H.; Kanaoka, Y. *Heterocycles* **1980**, *14*, 1255.

165. Hasegawa, T.; Miyata, K.; Ogawa, T.; Yoshihara, N.; Yoshioka, M. *J. Chem. Soc. Chem. Commun.* **1985**, 363.

166. Hasegawa, T.; Ogawa, T.; Miyata, K.; Karakizawa, A.; Komiyama, M.; Nishizawa; Yoshioka, M. *J. Chem. Soc. Perkin Trans. 1* **1990**, 901.

167. Hu, S.; Neckers, D. C. *Tetrahedron* **1997**, *53*, 2751.

168. Hasegawa, T.; Yamazaki, Y. *Tetrahedron* **1998**, *54*, 12223.

169. Kraus, G. A.; Chen, L. *Tetrahedron Lett.* **1991**, *32*, 7151.

170. Hu, S.; Neckers, D. C. *Tetrahedron* **1997**, *53*, 12771.

171. Hatanaka, Y.; Sato, Y.; Nakai, H.; Wada, M.; Mizoguchi, T.; Kanaoka, Y. *Liebigs Ann. Chem.* **1992**, 1113. Takechi, H.; Tateuchi, S.; Machida, M.; Nishibata, Y.; Aoe, K.; Sato, Y.; Kanaoka, Y. *Chem. Pharm. Bull.* **1986**, *34*, 3142. Sato, Y.; Nakai, H.; Wada, M.; Mizuguchi, T.; Hatanaka, Y.; Migita, Y.; Kanaoka, H. *Liebigs Ann. Chem.* **1985**, 1099. Sato, Y.; Nakai, H.; Wada, M.; Mizuguchi, T.; Hatanaka, Y.; Kanaoka, H. *Chem. Pharm. Bull.* **1992**, *40*, 3174. Sato, Y.; Nakai, H.; Mizoguchi, T.; Hatanaka, Y.; Kanaoka, Y. *J. Am. Chem. Soc.* **1976**, *98*, 2349.

172. Wada, M.; Nakai, H.; Aoe, K.; Kotera, K.; Sato, Y.; Hatanaka, Y.; Kanaoka, Y. *Tetrahedron* **1983**, *39*, 1273.

173. Griesbeck, A. G.; Henz, A.; Peters, K.; Peters, E. -M.; von Schnering, H. G. *Angew. Chem. Int. Ed. Engl.* **1995**, *34*, 474.

174. Griesbeck, A. G.; Henz, A.; Kramer, W.; Lex, J.; Nerowski, F.; Oelgemöller, M.; Peters, K.; Peters, E.-M. *Helvetica Chim. Acta*, **1997**, *80*, 912.

175. Griesbeck, A. G.; Nerowski, F.; Lex, J. *J. Org. Chem.* **1999**, *64*, 5213.

176. Griesbeck, A. G.; Kramer, W.; Oelgemöller, M. *Synlett* **1999**, 1169.

177. Seiler, M.; Schumacher, A.; Lindemann, U.; Barbosa, F.; Giese, B. *Synlett* **1999**, 1588.

178. Hu. S.; Neckers, D. C. *J. Chem. Soc., Perkin Trans. 2*, **1999**, 1771.

179. Griesbeck, A. G.; Heckroth, H. *Res. Chem. Intermediat* **1999**, *25*, 599.

180. Bach, T.; Bergmann, H,; Harms, K. *J. Am. Chem. Soc.* **1999**, *121*, 10650.

181. Adam, W.; Peters, K.; Peters, E.-M.; Stegmann, V. R. *J. Am. Chem. Soc.* **2000**, *122*, in print.

182. Wessig, P. *Synlett* **1999**, 1465.

183. Lindemann, U.; Neuburger, M.; Neuburger-Zehnder, M.; Wulff-Molder, Wessig, P. *J. Chem. Soc. Perkin Trans. 2*, **1999**, 2029.

184. Bertrand, S.; Glapski, C.; Hoffmann, N.; Pete, J.-P. *Tetrahedron Lett.* **1999**, *40*, 3169.

185. Bertrand, S.; Hoffmann, N.; Pete, J.-P. *Tetrahedron Lett.* **1999**, *40*, 3173.

186. Janz, K. M.; Scheffer, J. R. *Tetrahedron Lett.* **1999**, *40*, 8725.

187. Cheung, E.; Kang, T.; Raymond, J. R.; Scheffer, J. R.; Trotter, J. *Tetrahedron Lett.* **1999**, *40*, 8729.

188. Cheung, E.; Rademacher, K.; Scheffer, J. R.; Trotter, J. *Tetrahedron Lett.* **1999**, *40*, 8733.

189. Cheung, E.; Netherton, M. R.; Scheffer, J. R.; Trotter, J. *Tetrahedron Lett.* **1999**, *40*, 8737.

190. Giese, B.; Wettstein, P.; Stähelin, C.; Barbosa, F.; Neuburger, M.; Zehnder, M.; Wessig, P. *Angew. Chem. Int. Ed. Engl.* **1999**, *38*, 2586.

3

Photocycloadditions with Captodative Alkenes

Dietrich Döpp
Gerhard-Mercator-Universität Duisburg,
Duisburg, Germany

I. INTRODUCTION

During the last two decades the concept of synergistic stabilization of carbon-centered radicals by both an electron-withdrawing (captive, c) and an electron-donating (dative, d) substituent has been advanced; this extra stabilization is called the *captodative effect* [1]. Alternatively, the term *merostabilization* had been suggested [2] for the same phenomenon. Previous related investigations had been made on nitrogen-centered radicals [3], but the term *push–pull stabilization*, coined earlier in this connection [4], was not generally accepted. The magnitude of the captodative stabilization has been debated [5], including the experimental search [6] for solvent effects postulated [2] in this stabilization. Currently, synergetic radical stabilization energies of 17–21 kJ/mol [5b], 41 kJ/mol, [7], and 14 and 51.5 kJ/mol [8] are being envisaged. For theoretical treatments, see Refs. 8–10.

An alkene **1**, bearing both an electron donor d and an electron acceptor c at one terminus of the double bond, may be termed a captodative alkene. Addition of a radical R to its double bond will occur strictly regioselectively at the unsub-

---

This chapter is dedicated to Professor H.-G. Viehe on the occasion of his 70th birthday.

101

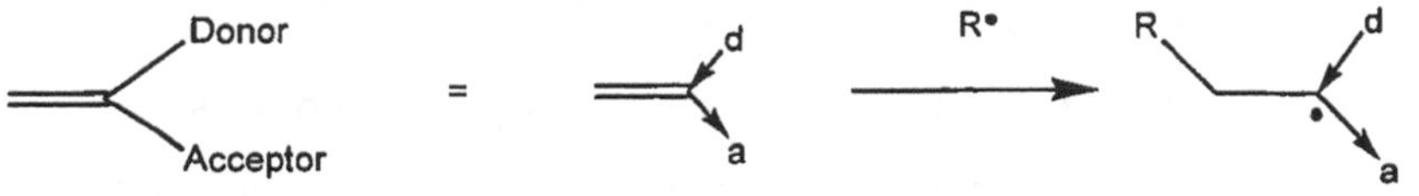

Scheme 1

stituted terminus and generate a captodatively (c,d)–stabilized adduct radical **2** [1] (Scheme 1).

Radicals **2** have dimerization and combination with other radicals as options for further reaction, disproportionation, and initiation of polymerization are not observed [1]. The unique features of c,d-alkenes have been investigated by photoelectron spectroscopy [11] and in many cycloadditions [1b,1d,12–14].

Since electronically excited states, especially n,π*-excited carbonyl groups, exhibit radical character, it was tempting to investigate whether the unique features of c,d-alkenes would influence the efficiency and/or the course of light-induced reactions involving the addition of an excited species to alkenes, as the Paternò–Büchi reaction or cyclobutane forming [$\pi^2 + \pi^2$] cycloadditions.

This account will summarize our findings in this field featuring a number of remarkably selective photocycloadditions.

Most reactions reported from our group will be cycloadditions involving α-cyanoenamines **3**. Any cycloadducts C derived from addition of the C=C double bond of these alkenes to other π systems are cyclic α-aminonitriles and thus hydrolyzable [15] to cyclic ketones K, the latter resembling adducts of ketene to said π systems. Thus, c,d-alkenes (and α-cyanoenamines in particular) are to be regarded as ketene equivalents in synthesis [14] (Scheme 2).

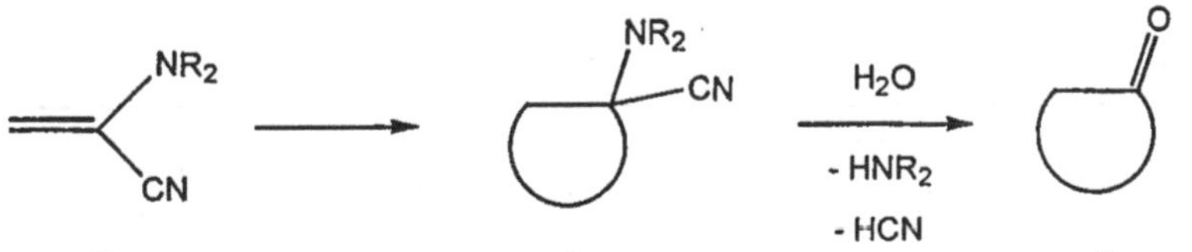

Scheme 2

II. [2 + 2] CYCLOADDITIONS TO VARIOUS π SYSTEMS

A. Light-Induced Cyclodimerizations of α-Cyanoenamines

Among typical captodative alkenes, 2-(alkylthio)propenenitriles and propenoic esters tend to form head-to-head [2 + 2] dimers spontaneously [16], whereas such reactivity is, with but one reported exception [17], generally not observed for 2-aminopropenenitriles 3 [1a]. The parent 2-aminopropenenitrile (3a) had been successfully dimerized, however, upon benzophenone sensitization to a mixture of *cis*- and *trans*-1,2-diaminocyclobutane-1,2-dicarbonitriles, i.e. of the head-to-head cyclodimers, in appreciable yield [18]. For 2-morpholinopropenenitrile (3b), benzophenone was inefficient in sensitizing [2 + 2] dimerization [19]. A systematic study revealed that the typical π,π* triplet sensitizers 9-fluorenone, 2-acetonaphthone, thioxanthone, or triphenylene in acetonitrile or benzene solution converted 3b to 4b in moderate yields [20] (Scheme 3).

Benzophenone, however, in its n,π*-T₁ state gave, besides small amounts of oxetane 5, a bicyclic dimer to which structure 9 was assigned on the basis of its ¹H and ¹³C NMR data and on the basis of such data for the hydrolysis product 10 [20]. Compound 9 is probably formed via initial hydrogen atom abstraction from 3b affording the α-amino radical 6, which is added to another molecule of 3b to form 7. The latter in turn undergoes intramolecular cyclization and hydro-

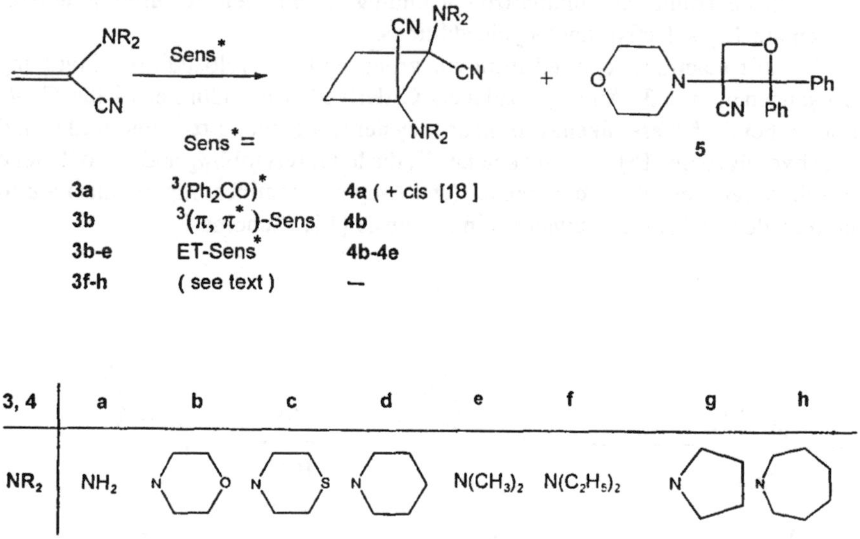

3, 4	a	b	c	d	e	f		g	h
NR₂	NH₂	(morpholine)	(thiomorpholine)	(piperidine)	N(CH₃)₂	N(C₂H₅)₂		(pyrrolidine)	(azepane)

Scheme 3

gen atom capture with formation of **9**. An analogous sequence is observed when **3d** is exposed to excited benzophenone in benzene solution [20] (Scheme 4).

For comparison, a look on the following early findings [21] seems justified. α-Acetoxypropenenitrile, albeit bearing a weaker donor than the α-cyanoenamines and vinyl sulfides discussed here, does have properties resembling a captodative alkene. Upon sensitization by acetophenone, it forms a diastereomeric mixture of the head-to-head [2 + 2] dimerization products [21]. Benzophenone sensitization in presence of butadiene gives rise to a 1:1 adduct (again as a mixture of diastereomers), namely, 1-acetoxy-2-vinylcyclobutane-1-carbonitrile, along with the formal Diels–Alder product and butadiene dimers [21]. The only property of 2-acetoxypropenenitrile, which is untypical for a c,d-alkene, is its tendency to keep up polymerization in forming a copolymer with butadiene in the benzophenone sensitized reaction [21]. Acetophenone and benzophenone sensitization seems to be more effective here than with α-cyanoenamines, in accord with a reduced electron richness of 2-acetoxypropenenitrile.

Electron-rich alkenes, on the other hand, may well be dimerized by processes involving initial electron transfer [22]. Efficient head-to-head photodimerization of 2-aminopropenenitriles **3b–d,g** is achieved using the typical electron transfer sensitizers 9-dicyanoanthracene (DCA) and 2,6,9,10-tetracyanoanthracene (TCA) in benzene as solvent [20].

All alkenes **3** used quench DCA* fluorescence at similar rates (e.g., **3b**: $k_q = 7.6 \times 10^9 \, M^{-1} \, s^{-1}$, **3e**: $k_q = 11.0 \times 10^9 \, M^{-1} \, s^{-1}$), a separate exciplex emis-

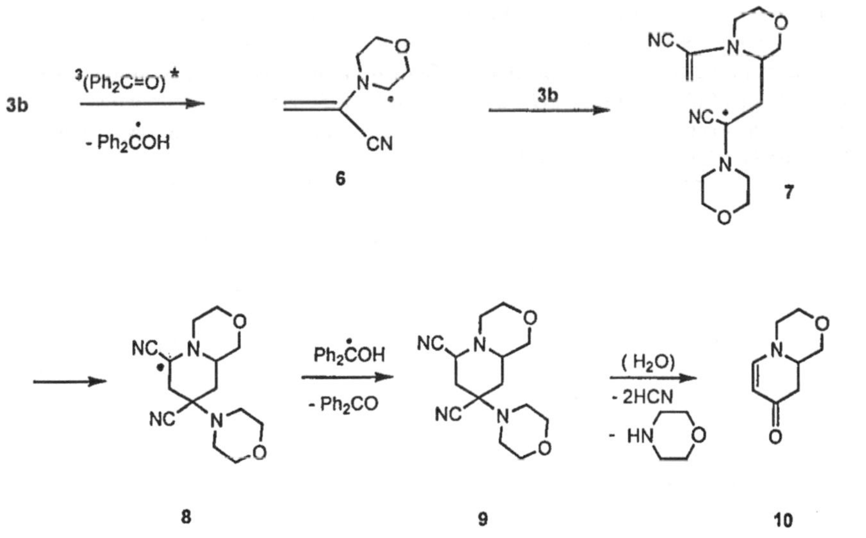

Scheme 4

sion, however, was not detected. Compounds **3f–h** could not be dimerized by this method and were recovered unchanged. At least for the dimerization of **3b**, 1,4-dicyanonaphthalene (DCN), 9-cyanoanthracene (CA), and 1,4-dicyanobenzene (DCB) were also effective (Table 1), whereas methyl-4-cyanobenzoate (MCB) was inefficient [20].

Table 1 also lists the reduction potentials [23] and singlet excitation energies of the aforementioned sensitizers, the estimate of ΔG for full electron transfer in benzene and acetonitrile solution according to the Rehm–Weller equation [24], the conversions of **3b** achieved and the preparative yields of **4b** obtained [20]. A clear trend such that the efficiency of dimerization follows ΔG_{ET} cannot be delineated from these data.

For oxidation potentials (in acetonitrile vs. SCE) of **3b-h**, see Table 2. The anodic oxidations are irreversible even at (CV)-scan rates of 1000 mVs^{-1}. The oxidation potentials of the successfully dimerized alkenes **3b–e** are $\geq$ 1.21 V vs. SCE, whereas **3f–h** (not dimerized by this method) have oxidation potentials below that value. As a consequence, calculated [23] ΔG values for full electron transfer to DCA* are more negative for **3f–h** than for **3b–e**.

While **3f–h** could not be dimerized, labile heterodimers by DCA* sensitization of common benzene solutions of alkenes **3g** (or **3h**) with **3b** could be obtained but not be isolated; thus, their ready decay (in the dark at ambient temperature) was monitored by ^{1}H NMR [20]. The same technique was employed to

Table 1 Variation of Sensitizer in the [2 + 2] Cyclodimerization of 2-Morpholinopropenenitrile (**3b**) in Benzene Solution (Half-Wave Reduction Potentials vs. SCE and Singlet Excitation Energies of Sensitizers in Acetonitrile Solution Taken from Ref. 23). Calculated [24] ΔG Values for Solvent-Separated Ion Formation in Benzene and Acetonitrile. Conversions of **3b** in Benzene, and Yields of **4b** Achieved (Based on Not Recovered Starting Material)

Acceptor	$E_{1/2}^{Red}$ (V)	$E_{0,0}^{S}$ (kJ mol^{-1})	ΔG (kJ mol^{-1}) Benzene	Acetonitrile	% Conversion of **3b**	% Yield of **4b**
TCA	−0.45[a]	280	−33	−109	47	91
DCA	−0.98[b]	270	+28	−49	81	88
DCN	−1.36[c]	333	+2	−75	62	93
CA	−1.58[b]	285	+70	−6	16	55
DCB	−1.69[c]	405	−39	−115	10	88
MCB	−1.79[c]	395	−20	−95	<3	—

[a] Ref. 23a.
[b] Ref. 23b.
[c] Taken from Ref. 23c and adjusted to value against SCE.

Table 2 Half-Wave Oxidation Potentials of Alkenes **3b–h** and Calculated [22c,24] ΔG Values for Electron Transfer in Both Benzene and Acetonitrile Using DCA ($E_{1/2}$, Red = 0.98 V [23b]) as Acceptor

Alkene	**3c**	**3b**	**3d**	**3e**	**3f**	**3g**	**3h**
$E_{1/2}$, ox (V)[a]	1.38	1.32	1.25	1.21	1.16	1.09	1.09
ΔG (kJ mol^{-1}) (C$_6$H$_6$)	+34	+28	+21	+17	+12	+6	+6
(CH$_3$CN)	−42.5	−48.5	−55.5	−59.5	−64.5	−70.5	−70.5

[a] Ref. 20. Determined vs. SCE in acetonitrile/Bu$_4$NBF$_4$, 25°C. Ar, atmosphere.

determine the activation parameters for [2 + 2] cleavage of **4b–e** (ΔH^* in kJ mol^{-1}/ΔS^* in J mol^{-1} K^{-1}): **4b**: 135.6 ± 3.1/87 ± 9; **4d**: 108.4 ± 4.0/16 ± 12; **4e**: 109.0 ± 5.1/28 ± 16.

At first glance, electron transfer–sensitized dimerization fails at higher ΔG values for solvent-separated radical ion formation (Marcus-inverted behavior [25]). However, since the same oxidation potential threshold for successful dimerization is also observed for sensitizers other than DCA and DCA* sensitization, solvent variation from methanol, acetonitrile, dichloromethane to benzene or toluene affects the dimerization efficiency only slightly (with a trend to more efficient dimerization in solvents of low dielectric constant). Therefore, other explanations need to be sought.

Electron transfer–sensitized dimerizations may proceed by addition of a second molecule of monomer to the radical cation formed in the electron transfer step, to form subsequently an open-chain dimer radical cation (or even a cyclized cyclobutane radical cation) prone to electron capture to generate a neutral biradical (or cyclobutane) [22]. It had been demonstrated earlier [26a] that in DCA*-sensitized dimerizations of alkenes in acetonitrile as solvent the radical cation/radical anion pair may be intercepted by a second molecule of alkene followed by back electron transfer to generate a biradical B and ground-state sensitizer [Scheme 5, Eqs. (1) and (2)]. Alternatively, with benzene as solvent, it has been suggested that the formation of an excited triplex composed of one molecule of

$$^1(\text{Sens} \text{---} 3)^* \quad \rightarrow \quad \overline{\text{Sens}^{\ominus} \ 3^{\oplus}} \tag{1}$$

$$\overline{\text{Sens}^{\ominus} \ 3^{\oplus}} \ + \ 3 \quad \rightarrow \quad \text{Sens} \ + \ \mathbf{B} \tag{2}$$

$$^1(\text{Sens} \text{---} 3)^* \ + \ 3 \quad \rightarrow \quad {}^1(\text{Sens} \text{---} 3 \text{---} 3)^* \tag{3}$$

$$^1(\text{Sens} \text{----} 3 \text{----} 3)^* \quad \rightarrow \quad \text{Sens} \ + \ \mathbf{B} \tag{4}$$

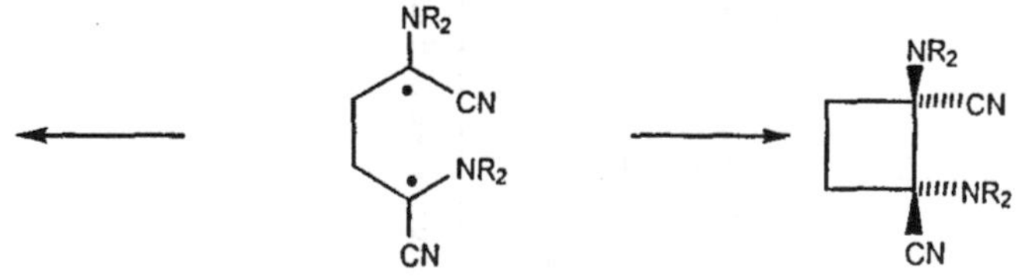

Scheme 5

DCA, benzene, and alkene is intercepted by a second molecule of the alkene, generating the cyclobutane [22a]. Furthermore, a triplex composed of two molecules of alkene and one molecule of sensitizer analogous to that suggested (sensitizer + alkene + diene) for the triplex Diels–Alder-reaction [26b] is feasible [see Eqs. (3) and (4) in Scheme 5]. Whatever pathway, the eventual intermediacy of a biradical B seems likely.

Alkenes 3 with low oxidation potential bear more effective donor groups which are better contributors in the captodative [1] stabilization at both termini of the tetramethylene 1,4 biradical B. The radical stabilization exerted by 1-pyrrolidinyl is estimated to exceed that of 1-piperidinyl by 16.7 kJ/mol [5b]. Thus, a line of more or less stabilized biradicals B may be conceived. The more stabilized ones may in turn experience a higher barrier to final ring closure than those bearing the less active donors, and return to starting materials 3 may well exceed cyclobutane formation.

B. Oxetane Formation

Oxetane formation from excited ketones and alkenes (the Paternò–Büchi reaction) has been amply reviewed with special emphasis on applications in synthesis [27]. Recently, it gained interest due to the facile stereoselective addition of N-acylenamines to electronically excited aldehydes opening an access to β-aminoalcohols [28].

The mechanism has been debated for some time and substantially refined [27], but at least for those cases where $^3(n,\pi^*)$-excited aldehydes or ketones act on normal or electron-rich alkenes, a two-step reaction passing through an intermediate 2-oxa-1,4 diradical has been proposed [27]. This pathway should be especially favorable with captodative alkenes, since one terminus of that diradical formed by attack of the carbonyl oxygen on the unsubstituted carbon atom should experience captodative stabilization. A different approach to explain the regioselectivity of the Paternò–Büchi reaction based on the concept of local softness has been presented [29] (Scheme 6).

Whereas benzophenone, as mentioned in the previous section, only sluggishly forms oxetanes with alkenes 3a [18] and 3b [19], both symmetrical (11) and unsymmetrical $^3(n,\pi^*)$ 1,2-diarylethanediones (13) undergo highly regio- and stereoselective head-to-head additions to various 2-aminopropenenitriles (3b, d,

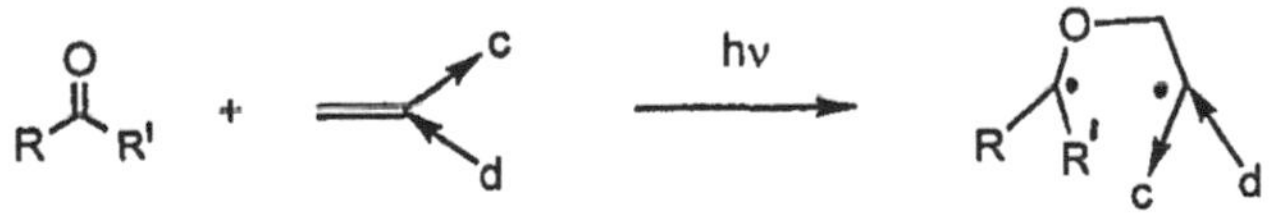

Scheme 6

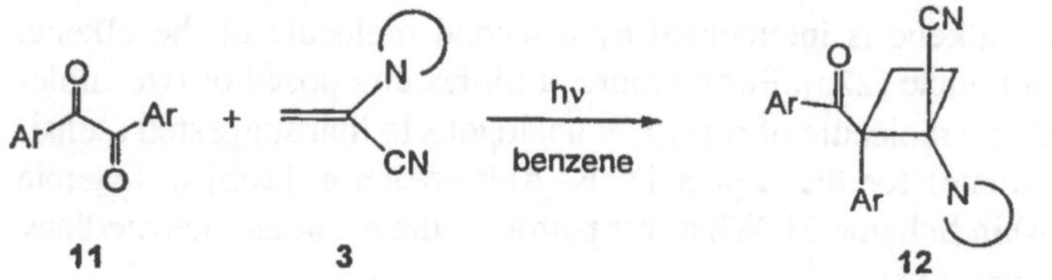

Ar = C$_6$H$_5$, 4-H$_3$CO-C$_6$H$_4$, 4-H$_3$C-C$_6$H$_4$, 4-Cl-C$_6$H$_4$, 4-F$_3$C-C$_6$H$_4$, 2-C$_{10}$H$_7$

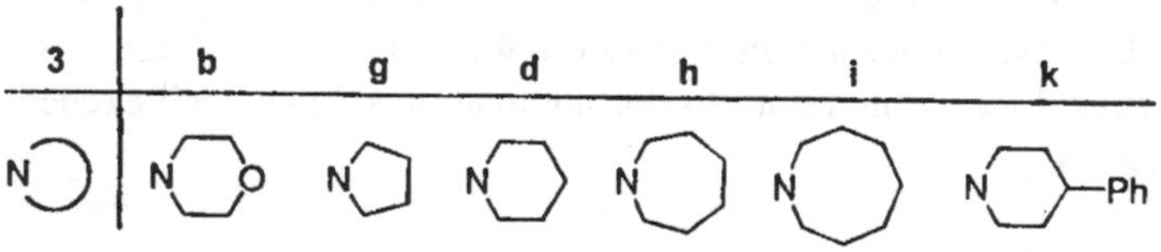

Scheme 7

g, h, i, k) forming oxetanes **12, 14**, and **15** in moderate to good yield [30–32] (Schemes 7 and 8).

Only one stereoisomer of one of the a priori conceivable regioisomers is formed in each case. The connectivity and the relative configuration of the products, derived largely from ^{1}H NMR data, including nuclear Overhauser intensity enhancement studies [30], reflect the geometry of the most easily accessible 1,4-diradical intermediate and the preferred mode of ring closure [27a,33,34]. In all cases, the cyclic amino group is oriented cis to the aryl moiety of the "participating" aroyl group, and the nonparticipating aroyl group is placed cis to the nitrile function. This relative configuration has also been established by x-ray crystal-

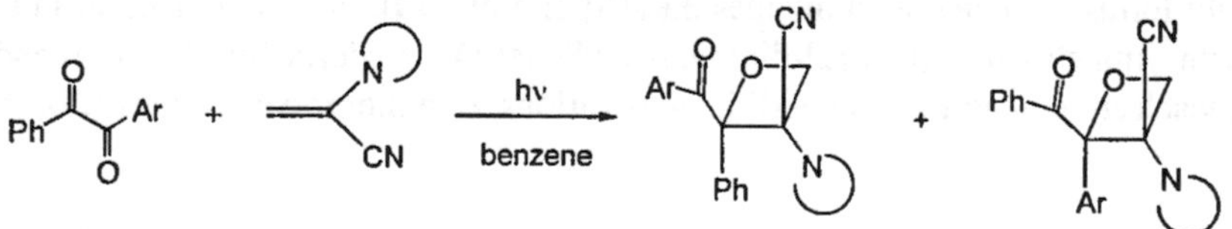

Ar = 4-Br-C$_6$H$_4$, 4-H$_3$C-C$_6$H$_4$, 4-H$_3$CO-C$_6$H$_4$, 2-naphthyl

Scheme 8

lographic structure determinations of two oxetanes **14** (Ar = 4-Br-C$_6$H$_4$ and 4-H$_3$C-C$_6$H$_4$) [32] and in the addition of **3b,h** to the keto carbonyl of methyl phenylglyoxalate [35]. Similar pronounced preferences have been found in the oxetane formation with related electron-rich alkenes, namely, enol acetates [36] and *N*-acylenamines [28] to benzaldehyde.

C. [2 + 2] Photocycloadditions to the Hetero Ring of Indoles and Benzo[b]thiophenes

A number of highly regio- but sometimes stereounselective cycloadditions of α-cyanoenamines and 2-(*tert*-butylthio)propenenitrile (**16**) to benzoannellated heterocycles have been found. (It should be noted that the term "regioselective" is used in this chapter following common usage while, strictly speaking, "selective" refers to the *direction of addition*.)

Photoexcited 3-acetylbenzo[b]thiophene (**17**) added 2-morpholinopropenenitrile (**3b**) regioselectively with formation of stereoisomeric cyclobuta[b][1]-benzothiophenes **18** and **19** in a 4:1 ratio [37]. Likewise, alkene **16** is added to the C-2/C-3 bond of **17** to give a 3:1 mixture of adducts, the major one of which could be separated off by crystallization from the mixture [37].

2-Acetylbenzo[b]thiophene (**22**), when excited in the presence of **3b**, gave **23** as the sole product. Again, the structure of the latter had been confirmed by a single-crystal x-ray structural analysis [37] (Scheme 9).

The following facets seem noteworthy: 1) The main products always bear the donor group endo (pointing to the benzoannellated bridge). This feature will also be encountered in most cycloadditions described below. 2) If one restricts discussion to the enone subunit of **17** and **22**, all cycloadditions mentioned here are highly regioselective and strictly head-to-tail. 3) The product pattern generally suggests the intermediacy of the most stable 1,4 biradical. While 2-acetyl-benzo[b]thiophene (**22**) itself may be regarded as captodative alkene (C-2 is substituted by a donor and an acceptor), benzylic stabilization of a radical center at C-3 seems to be overriding any c,d stabilization to be gained by attack at C-3 of **22**.

When ethyl benzo[b]thiophene-2-carboxylate (**24a**) is substituted for **22** in the photocycloaddition reaction to **3b**, the regioselectivity is reverted and the stereoselectivity (endo/exo) is lowered: The main product (52% yield at 90% conversion of **24a**) is the donor-endo head-to-head adduct **29a** followed by the donor-endo head-to-tail adduct **27a** (20%), the donor-exo head-to-head adduct **30a** (12%) and a small amount (2%) of the donor-exo head-to-tail adduct **28a** with 10% of **24a** recovered unchanged [38]. This means that the two regioisomers are formed in a 64:22 ratio and the main regioisomer is exactly that one not being formed from the corresponding methyl ketone **22**. Thus, the ketone **22** and

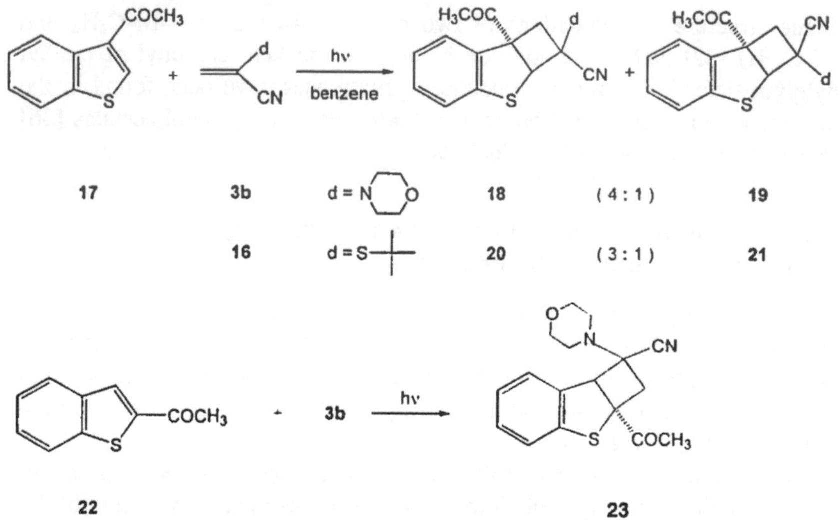

17 3b d = N⟶O 18 (4 : 1) 19

16 d = S⟶ 20 (3 : 1) 21

22 23

Scheme 9

the corresponding ester **24a** show opposite regioselectivities in the addition to the same alkene **3b** [38] (Scheme 10).

N-Acetylindole-2-carboxylic acid ethyl ester (**24b**) undergoes a stereoselective but regiounselective [2+2] photoaddition of **3b** with formation of the donor-endo products **27b** (head-to-tail) and **29b** (head-to-head). In addition, 7% of **28b** (donor-exo, head-to-tail) is formed [38]. The structures of all three products have been unequivocally confirmed by single-crystal x-ray structural analyses [38].

Thus a total of 59% of products (**27b**, **28b**) probably have been formed from biradical **25b** arising from attack of C-2 of indole **24b** on C-3 of **3b**, whereas the minor products **29b**, **30b** may stem from biradical **26b**. It ought to be pointed out, though, that statements on the relative importance of 1,4 biradicals should not be based solely on product distributions, since nothing is known at present about the rate constants of ring closure and back reaction (to starting materials) of biradicals **25** and **26**.

According to Hastings and Weedon [39], triplet *N*-benzoylindole, with its carbon atom 2, attacks alkenes at that terminus, which is less capable of stabilizing a radical center. This trend is possibly also present in the reactions discussed here, although an additional substituent at C-2 may render attack on olefins with carbon atom 3 more competitive. The same considerations should hold for

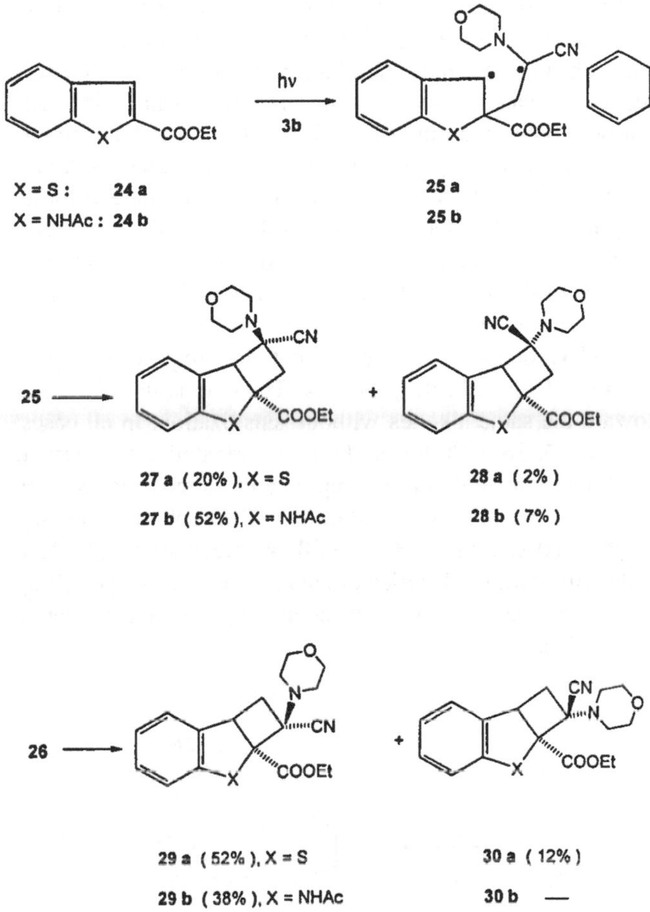

X = S :   24 a
X = NHAc : 24 b

25 a   26 a
25 b   26 b

25 ⟶

27 a (20%), X = S   28 a (2%)
27 b (52%), X = NHAc   28 b (7%)

26 ⟶

29 a (52%), X = S   30 a (12%)
29 b (38%), X = NHAc   30 b   —

Scheme 10

the above-mentioned [2 + 2] photocycloadditions of benzothiophenes **17, 22,** and **24a.**

D. [2 + 2] Photocycloadditions to Coumarins and Thiocoumarins

Various [2 + 2] cycloadducts have recently become available from additions of α-cyanoenamines and 2-(*tert*-butylthio)propenitrile to the C-3, C-4 double bond of coumarins and thiocoumarin.

The photochemistry of coumarin (**31**) has been well investigated. Upon direct irradiation in nonpolar solvents, self-quenching dominates; small amounts of the cis–syn–cis head-to-head dimer may, in addition, be formed in polar solvents like ethanol [40]. Benzophenone sensitization affords cis–anti–cis head-to-head dimerization in both polar and nonpolar solvents [40]. The tendency to dimerization is increased with increasing polarity of the solvent used [41], and the cis–syn–cis head-to-tail dimer is also formed [41]. Thiocoumarin (**32**) forms all four possible [2 + 2] dimers [42] when irradiated in the crystalline state with light of λ > 340 nm. Dimerizations and [2 + 2] cycloadditions of thiocoumarins have been recently reviewed [43], showing the same trends with respect to regioselectivity as observed in our studies.

The parent compound **31**, being weakly fluorescent, does require benzophenone sensitization to form adducts **33b, d, g**, and **34**. Thiocoumarin (**32**), however, is photoactive toward the same alkenes without sensitization. In all cases, solely head-to-tail products **35, 36** with donor (D) endo orientation are formed [44,45]; thus, these additions are both highly regio- and stereoselective (see Scheme 11). This statement also applies to the addition of 2-morpholinopropenenitrile (**3b**) to the 4-methylated coumarins **37** and **38**. Whereas the highly fluorescent dye **38** reacts directly with (and under efficient fluorescence quenching by) **3b**, the successful conversion of **37** into **39** again requires benzophenone sensitization [44] (Scheme 12).

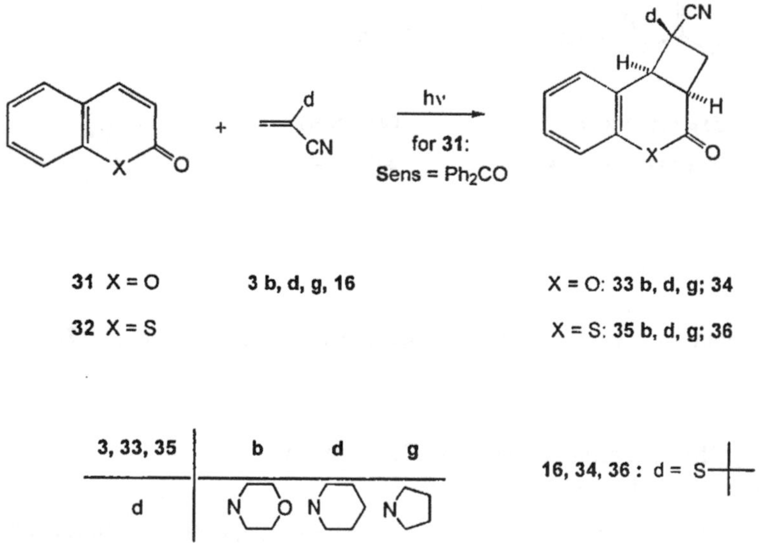

31 X = O   3 b, d, g, 16   X = O: 33 b, d, g; 34

32 X = S   X = S: 35 b, d, g; 36

3, 33, 35	b	d	g
d	N⟩O	N⟩	N⟩

16, 34, 36 : d = S—

Scheme 11

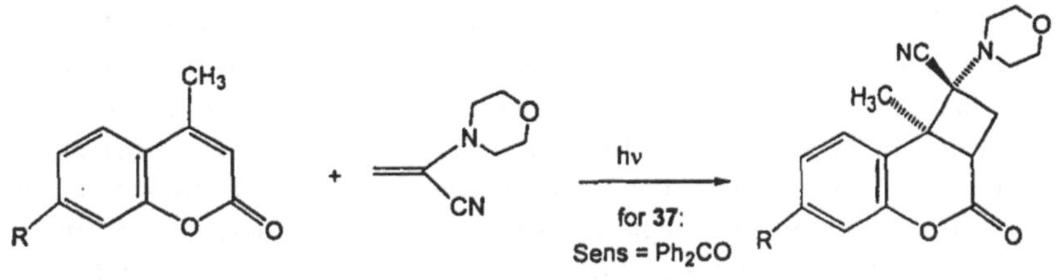

37 : R = H   3b   39 : R = H
38 : R = NEt₂   40 : R = NEt₂

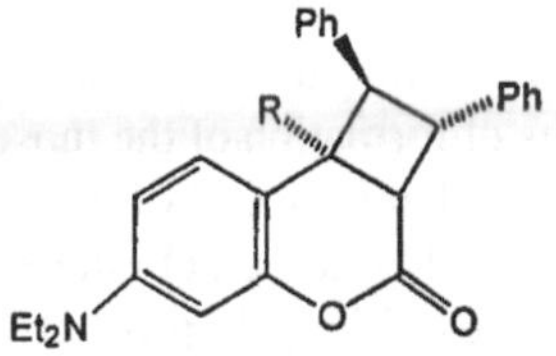

41 (R = H, CH₃, O(CH₂-CH₂)₂N)

Scheme 12

Whereas the 4-unsubstituted coumarins give rise to the donor-endo products 33–36, the relative configuration of 39 and 40 is donor-exo, as clearly corroborated for 40 by a single-crystal x-ray structural analysis [44].

While the geometry (donor-endo) of products 33–36 may be predetermined by either a preceding exciplex and/or attractive secondary interactions between the positivated donor and the arene moiety stabilizing a donor-endo conformation of the biradical, the final ring closure has to be fast to preserve this kind of preorientation. An additional methyl group, as in 37, 38, may slow down this final ring closure and thus allow for conformational relaxation of the biradical with donor-exo geometry finally closing to the donor-exo products 39, 40. An attempt to explain the otherwise general tendency towards formation of donor-endo adducts in the addition of captodative alkenes to π systems will be made later.

It had earlier been observed [46] that photoexcited coumarin 38 adds both acrylonitrile and butylvinyl ether across the C-3, C-4 bond regioselectively to form diastereomeric mixtures (butoxy and cyano groups endo and exo) of one regioisomer corresponding to 40 in either case. Any products of photodimerization of 38 have not been detected [46]. The results had been interpreted in terms of a highly regioselective and nonconcerted addition of C-2 of the alkene to C-

3 of **38** in its fluorescent singlet state [47,48]. Three different 4-R, 7-diethylamino coumarins (with R = H, CH_3, morpholino) upon excitation add *trans*-stilbene uniformly to products of type **41** [49]. This result, however, bears no resemblance to the stereoselectivities observed with alkenes of type **3** or **16**, since these cannot exist as E and Z isomers. The highly fluorescent 3-(2-benzthiazolyl)coumarin (**42**) adds **3b, d, g** with the same regioselectivity as before but stereounselectively to donor-endo/donor-exo mixtures of adducts **43** (see Scheme 13). 2-(*tert*-Butyl-thio)propenenitrile (**16**) does not form an adduct. All c,d-alkenes tested, including **16**, do effectively quench the fluorescence of **42** in benzene and acetonitrile solution [44,45]. More detailed investigations aimed at separating dynamic and static quenching are needed.

The results obtained so far in the photocycloadditions of c,d-alkenes may be interpreted in the following way. The regioselectivities may be explained in terms of formation of the most stable intermediate 1,4-biradical formed via initial bonding between C-3 of the alkene and C-3 of the coumarin. While such a biradi-cal would be a logical intermediate in the cases of triplet sensitization, it seems to be less likely if product formation appears to be part of the fluorescence quenching of **38** or analogous dyes.

The thioxo analogs of coumarin and thiocoumarin, namely, **44** and **45**, pre-fer [2 + 2] addition to the C=S bond as delineated from the alkenes **48, 49** formed (as mixtures of E and Z isomers) [45]. Intermediates **46** and **47** could not be isolated (Scheme 14).

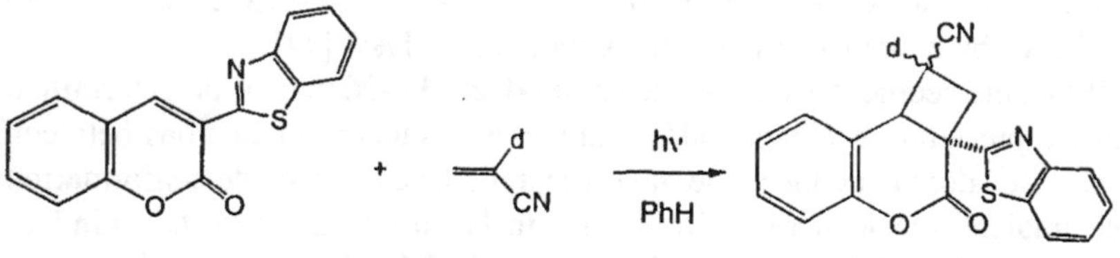

3, 43	b	d	g
d	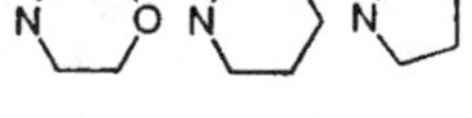 N−morpholino	N−piperidino	N−pyrrolidino
d-endo/ d-exo	2	2.6	1.8

Scheme 13

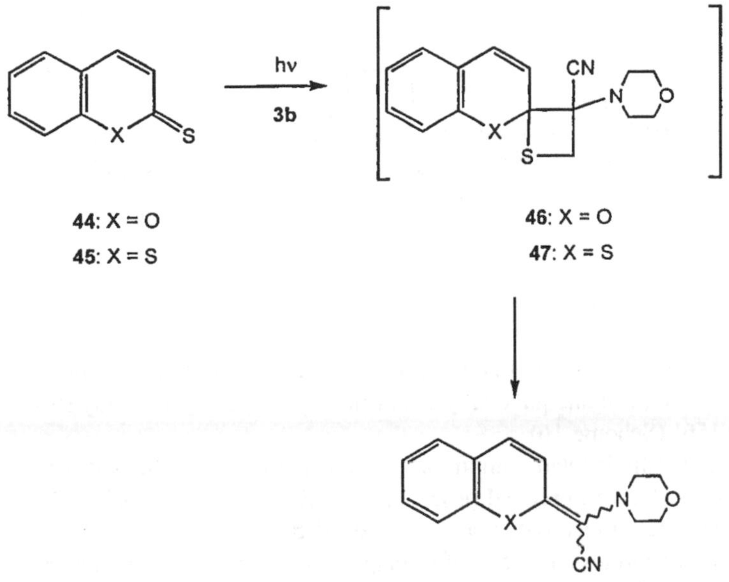

44: X = O
45: X = S

46: X = O
47: X = S

48: X = O, 52%; E/Z = 1:1
49: X = S, 42%; E/Z = 14:1

Scheme 14

E. [2 + 2] Photocycloadditions to Acenaphthylene

As a consequence of the rigid arrangement of its C-1, C-2 double bond, acenaph-thylene (50) tends to undergo homoaddition to its cis–syn–cis dimer via S_1^* and to the cis–anti–cis dimer via T_1^* [50]. External heavy-atom effects brought in by halogenated solvents promote the intersystem crossing to T_1^* ($\phi \rightarrow 1$) and at the same time an increase in the yield of dimers, with the cis–anti–cis isomer prevailing. A biradical is therefore invoked as a logical intermediate [50]. The [2 + 2] heteroadditions have been reported with maleic anhydride [51], acryloni-trile in dibromomethane [52], pentadiene (cis and trans) and cyclopentadiene [53]. The addition reported here comes closest to the previously investigated addition of 2-chloropropenenitrile to 50 in bromoethane as solvent giving rise to a 1.3:1 mixture of diastereomeric adducts 51 and 52 [54] (Scheme 15).

The photoaddition of 2-morpholinopropenenitrile (3b) to 50 in cyclohexane as solvent (bromoethane would react with the cyanoenamine) gave, besides 14% of cis–syn–cis and 5% of cis–anti–cis dimers, 38% of endo adduct 53 and 12% of exo adduct 54. If oxygen is not rigorously excluded, 5% of acenaphthenone

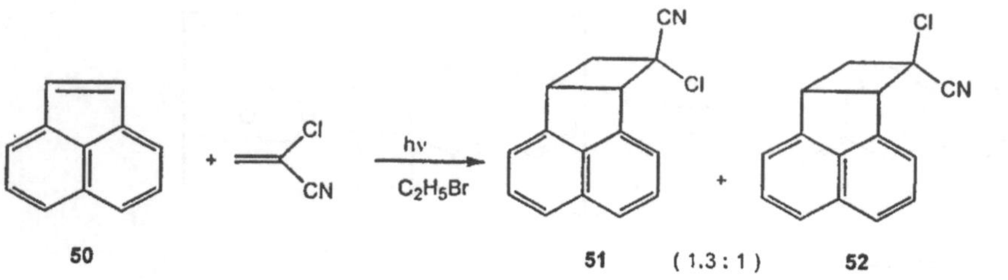

Scheme 15

and 6% of cyanocarbonylmorpholine are also formed [55,56]. The structures of both **53** and **54** have been unequivocally confirmed by single-crystal x-ray structural analyses [56] (Scheme 16).

When the chirally labeled 2-aminopropenenitrile (S)-**31** (for the first preparation, see Ref. 57; for an improved procedure, see Ref. 56) was substituted for **3b**, a moderate diastereoselectivity was observed [55,56]. In addition to 3% anti- and 11% syn-dimer and less than 5% of acenaphthenone, the diastereomers **56–59** have been formed in the percentages given as determined by ¹H NMR analysis [56]. (S)-**31** may be added to photoexcited **50** with either its Re or its Si face. In either option, the cyclic amino residue may be oriented endo or exo to the arene moiety (Scheme 17).

Separation of products is severely hampered by their sensitivity to hydrolysis and epimerization (endo → exo) at the aminonitrile carbon atoms under the influence of traces of acid (e.g., present in CDCl₃). On the other hand, this unidirectional epimerization can be used as criterion for structural assignment.

The donor-exo products **56** and **57** may together be separated from all other products by dry column chromatography, whereas the donor-endo products **58, 59** escape every trial of separation by ready hydrolysis of the aminonitrile function to the oxo group [56] upon chromatography. Thus, the structural assignment always rests on a set of criteria: 1) chemical shift differences for donor-endo and donor-exo isomers, 2) hydrolyzability, 3) epimerizability, 4) NOE signal enhancements.

Since hydrolysis of **53** (or **54**) gives rise to the ketone **55**, hydrolysis of the total product mixture **56–59** or a part thereof, besides regenerating the auxiliary (S)-**31**, would provide more or less enantiomerically enriched samples of ketone **55**.

F. [2 + 2] Photocycloadditions to Naphthalenes and Phenanthrenes

Photocycloadditions to aromatic compounds have been [58] and continue to be [59] reviewed.

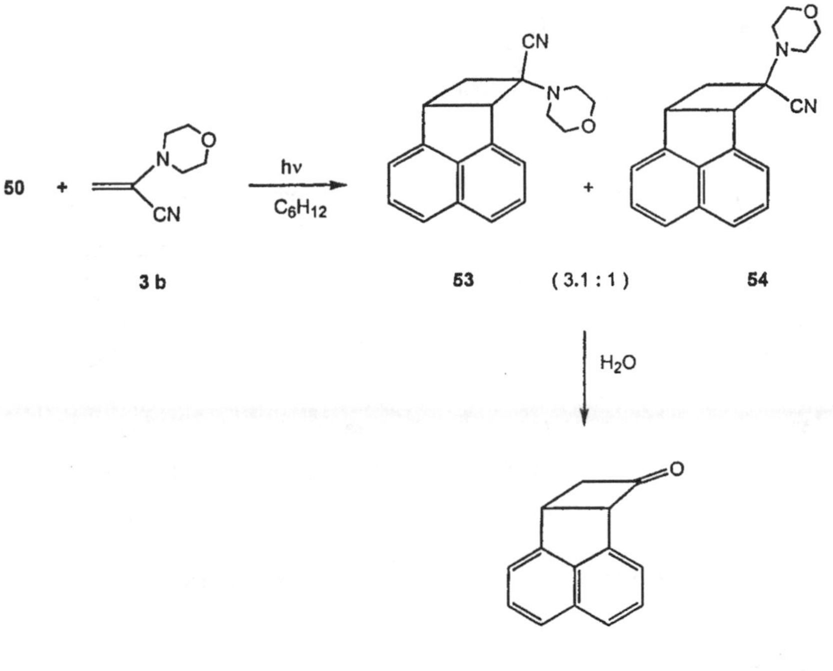

Scheme 16

Both the parent hydrocarbons as well as cyano-, acetyl- and alkoxycarbo-nyl-substituted derivatives have been found to be capable of photoaddition of various captodative alkenes, predominantly α-cyanoenamines, with formation of [2 + 2] adducts at skeletal double bonds. The [2 + 2] photocycloadditions of c,d-alkenes to various acylnaphthalenes and acylquinolines will be treated in another context in Section III.

Naphthalene itself slowly and in very modest yield forms two morpholino-endo adducts **60** and **61** with 2-morpholinopropenenitrile (**3b**) along with some *N*-cyanocarbonylmorpholine [56]. Chromatography on air dried silica gel brings about hydrolysis of the aminonitrile function giving rise to the ketones **62** and **63** [56]; see Scheme 18.

The regio- and stereoselectivity observed in the addition of **3b** to naphtha-lene resembles that of [2 + 2] photoaddition of 2,3-dihydrofuran [60] and acrylo-nitrile [61] to the same hydrocarbon.

Photoexcited 1-naphthalenecarbonitrile **64** also adds **3b** in the [2 + 2] mode unidirectionally both at the C-1, C-2 and the C-7, C-8 bond and with formation

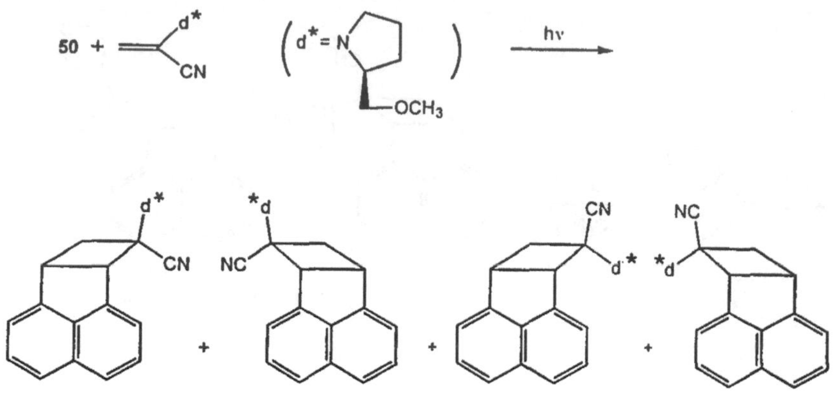

56 (*Re* - *exo*) 10%

57 (*Si* - *exo*) 15%

58 (*Si* - *endo*) 13%

59 (*Re* – *endo*) 17%

Re : *Si* = 1.9 : 1

Re : *Si* = 1.3 : 1

d*(*endo*) : d*(*exo*) = 2 : 1

Scheme 17

of only one each (**65, 66**) of the two a priori possible diastereomers in either case [62]. The structure of compound **65** again has been unequivocally confirmed by a single-crystal x-ray crystallographic structure determination. Structures other than **65** and **66** and thus any alternative structures for **66** could be ruled out on the basis of ^{1}H NMR data. The morpholino-endo assignment for **66** is based on NOE experiments: Saturation of the morpholino $N(CH_2)_2$ signals enhances the intensity of the 3H signal in both **65** and **66**. That **66** indeed is also a cyclobutanaphthalene is also supported by a comparison of the UV spectra of **66** and ketone **67** with that of **65** (Scheme 19).

It should be noted that the direction of addition of the cyanoenamine in the addition to the 1,2 bond of **64** is opposite to the direction of addition of various vinyl ethers to **64** [63] and gives only one stereoisomer. However, it resembles both the regioselectivity and the stereocontrol of the cycloadditions of 2-chloropropenenitrile and α-methylstyrene to the C-1, C-2 bond of 4-methoxy-1-naphthonitrile [64].

Like in the additions to 1-naphthalenecarbonitrile (**64**) reported here, the same high stereoselectivity is observed in the addition of **3b** to the 9, 10 bond

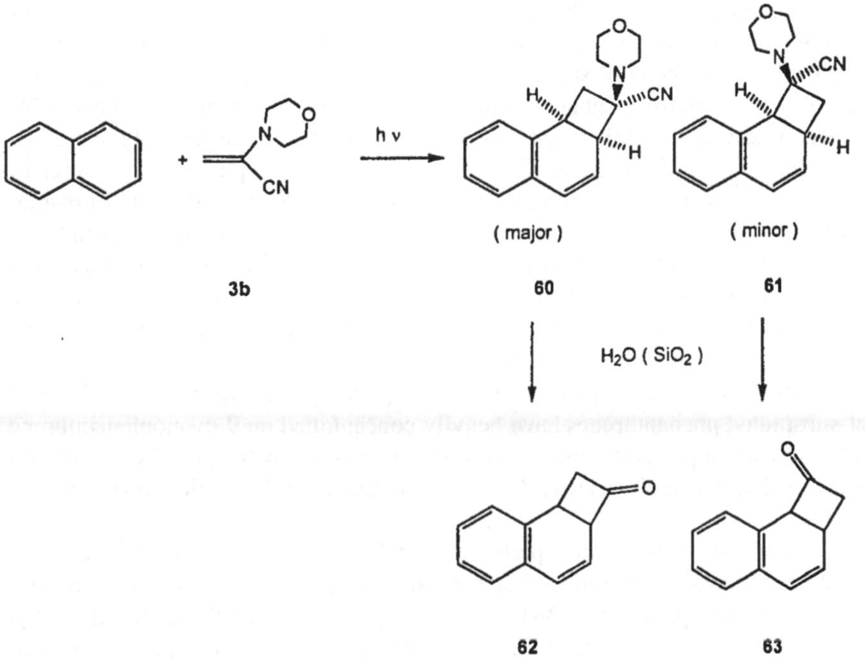

Scheme 18

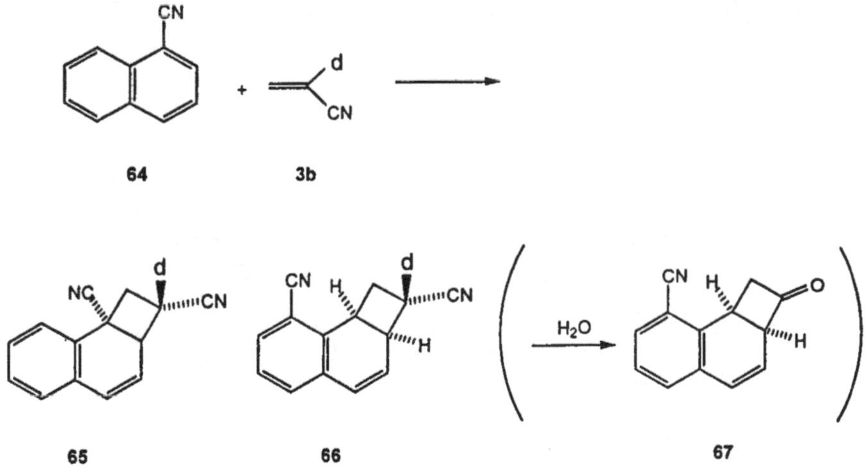

Scheme 19

of phenanthrene: The morpholino-endo adduct **68** is obtained in 47% yield and readily converted to the ketone **69** [56] (Scheme 20). The structural assignment of **68** again rests on NOE studies. Irradiation into the morpholino $N(CH_{ax})_2$ signals effects an intensity enhancement of the aromatic 4H signal [56]. Other NOE effects observed with this compound also support the structural assignment.

Earlier investigations of [2 + 2] photoadditions to phenanthrene refer to the reactions with 1,2-dichloroethene and acrylonitrile [65], dimethylfumarate and/or maleate [65–67], and fumarodinitrile [68]. The participation of exciplexes in [2 + 2] photocycloadditions to this hydrocarbon has been strongly supported [66,68] on the basis of emission studies including exciplex quenching and kinetic data. An alternative singlet biradical pathway has been suggested, also questioning the participation of exciplexes [67].

Mechanistic investigations of [2 + 2] photocycloadditions to the skeleton of substituted phenanthrenes have heavily concentrated on 9-cyanophenanthrene [69–74] and on phenanthrene-9-carboxylates [75–78], with special emphasis on participation of exciplexes [69,71–74,78] and Lewis acid modification of reactivity [75].

According to our own experience, with benzene as solvent the fluorescence of phenanthrene-9-carbonitrile (**70**, 3×10^{-5} M, $\lambda_{exc} = 320$ nm) is quenched by 2-morpholinopropenenitrile (**3b**) and *tert*-butyl vinylsulfide (**76**), that of methyl phenanthrene-9-carboxylate (**73**, 2×10^{-5} M, $\lambda_{exc} = 306$ nm) by **3b**, 2-(1-piperi-

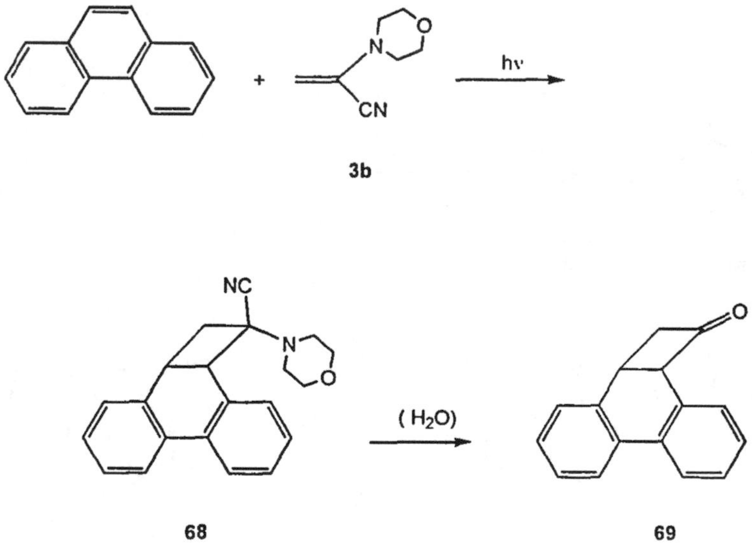

3b

68

(H_2O)

69

Scheme 20

dinyl)propenenitrile (**3b**), 2-(hexamethyleneimino)propenenitrile (**3h**), 2-(*tert*-butylthio)-propenenitrile (**16**) and **76**. For nitrile **70**, linear plots of l_0/l vs. **3b** or **76**, respectively, have been obtained (with a very slight upward curvature in the plot for quenching by **76**), whereas for **73** and alkenes **3b,d,h** all plots of l_0/l vs. [quencher] are markedly curved upward, indicating the participation of a static component in the fluorescence quenching, especially effective at higher concentrations of quencher [79].

The different effects of **16** and **76** on the fluorescence of the ester **73** are especially striking: While a rate constant for dynamic quenching $k_q = 1.1 \times 10^9$ $M^{-1} s^{-1}$ can be extracted from the linear plot of l_0/l vs. concentration of **76**, the strongly upward-curved plot for quenching by **16** (the electron poorer of the two olefins! **16**: first IP 9.12 eV [11b], **76**: first IP 8.07 eV [80]) suggests strong complexation. By treating the data according to accepted procedures to separate the dynamic and static contributions [81], a rough estimate of the equilibrium constant K for complexation can be made. For the pair **16/73**, K would come out as high as 1700 assuming a 1:1 stoichiometry [79]. Likewise, ground-state complex formation of **73** is also observed with α-cyanoenamines, $K = 25$ (**3b**), 169 (**3d**), 944 (**3h**) [79].

Triplet-detection and quenching experiments [82] with methyl-9-phenanthroate (**73**, $E_T = 289$ kJ mol^{-1}) had also been carried out. Excitation of benzene solutions of **73** at 354 nm with 15-ns pulse duration and time-delayed absorption spectroscopy of the transient between 400 and 700 nm revealed a maximum at 490 nm which was assigned to the lowest triplet excited state of **73**, $\tau_T = 5$ μs (under argon), and 65 ns upon saturation with oxygen, $k(O_2) = 2 \times 10^9$ $M^{-1} s^{-1}$. Alkenes **3b**, **16**, and **76** (listed in the order of decreasing efficiency) reduce the lifetime of the transient, $k_q = 3.5 \times 10^8$ (**3b**), 7×10^7 (**16**), and 8.7×10^4 (**76**) $M^{-1} s^{-1}$ [82].

The larger the ground-state interaction, the lower the tendency to undergo cycloaddition. Again, with benzene as solvent, alkene **3b** is smoothly added to excited **70** with formation of a 4:3 mixture of the head-to-tail adducts **71** (donor-endo) and **72** (donor-exo), in a combined yield of 79% at 29% conversion of **70**. In addition, 21% of the head-to-tail syn-dimer **79** are obtained. While the regioselectivity is high, the stereoselectivity is not; there is only a slight preference for the donor-endo adduct **71**. The structures of **71** and **72** have been assigned on the basis of ^{1}H NMR data and a single-crystal x-ray structural analysis of **72** [79] (Scheme 21). However, whereas 2-(*tert*-butylthio)propenenitrile (**16**) (strongly complexing **73**!) does not show any cycloaddition to **70** at all, *tert*-butyl vinylsulfide (**76**) is added smoothly with formation of a single adduct **77** in 48% yield (the structure is again assigned on the basis of a single-crystal x-ray structural analysis) at 90% conversion of **70**. It is noteworthy that both the stereoselectivity is high and the direction of addition is opposite to that of **3b**, which suggests that different pathways are followed in both additions [79,83].

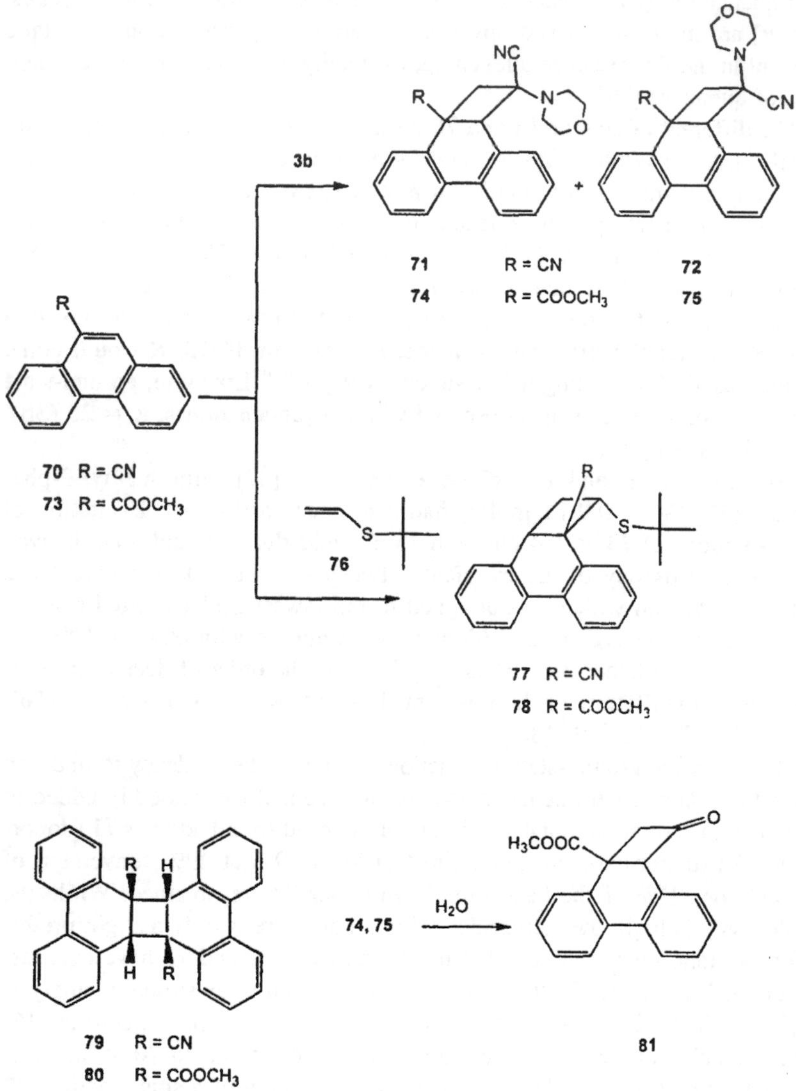
3b
NC
N
O
R
CN
+
71 R = CN 72
74 R = COOCH3 75
70 R = CN
73 R = COOCH3
S
76
77 R = CN
78 R = COOCH3
R H
H R
79 R = CN
80 R = COOCH3
74, 75
H2O
H3COOC
O
81

Scheme 21

Methyl phenanthrene-9-carboxylate (73), upon photoexcitation in benzene, adds 3b with formation of the stereoisomeric head-to-tail adducts 74 (46%, donor-endo) and 75 (33%, donor-exo), in addition to 21% of dimer 80, at 29% conversion of 73. The structural assignment is based on a crystal structure analysis of 75 and characteristically different ^{1}H chemical shifts for both isomers [79].

In conventional quenching experiments [79,82,83], the triplet quencher octafluoronaphthalene (E_T = 237 kJ mol^{-1}) [84] had almost no influence on the photoaddition of tert-butyl vinylsulfide (76) to 73. This suggests that 76 is added to the excited singlet state of 73. In the addition of 3b to 73, however, the formation of the donor-exo adduct 75 (see Scheme 21) is suppressed markedly, the formation of the donor-endo adduct 74 to a smaller extent. At the same time, the proportion of dimer 80 is increasing with the increase of the quencher concentration. This suggests that while the triplet reaction leading to both 74 and 75 is quenched, more starting material depletion is achieved through the dimerization of the singlet excited state of 73. A very similar result has been obtained with 3,3,4,4-tetramethyl 1,2-diazetine-1,1-dioxide (TMDD) [85] as triplet quencher (E_T = 147 kJ mol^{-1} [86]).

When 2-(1-piperidinyl)- (3d) and 2-(1-pyrrolidinyl)propenenitrile (3g) were substituted for 3b in the photoaddition to 73 in dichloromethane as solvent, only the donor-exo stereoisomers 82 (60%) and 83 (47%) were isolated as such together with appreciable amounts of the hydrolysis product 81 and dimer 80 [79,83]. It appears that the increased donor character in the cyclic amino group and the more hindered endo orientation of the donor promote sensitivity to hydrolysis [79,83] (Scheme 22).

Addition of the chirally labeled alkene (S) 31 to excited 73 gave a mixture of [2 + 2] adducts too sensitive to hydrolysis to allow for chromatographic sepa-

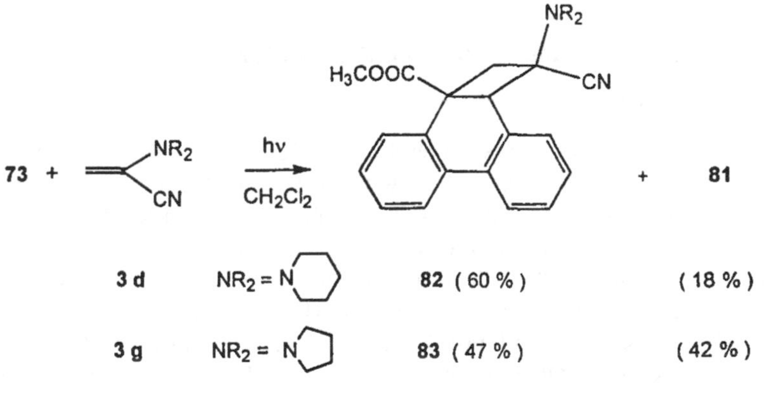

Scheme 22

ration. Thus, this mixture was subjected to hydrolysis [15] and ketone **81** (56%) was separated from dimer **80** (44%). The ketone sample had $[\alpha]^{20}_D = 6.06$ (CH$_2$Cl$_2$, c = 1) and consisted of a mixture of enantiomers in the ratio 1:2.6 as determined using a chiral NMR shift reagent. From this it follows that a double-facial selection had been operative, the details of which await further clarification. After hydrolysis, optically active ketone **81** with 44% enantiomeric excess has been obtained [79].

III. [4 + 2] PHOTOCYCLOADDITIONS TO CONDENSED ACYL ARENES SUCCESSFULLY COMPETING WITH [2 + 2] PHOTOCYCLOADDITIONS

A. General Remarks

In contrast to the wealth of known light-induced [2 + 2] photocycloadditions to naphthalenes and phenanthrenes [58,59], [4 + 2] photocycloadditions to the naphthalene skeleton are comparatively rare [87–94], as are formal [4 + 4] photo-additions [58,95] and [4 + 4] dimerizations [58,96,97]. A novel 1,8 photoaddition of naphthalene 1,4-dicarbonitrile, the corresponding 4-cyano-1-methyl ester, and dimethylnaphthalene-1,4-dicarboxylate to various alkenes has recently also been reported [98].

For [4 + 2] ("photo–Diels–Alder") and [4 + 4] additions to anthracene, see Refs. 58 and 99.

Photo–Diels–Alder reactions have theoretically been treated with emphasis on stereocontrol and mechanism [100].

In the following chapters, such [4 + 2] photoadditions of c,d-alkenes to 1-acylnaphthalenes and acylquinolines, as well as to anthracenes and benzoannel-lated anthracenes, will be reviewed.

B. Acylnaphthalenes and Naphthoic Acid Esters

1. Scope and Limitations

By far the largest number of successful photocycloadditions of c,d-alkenes to the naphthalene nucleus have been carried out with 1-acetonaphthone (**84**) [57,101–113]. With alkenes **3b–d, g–p** (i.e., those two aminopropenenitriles **3**, in which the amino nitrogen is part of a ring), the main (and often the only isolated) product is the donor-endo 1,4 adduct **85** as the result of a highly regio- and stereoselective formal [4 + 2] photoaddition (see Scheme 24). Thus, crystalline products **85b** [101,102,105,106], **85c, k, n–p** [106], **85d, g** [103,111], **85h, i** [111], **85l** [57], **85m** [103,110] may be obtained in moderate to agreeable yields (Scheme 23). Often these products directly separate from the photolyzed solutions, so conver-

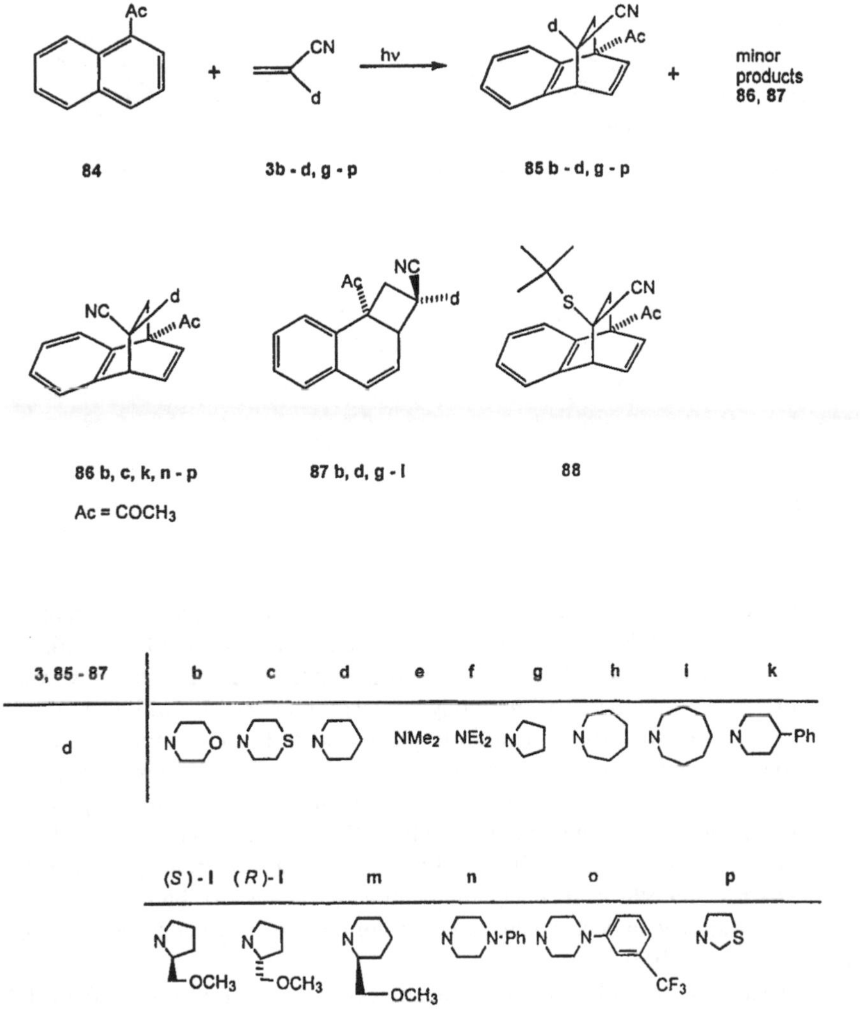

84   3b - d, g - p   85 b - d, g - p

86 b, c, k, n - p   87 b, d, g - l   88

Ac = COCH₃

3, 85 - 87	b	c	d	e	f	g	h	i	k
d	N◯O	N◯S	N◯	NMe₂	NEt₂	N◯	N◯	N◯	N◯-Ph

	(S)-l	(R)-l	m	n	o	p
	N◯-OCH₃	N◯-OCH₃	N◯-OCH₃	N◯N-Ph	N◯N-◯CF₃	N◯S

Scheme 23

sions may be driven to 70–80%. There is only little dependence on solvent polarity.

Occasionally, the 9-epimeric 1,4 adducts **86** are also detected in the mother liquor, and from ¹H NMR integration ratios of formation close to 94:6 in the cases (85/86)b,c,k,n,o or 89:11 for 85p/86p may be determined [106].

Preliminary results indicate that with noncyclic amino donors, i.e., in the

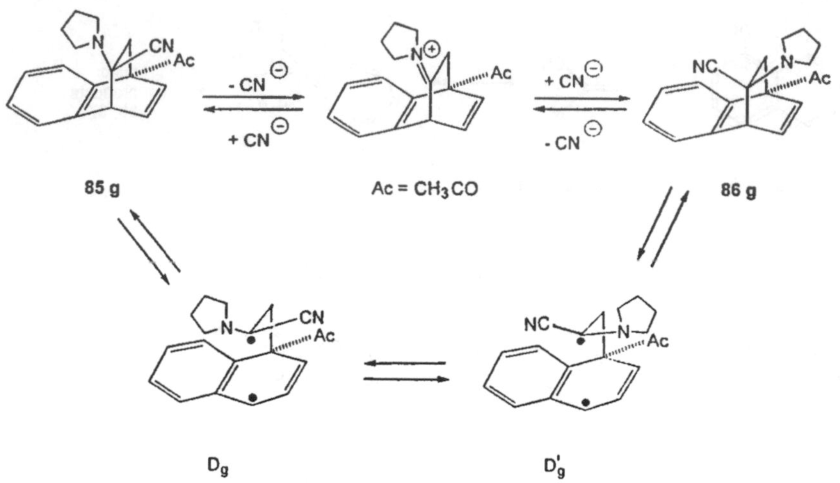

Scheme 24

cases of photoaddition of alkenes **3e,f** or their homologs with higher dialkylamino donors, the donor-endo- and donor-exo isomeric adducts of type **85** and **86**, respectively, may be present in close to equal amounts [110,112]. One has to be aware of artefacts, though. The thermodynamically more stable (while less sterically hindered) donor-exo isomers **86** might originate from the more sterically hindered donor-endo isomers **85** via either C-4/C-9 homolysis and reclosure (via Dg and D′g) or by cyanide release, to generate an iminium ion, and recapture of cyanide, as outlined for compound **85g** [103] (Scheme 24). Indeed, the adduct of 2-(1-pyrrolidinyl)propenenitrile (**3g**) to 1-acetonaphthone, namely **85g**, or the analogous product **89g** from 1-naphthophenone (Scheme 25), build up the C-9 epimer upon storage in deuteriochloroform solution, probably under the influence of traces of acid [103].

Other 1-acylnaphthalenes have been less successfully converted to 1,4 adducts. Compared to **84**, 1-naphthophenone undergoes sluggish additions to **3b,c,g**, forming **89b,c,g** in moderate to low yields [19,103] (Scheme 25). The 1,4 adduct of 2-(*tert*-butylthio)propenenitrile (**16**) to 1-naphthaldehyde was found to be unstable [19] compared to the stable adduct **88** obtained from **84** and the same olefin. The same aldehyde, on the other hand, underwent smooth addition to **3b** to give a 59% yield of **90** [101]. The isomeric 2-naphthaldehyde, however, did not undergo photoadditions to 2-aminopropenenitriles but to **16**, forming **91** in 27% yield [19].

As mentioned above, the detection of small amounts of 1,2 adducts **87** is noteworthy, although the isolable yields may be just a few percentage points

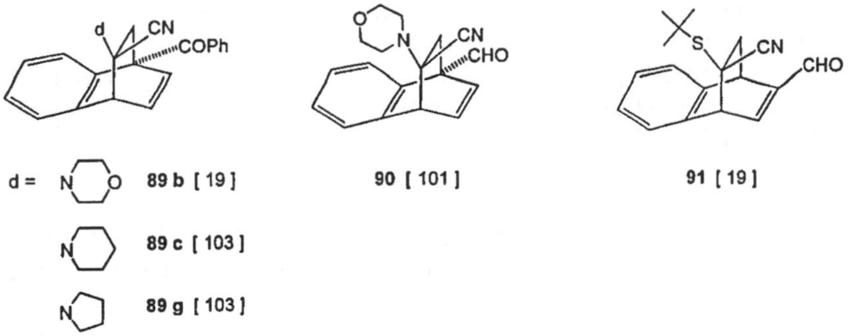

d = N☐O   89 b [19]

     N☐   89 c [103]

     N☐   89 g [103]

Scheme 25

or less. From the two C-2 epimers possible a priori, the donor-exo isomers **87** as shown are clearly dominating (**87b** [105], **87d** [111], **87g** [103,111], **87h** [111,113], **87i** [111]. C-2 epimerizations have not been detected. The structural assignments rest on the following:

1. ^{1}H NMR data, especially those of the AB and ABX systems, in comparison with those for the isomeric 1,4 adducts,
2. An x-ray crystal structure analysis of **87h** [113], and
3. The UV spectrum of **87b** (in chloroform) featuring shoulders (log ϵ) at 320 (2.94), 302 (3.34), and 290 (3.41) nm, in addition to two maxima at 276 and 266 nm (3.83). In contrast, the UV spectrum of **85b**, due to lack of conjugation between the benzenoid ring and the C-2/C-3 double bond, has absorptions starting with a broad plateau at 290 nm (1.36) and a structured absorption peaking at 261 nm (2.38) in the same solvent [19].

Interestingly, compound **87b** could not be converted to the isomeric 1,4 adduct **85b** by either 313 nm irradiation or thermally in solution; in both cases, cleavage into starting materials **84** and **3b** was the sole reaction [105]. From this it may be delineated that **85b** and **87b** cannot be interconverted by either a concerted [1,3] shift (the exo product **87b** would lead to the exo product **86b** instead of the endo product **85b**) or stepwise via a biradical intermediate.

UV irradiation ($\lambda \geq 280$ nm) of common solutions of **84** and either one of the alkenes **3d,g,h,i** in perdeuterated benzene and careful following of the changes in the ^{1}H NMR spectra of these solutions within suitable time intervals revealed that at the beginning of the experiment, the donor-exo 1,2 adducts **87** were built up more rapidly than any 1,4 adduct. At approximately 30–40% conversion of starting material the 1,2 adducts were degraded, again giving way to more 1,4 adduct **85** [111]. The latter accumulated due to its stability under the irradiation conditions. From this it may be concluded that the [2 + 2] photoaddition (leading to type **87** products) and the [4 + 2] photoaddition (to generate **85**, **86**) are two independent competing reactions, and no common intermediate would exist (Scheme 26).

2. Mechanistic Details

Some comments are necessary with respect to the proposed biradical D (which had also been invoked above (Scheme 24) in the isomerization of compound **85g** into its epimer **86g**):

1. Since the [4 + 2] 1,4-addition of **3b** to **84** on a preparative scale can be quenched [101] by 3,3,4,4-tetramethyldiazetine-1,2-dioxide (TMDD) [85,86], and a selection of 2-donor-substituted propenenitriles quenches π,π^*-triplet 1-acetonaphthone (E_T = 236 kJ mol^{-1} [114]) with rate constants around 10^8–10^9 M^{-1} s^{-1} [115], the observed 1,4 additions are envisaged as originating from the (lowest excited) $^3(\pi,\pi)^*$ triplet state of **84**. This state has been observed by triplet–

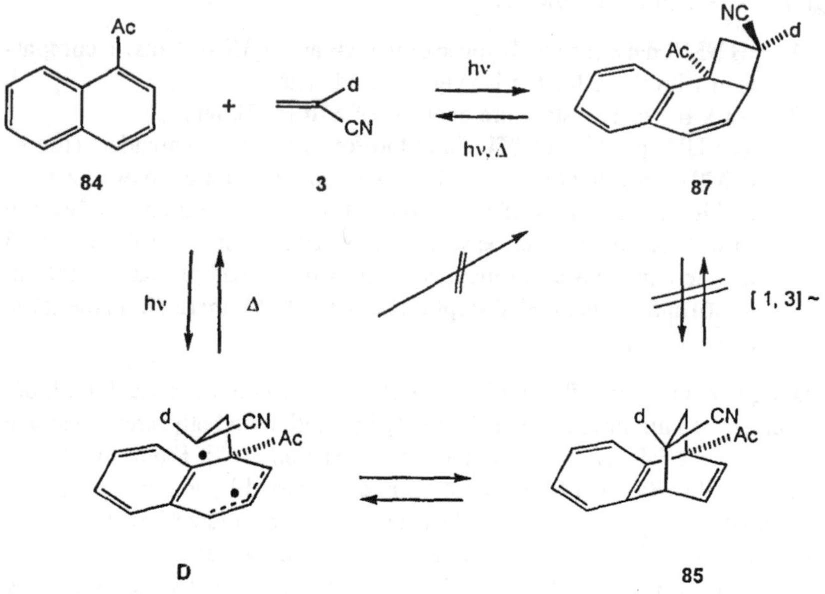

Ac = COCH$_3$ d: see previous schemes

Scheme 26

triplet absorption spectroscopy (λ_{max} = 480 nm) in methanol solution. A quenching rate constant k_q = 0.5 $\times$ 10^{10} M^{-1} s^{-1} for TMDD has been determined [115]. A biradical (first in its triplet state) would be the logical intermediate in the cycloaddition of an alkene to a triplet arene. Such a species, however, could not be detected by transient spectroscopy or by reaction with methylviologen.

2. Captodative [1] alkenes are most predestined to form biradicals in such processes. At one terminus, the biradical derived from **84** is resonance-stabilized; at the other, it is stabilized by the synergistic captodative [1] effect.

3. The preferred endo orientation of the donor group in the 1,4 adducts may originate from *either* a preferred orientation of the two reactants in an exciplex (prior to any bond forming event) which is retained in the subsequent steps (no proof can be presented so far for this possibility) *or* from the preferred geometry of the intermediate biradical D in its U-shaped conformers (note that extended conformations may exist as well but may be more prone to reversal to starting materials). The implications of two levels of selection have been discussed by Scharf and co-workers [116]. The underlying principles worked out by these authors, i.e., an entropy control in the first and an enthalpy control in the second level of selection, may also be operative here.

4. The captodative effect of radical stabilization [1] implies that the dative substituent donates charge to the radical center and the captive substituent withdraws charge. By this transfer, the dative substituent in D with its partial positive charge may experience attraction by the electron-rich π system of the benzenoid ring, while a negatively charged captor would experience repulsion. Thus, the additionally stabilized donor-endo conformer D_n may preferentially cyclize to product whereas the destabilized donor-exo conformer D_x may prefer to return to starting materials (Scheme 27).

5. In accordance with this explanation is the only moderate quantum yield for product **85b** formation ϕ_p = (2.16 $\pm$ 0.15) $\times$ 10^{-2} (in methanol) [105]. A low quantum yield may well be a prerequisite for high selectivity.

6. Since the quantum yield of **85b** formation is only slightly lower in cyclohexane, benzene, and acetonitrile compared to that in methanol [105], the formation of intermediate net charged species or zwitterions is of little, if any, kinetic significance.

3. Diastereoface Differentiating Photoadditions

If entropy- and/or enthalpy-controlled stereoselection mechanisms [116] are operative, the reaction should be sensitive to diastereofacial selection.

When either enantiomer of 2-(2-methoxymethyl-1-pyrrolidinyl)propenenitrile, namely, (*S*)-(−)-**3l** and (*R*)-(+)-**3l** [57,103], respectively, were applied in the photocycloaddition to **84** in cyclohexane, only one single diastereomeric adduct, namely, (+)-**85l** or (−)-**85l**, respectively, was obtained [57,103] (Scheme 28). Both adducts showed opposite rotations and were converted by copper-ion

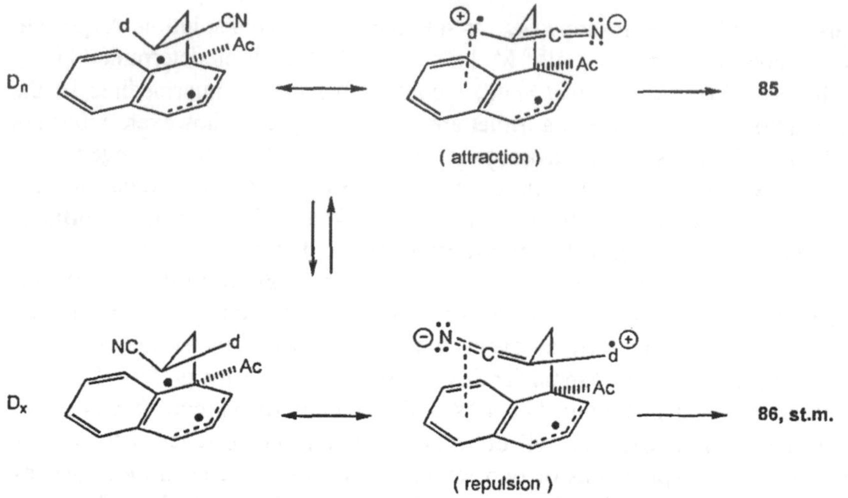

Scheme 27

assisted hydrolysis of the aminonitrile function [15] into the diketones (−)-**92** and (+)-**92**, respectively. The absolute configuration of (−)-**92** and thus also at C-1 and C-4 of (+)-**85l** could be delineated from the circular dichroism of its 2,3-dihydro derivative (+)-**93**. The latter showed a positive Cotton effect ($\Delta\epsilon$ = +12.87 at 306.8 nm and +12.94 at 296.8 nm) [103], resembling the same positive Cotton effect observed for (1R,4R)-bicyclo[2.2.2]oct-5-en-2-one [117] and as to be expected for this absolute arrangement of homoconjugated double bond and carbonyl group in β,γ-unsaturated ketones [118].

Thus, the absolute configuration of (+)-**93** (1S,4R) and therefore also for (−)-**92** (1R,4R) was established as shown in Scheme 28. Since the donor-endo geometry of (+)-**85l** was evident from the ^{1}H NMR spectrum (see above), the absolute configuration of (+)-**85l** was derived as 1R,4R,9R,2'S as shown.

Surprisingly, when alkene (S)-(+)-**3m**, carrying the higher ring homologous auxiliary compared to **3l**, was used in the same photoaddition, the cycloaddition was less diastereoselective [103,110]. Furthermore, in the main adduct (−)-**85m**, as unambiguously demonstrated by an x-ray crystal structure determination, the new stereogenic centers created had the *opposite chirality* as that in (+)-**85l** [110]. As a consequence, hydrolytic destruction of the C-9 stereocenter led to the diketone (+)-**92**. The structure of the minor isomer (not isolated from mother liquors) awaits clarification.

Whereas with (S)-(−)-**3l** the diastereofacial selection was ketone-*Si*/alkene-*R* [103], exchange of the five-membered auxiliary against its six-membered ho-

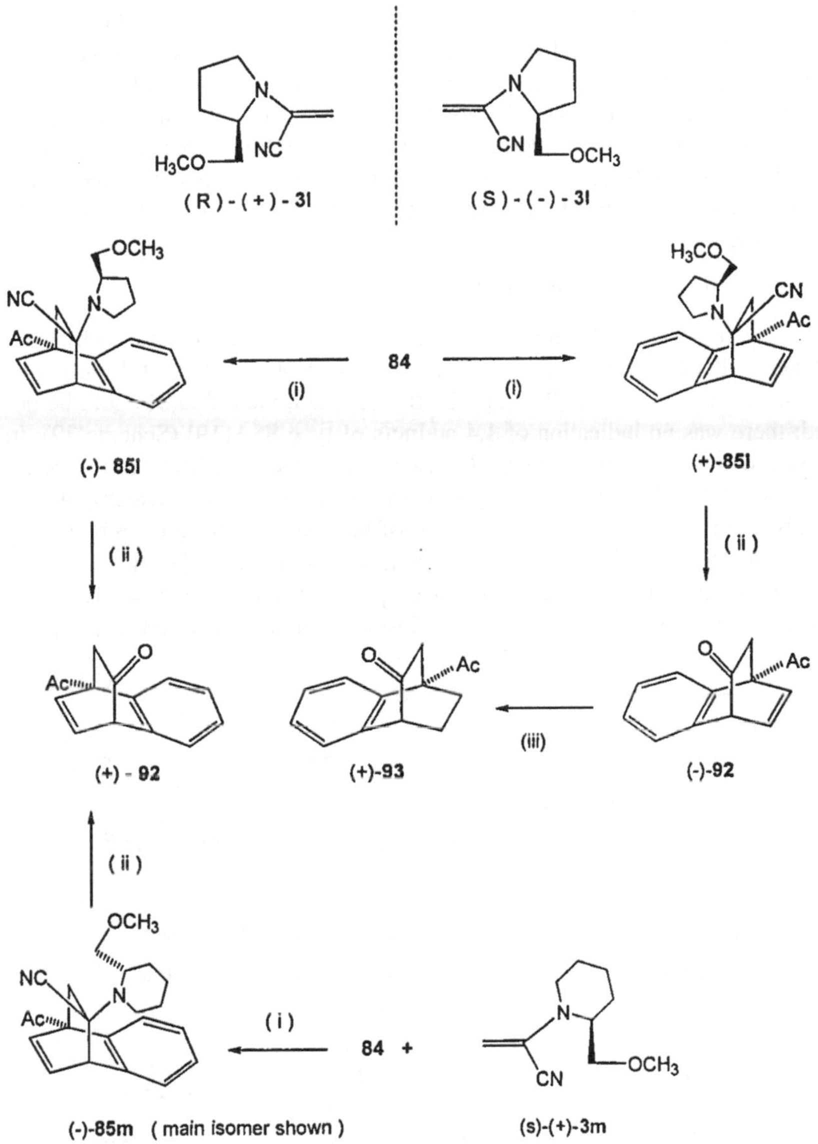

(R)-(+)-3l   (S)-(-)-3l

(-)-85l   84   (+)-85l

(+)-92   (+)-93   (-)-92

(-)-85m (main isomer shown)   (s)-(+)-3m

(i) hv, >280 nm, C$_6$H$_{12}$ − (ii) CuSO$_4$/H$_2$O/Na$_2$HPO$_4$, room temp. − (iii) EtOAc, 1 atm H$_2$/Pt, room temp. − Ac = COCH$_3$

Scheme 28

molog of the same chirality led to a reversal in the preferred orientation, i.e., ketone-*Re*/alkene-*Si* as found for the addition of (*R*)-(+)-**3l** to ketone **84** [110]. This finding demonstrates the importance of steric effects either in the phase of reactants approaching or in the selection between "successful" and "unsuccessful" biradicals. It is conceivable that the preferred conformations in the participating alkenes (**3l** or **3m**) are of decisive influence. However, more detailed investigations, including temperature dependence studies [116], are needed to clarify these findings.

4. Ring-Substituted 1-Acetonaphthones

When 4-cyano-(**94a**) and 4-methoxy-1-acetophenone (**94b**) (0.1 M each) were irradiated in dry benzene in the presence of an equimolar amount of 2-morpholi-nopropenenitrile (**3b**), only [2 + 2] 1,2-adducts (**95a,b** and **96a,b**) could be detected; there was no indication of 1,4 adducts of type **85** [119] (Scheme 29). In both cases, the donor-exo isomers (**95a,b**) dominated (exo/endo ratios determined by ¹H NMR integration, major isomer isolated by chromatography and crystallization). The relative configuration of **95b** has also been unambiguously corroborated by a single-crystal x-ray structure determination [119].

On the other hand, 2-methoxy-1-acetophenone and **3b** gave solely a 83% yield of the 1,4 adduct **97** [119]. These findings support the interpretation that

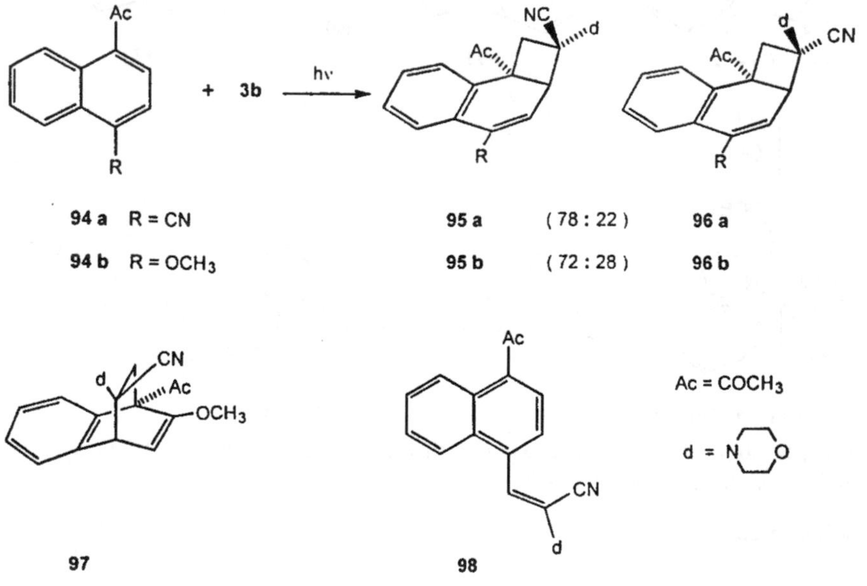

94 a R = CN	**95 a** (78 : 22)	**96 a**
94 b R = OCH₃	**95 b** (72 : 28)	**96 b**

Ac = COCH₃

d = N⌒O

97 **98**

Scheme 29

any intermediate biradicals avoid ring closure at additionally substituted centers.

A totally different course of reaction was observed for the light-induced interaction of **3b** with 4-bromo- [119,120] and 4-chloro-1-acetonaphthone [120]. From the former, substantial amounts of morpholinium bromide and the substitution product **98**, clearly originating from a multistep sequence, were formed [119]. Again, the structure of the latter was confirmed by a single-crystal x-ray structure determination [120]. At present, an electron transfer–initiated release of bromide from C-4, followed by attack of the 4-acetyl-1-naphthyl radical on **3b**, seem to be decisive steps in the formation of **98**.

5. Photocycloadditions to Naphthoic Esters

Six captodative 2-aminopropenenitriles have been successfully added to methyl-1-naphthoate (**99**) [107,108]. One or two minor byproducts, detected by additional weak signals in the ^{1}H NMR spectra of the total photolysates, accompany the main and solely isolated [4 + 2] 1,4-adducts **100b,c,k,n-p**. The endo orientation of the cyclic donor groups has been corroborated by NOE intensity enhancement studies [108] in the cases of **100c,p** and in the other cases from analogy with **100c,p** as evident from the similarities in the 2H, 3H, 4H ABX chemical shifts (Scheme 30).

Since 0.025 M 3,3,4,4-tetramethyl 1,2-diazetine-1,2-dioxide [85,86] completely quenches the 1,4 addition of **3b** to **99** [107], π,π^*-triplet reactivity is

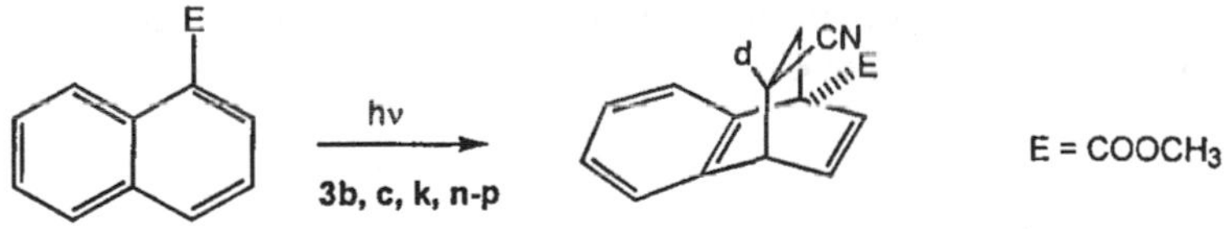

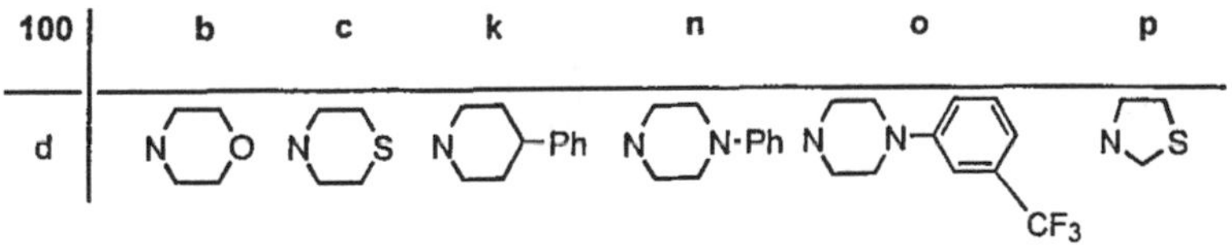

Scheme 30

envisaged for **99** as it is for 1-acetonaphthone (**84**). The relatively sluggish photo-addition compared to **84** is attributed to the lower singlet–triplet intersystem crossing efficiency [121] of **99** [108.]

In contrast to this behavior, enol ethers and *tert*-butyl vinylsulfide are readily added to C-1, C-2 of **99** in a [2 + 2] mode [122]. This means that **99** adds captodative 2-aminoacrylonitriles via the triplet state in a [4 + 2] 1,4-mode and electron-rich alkenes of the type mentioned in a (singlet?) [2 + 2] 1,2-mode.

Methyl-2-naphthoate does not add **3b** upon excitation but forms the known [96b] dimer instead [106].

Methyl-4-bromo- (**101a**) and methyl 4-chloro-1-naphthoate (**101b**), in analogy to 4-bromo- and 4-chloro-1-acetonaphthone (see above), undergo dis-placement of the halogen atom when irradiated in cyclohexane solution in the presence of **3b** with formation of esters **102** [122] (Scheme 31). Very likely the same reaction course is effective in this case as with the halogenated 1-acetonaph-thones.

While dimethyl naphthalene-1,4-dicarboxylate (**103**) and related com-pounds undergo a stepwise 1,8 photoaddition of isobutene [98b], no such addition is observed with 2-morpholinopropenenitrile (**3b**). Instead, a sensitized [2 + 2] dimerization of the latter to the previously described [19,20a] cyclobutane **4b** (Scheme 32) occurred [123] when the irradiation was carried out in diethyl ether. In contrast, in acetonitrile solution, [2 + 2] 1,2-addition dominated, yielding product **104** [124].

The photochemistry of carboxylic acids so far has received little attention in comparison with the wealth of investigations with carboxylic esters and, above all, with ketones. This may have practical reasons, but the potential for decarbox-ylation also needs to be considered.

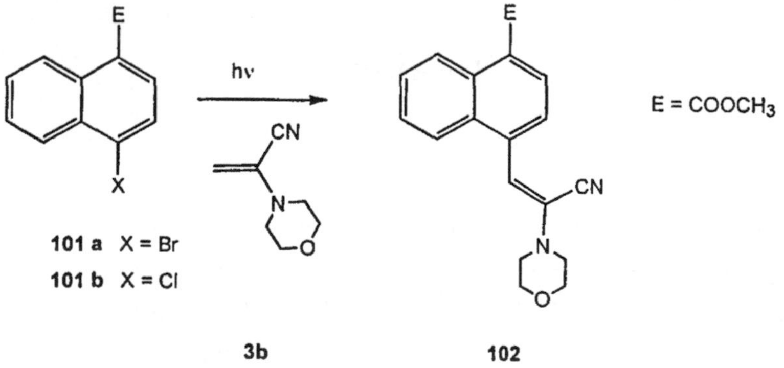

101 a X = Br
101 b X = Cl

E = COOCH$_3$

3b   102

Scheme 31

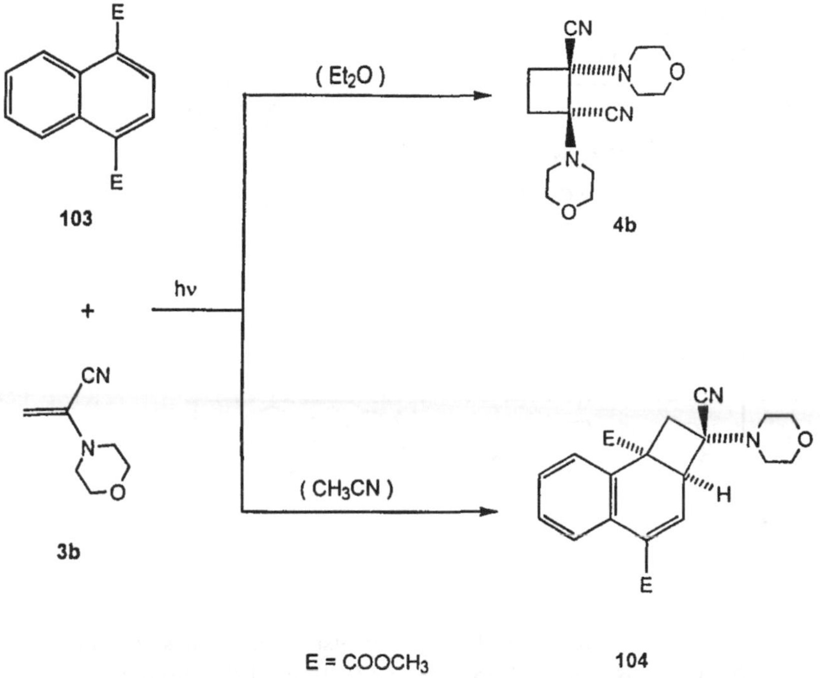

E = COOCH₃

Scheme 32

Therefore, it may be of interest that the carboxylic acids corresponding to esters **99** and **103**, namely, naphthalene-1-carboxylic and naphthalene-1,4-dicarboxylic acid, were found to undergo the same types of photoadditions ([4 + 2] and [2 + 2] in the case of naphthalene-1-carboxylic acid, and [2 + 2] only in the case of the 1,4-diacid) with the same preferences with respect to product configurations [124]. Naphthalene-2-carboxylic acid proved to be unreactive under the same conditions.

6. Di-π-methane Rearrangements of Cycloadducts

The 1,4 adducts **85** and **100** are in fact dihydrobenzobarrelenes and as such rigid di-π-methanes [125]. The various functionalities attached to the skeleton of the methyl ketones **85** do not interfere in any way with di-π-methane reactivity: Upon 254 nm excitation in acetonitrile **85b** is transformed smoothly into the dihydrobenzosemibullvalene **105b** in 80% yield [105]. In the same way, the 1,4 cycloadducts **85d,g–i** are transformed into **105d,g–i** (Scheme 33). In some cases, small fractions of the starting materials **85** are cleaved into the components **84** and **3**. Duplication of the ¹H NMR signal sets suggests further that C-9 epimers of **85g–i**

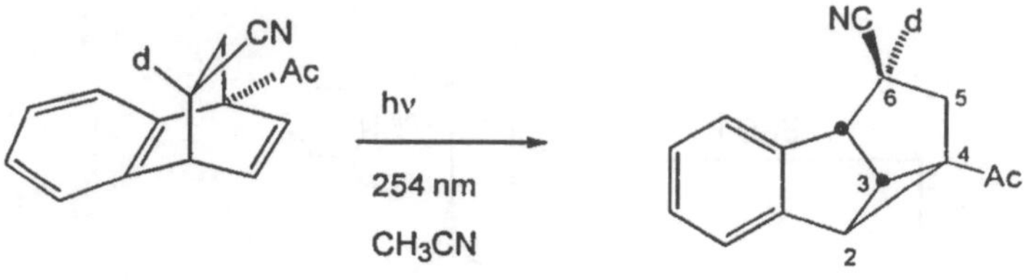

85, 105	b	d	g	h	i
d	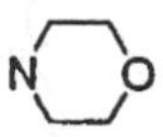 N⟨ ⟩O	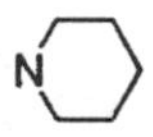 N⟨ ⟩	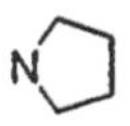 N⟨ ⟩	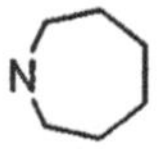 N⟨ ⟩	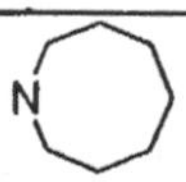 N⟨ ⟩

Scheme 33

and consequently C-6 epimers of **105g–i** are also formed in small amounts [111]. An unambiguous structural proof is available for **105d** through a single-crystal structural analysis [111].

While the methyl ketone bicyclics **85b,d,g–i** show di-π-methane reactivity, the methyl ester **100b**, when subjected to otherwise the same irradiation conditions, did not show any semibullvalene type of product. Instead, retrocleavage into **3b** and methyl 1-naphthoate (**99**) predominated. In addition, some intractable product of higher molecular weight was produced [120b].

The di-π-methane rearrangement is governed by multiplicity effects. As a rule of thumb it is accepted that conformationally flexible acyclic reactants react via the excited singlet state, whereas cyclic and bicyclic reactants tend to react via the triplet excited state [125a,b]. Since the successful rearrangements reported for **85b,d,g–i** were carried out without external sensitization, as was the unsuccessful trial with **100b**, it may be concluded that the excited methyl ketones **85** underwent intersystem crossing to a reactive triplet state sufficiently rapidly, whereas the ester **100b** did not. For a rationalization of the rearrangement, Ref. 125 or the cited original literature containing ample precedence should be consulted.

C. Acyl Quinolines

In the preceding chapter the hypothesis was put forward that an attractive interaction between the partly positively charged donor group and the benzenoid ring in the U-shaped donor-endo-oriented biradical D_n was responsible for the preferential formation of type **85** products (see Scheme 28).

If the benzenoid ring in D_n were replaced by a (electron poorer) pyridine ring, the attractive interaction should decrease to some extent. Thus, six selected acylquinolines (**106a–c, 107a–c**) were tested as addends for 2-morpholinopropenenitrile (**3b**), 2-(*tert*-butylthio)propenenitrile (**16**), and *tert*-butylvinylsulfide (**76**).

In all cases, a decreased reactivity was observed compared to analogous reactions with 1-acylnaphthalenes, attributed to the lower electron density available in the quinolines. This aside, there is little change in regio- and stereoselectivity [126,127] (Scheme 34).

The conversion and yields were generally lower than with the corresponding naphthalenes. At best, 33% of endo-1,4 adduct (here **109a**) and 13% of 1,2 adduct (here **108a**, one isomer only) were obtained. In the other cases, yields were even lower. The benzoylated compounds **106b** and **107b** did not give rise to any 1,2 adducts, and only traces at best of donor-exo 1,4 adducts (**111b** or **113b**, respectively) could be observed besides the endo epimers **110b** and **112b**. Sulfides **16** and **76** could successfully be added only to methylquinoline-8-carboxylate (**106c**), giving rise to adducts **114** and **115** [126,127]. Hydrolysis of donor-endo **110a,c** furnished the dicarbonyl compounds **112** and **113** [127].

Upon ≥280 nm irradiation in hexadeuterobenzene, **108a** was cleaved into the components **106** and **3b**. This suggests that, as in the case of the acylnaphthalenes, the 1,2 adducts are built up as long as they are screened by starting materials **106** and **107** and eventually are degraded by the same UV wavelength range that builds up all photoproducts and to which the 1,4 adducts are stable [127].

D. Anthracenes and Benzanthracenes

1. 1-(9-Anthyl)ethanone

In 0.1 M benzene solution and in the presence of an equimolar amount of 2-(1-piperidinyl)propenenitrile, the title compound **116** was transformed into the head-to-tail [4 + 2] 9,10-adduct **117** [103]. Large amounts of the well-known head-to-tail dimer **118** [128] impeded the isolation of **117**, so the dimer was cleaved by 254-nm irradiation of the crude photolysate. Filtration and concentration allowed precipitation of **117** and recovery of the starting material. Based on nonrecovered starting material **116**, the yield of **117** was 62% [103] (Scheme 35).

2. Addition to the Hydrocarbons: Anthracene and Anellated Anthracenes

Light-induced cycloadditions of alkenes to anthracene occur at the carbon atoms 9 and 10 in a stepwise fashion [99] via intermediate biradicals. For this reason

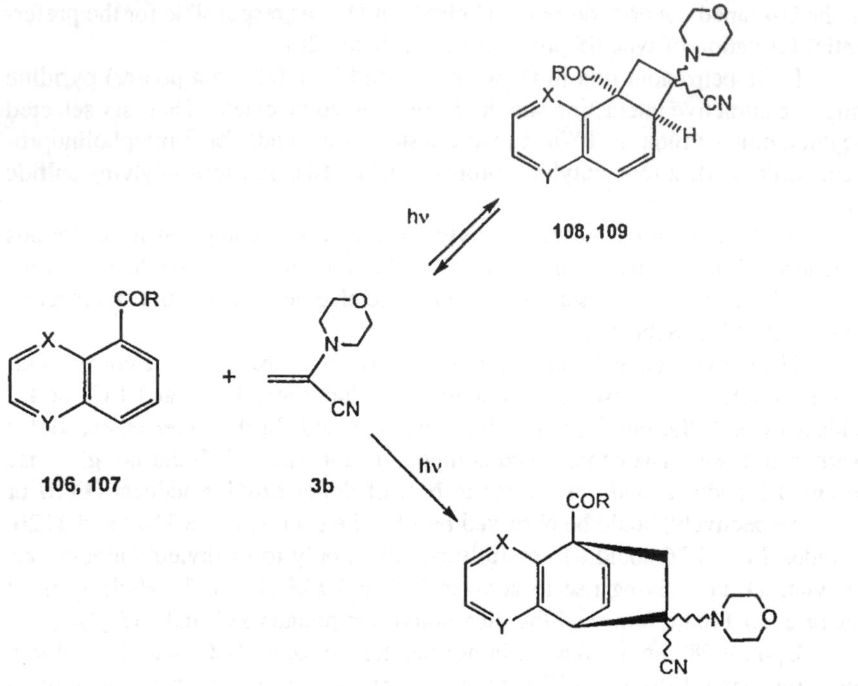

108, 109

106, 107   **3b**

110, 111

106, 108, 110 : X = N,Y = CH.- 107, 109, 111 : X = CH. Y = N

106 - 111 : a : R = CH₃; b : R = C₆H₅ ; c : R = OCH₃

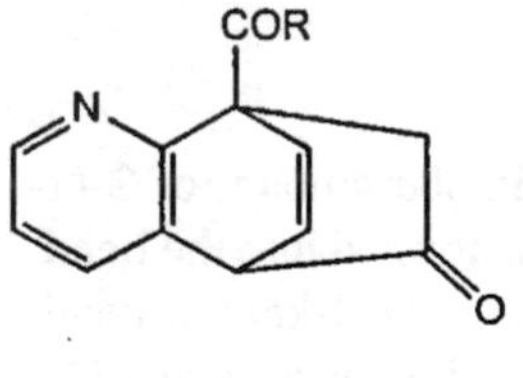

112 : R = CH₃
113 : R = OCH₃

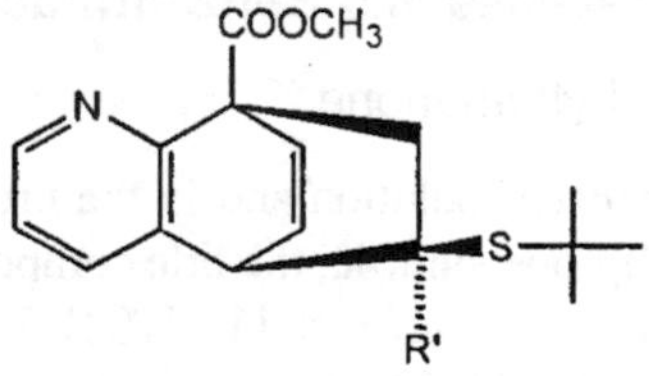

114 : R' = CN
115 : R' = H

Scheme 34

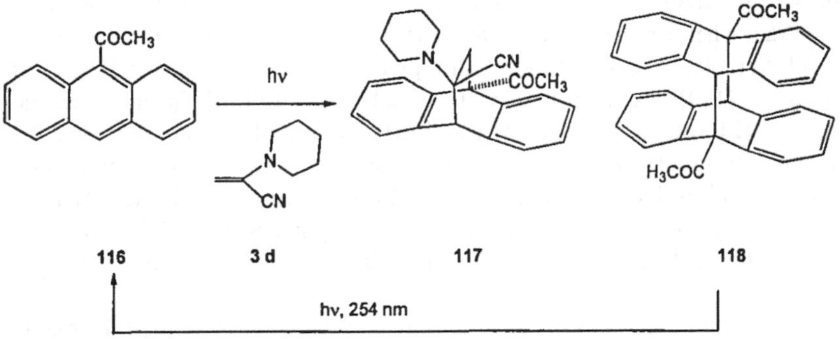

Scheme 35

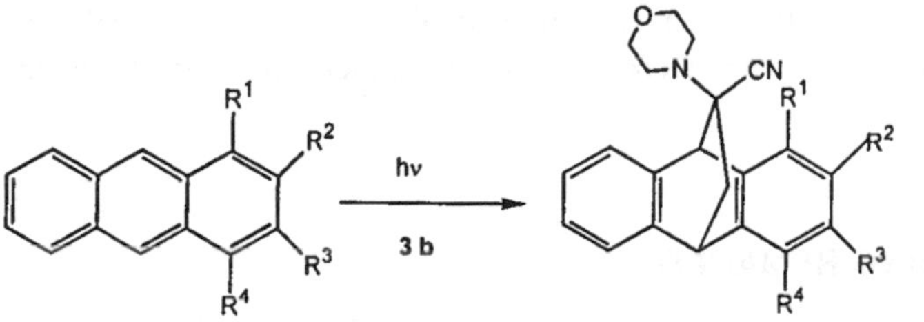

119 a $R^1 - R^4 = H$

    b $R^1, R^2 = X; R^3, R^4 = H$

    d $R^1, R^2 = X; R^3, R^4 = H$

X = $-HC = C - C = CH-$
            | |
            H   H

120 a $R^1 - R^4 = H$

    b $R^1, R^2 = X; R^3, R^4 = H$

    c $R^1, R^2 = H; R^3, R^4 = X$

    d $R^1, R^2 = X; R^3, R^4 = H$
       (*endo* and *exo*)

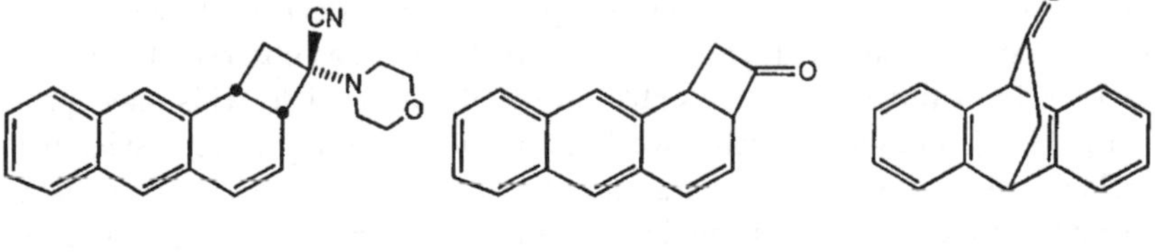

121   122   123

Scheme 36

captodative [1] alkenes should be especially suitable for such additions. Anthracene (119a), when irradiated ($\lambda \geq$ 280 nm) in benzene solution in the presence of a fivefold molar excess of 2-morpholinopropenenitrile (3b), gave a 35% yield of the dimer [129] besides a 53% yield of adduct 120a [56]. In one case, the 1,2 adduct 121 was also formed (12%) [56]. By hydrolysis on silica gel, the latter could be transformed into the cyclobutanone 122 and the former into the symmetrical bridged ketone 123 [56]. The structures of 121 and 122 have been derived by comparison with the corresponding naphthalene 1,2 adducts 60 and 62 (Scheme 18), the latter being a well-characterized [130] compound (Scheme 36).

Benz[a]anthracene (119b) gave a total of 56% yield of adducts 120b,c in a 1:4 ratio (one stereoisomer in either case, with the morpholino group oriented toward the benzenoid ring and away from the naphthalene moiety [131]. In contrast, with dibenz[a,c]anthracene (119d), both stereoisomers of adduct 120d (50% endo and 40% exo) were collected. The additions in all cases occurred at the sites which are "anthracene-9,10-like." The structural assignments were made on the basis of ^{1}H NOE intensity enhancement studies showing the spatial proximity of morpholine aminomethylene protons and either benzo or phenanthro protons [131].

IV. CONCLUDING REMARKS

Although much work has yet to be done on the reactions described herein to unravel their full potential and to arrive at a thorough understanding of the selectivities observed, it should be evident from the body of work accumulated that the ketene equivalency of the 2-aminopropenenitriles opens access to introduction of functionalized anellated cyclobutanes and to ethano bridges bearing keto group precursors as α-aminonitrile functions with high regio- and stereoselectivity. The captodative nature of the olefins employed is certainly of help in stabilizing biradical intermediates and thus promoting stepwise cycloadditions.

Aside from this, 2-aminopropenenitriles (and perhaps to a smaller extent also the analogous vinyl sulfides) may have good complexing properties. This kind of behavior may also be present in the captodatively substituted radical termini of intermediate biradicals and thus participate in steering stereoselectivity. It appears that certainly all photo–Diels–Alder additions observed follow the principle of formation of the most stable biradical intermediate. This may apply also to [2 + 2] additions, which have been found to be highly regioselective. Two related cases with biradical character in the rate-determining step for cyclization have been found in the *intra*molecular photoaddition of enol ether moieties tethered to triplet excited acetonaphthones [132].

ACKNOWLEDGMENTS

The work carried out at Duisburg has been supported by the provincial government of Nordrhein-Westfalen, Deutsche Forschungsgemeinschaft, Alexander-von-Humboldt-Stiftung and Fonds der Chemischen Industrie. The author thanks these institutions. Technical assistance in preparing the manuscript by J. Leven is gratefully acknowledged.

REFERENCES

1. Reviews: (a) Viehe, H. G., Merényi, R.; Stella, L.; Janousek, Z. *Angew. Chem.*, **1979**, *91*, 982–997; *Angew. Chem. Int. Ed. Engl.* **1979**, *18*, 917. (b) Viehe, H. G.; Janousek, Z.; Merényi, R.; Stella, L. *Acc. Chem. Res.* **1985**, *18*, 148–154. (c) Viehe, H. G.; Merényi, R.; Janousek, Z. *Pure Appl. Chem.* **1988**, *60*, 1635–1644. (d) Viehe, H. G.; Janousek, Z.; Merényi, R. *Free Radicals in Synthesis and Biology*, Minisci, F.; ed., **1989**, Kluwer Academic Publishers, Dordrecht, 1–26. (e) Sustmann, R.; Korth, H.-G. *Adv. Phys. Org. Chem.* **1990**, *26*, 131–178.
2. Katritzky, A. R.; Zerner, M. C.; Karelson, M. M. *J. Am. Chem. Soc.* **1986**, *108*, 7213–7214 and references cited therein.
3. Balaban, A. T.; Bologa, U.; Caproiu, M. T.; Grecu, N.; Negoita, N.; Walter, R. I. *J. Chem. Research* (S) **1988**, 274–275.
4. Balaban, A. T. *Rev. Roum. Chim.* **1971**, *16*, 725–737.
5. (a) Zamkanei, M.; Kaiser, J. H.; Birkhofer, H.; Beckhaus, H.-D., Rüchardt, C. *Chem. Ber.* **1983** *116*, 3216–3234. (b) Bordwell, F. G.; Lynch, T.-Y. *J. Am. Chem. Soc.* **1989**, *111*, 7558–7562. (c) Birkhofer, H.; Beckhaus, H.-D.; Rüchardt, C. *Substituent Effects in Radical Chemistry*, Viehe, H. G.; Janousek, Z.; Merényi, R. Eds., D. Reidel Publishing Co: Dordrecht, 1986, 199–218.
6. (a) Beckhaus, H.-D.; Rüchardt, C. *Angew. Chem.* **1987**, *99*, 807–808. *Angew. Chem. Int. Ed. Engl.* **1987**, *26*, 770–771. (b) Olson, J. B.; Koch. T. H. *J. Am. Chem. Soc.* **1986**, *108*, 756–761.
7. Welle, F.; Verevkin, S. P.; Keller, M.; Beckhaus, H.-D.; Rüchardt, C. *Chem. Ber.*, **1994**, *127*, 697–710.
8. Crans, D.; Clark, T.; v. Ragué-Schleyer, P. *Tetrahedron Lett.* **1980**, *21*, 3681–3684.
9. Pasto, D. J. *J. Am. Chem. Soc.* **1988**, *110*, 8164–8175.
10. Leroy, G.; Peeters, D.; Sana, M.; Wilante, C. *Substituent Effects in Radical Chemistry*, Viehe, H. G.; Janousek, Z.; Merényi, R., Eds., D. Reidel Publishing Co: Dordrecht, **1986**, 1–48.
11. (a) Riga, J.; Verbist, J. J.; Mignani, S., Janousek, Z.; Merényi, R.; Viehe, H. G. *J. Chem. Soc. Perkin Trans. II* **1985**, 883–885. (b) Sustmann, R.; Müller, W.; Mignani, S.; Merényi, R.; Janousek, Z.; Viehe, H. G. *New J. Chem.* **1989**, *13*, 557–563.
12. Stella, L. *Substituent Effects in Radical Chemistry*, Viehe, H. G.; Janousek, Z.; Merényi, R., Eds., D. Reidel Publishing Co: Dordrecht, **1986**, 361–370.
13. De Cock, C.; Piettre, S.; Lahousse, F.; Janousek, Z.; Merényi, R.; Viehe, H. G. *Tetrahedron* **1985**, *41*, 4183–4193.

14. Boucher, J.-L.; Stella, L. *Tetrahedron* **1985**, *41*, 875–887.
15. Büchi, G.; Liang, P. H.; Wüest, H. *Tetrahedron Lett.* **1978**, 2763–2764.
16. (a) Gundermann, K.-D. *Intrascience Chem. Reports* **1972**, *6*, 91–112. (b) Gundermann, K.-D.; Huchting, R. *Chem. Ber.* **1959**, 415–424.
17. Toye, J. *Thèse*, Université Catholique de Louvain, Louvain-la-Neuve, 1977.
18. Ksander, G.; Bold, G. H.; Lattmann, R.; Lehmann, C.; Früh, T.; Xiang, Y.; Inomata, K.; Buser, H. P.; Schreiber, J.; Zass, E.; Eschenmoser, A. *Helv. Chim. Acta* **1987**, *70*, 1115–1172.
19. Memarian, H. R. *Doctoral Thesis*, University of Duisburg, 1986.
20. (a) Döpp, D.; Bredehorn, J. *J. Inf. Recording* **1996**, *22*, 401–407. (b) Döpp, D.; Bredehorn, J. *unpublished*. Bredehorn, J. *Doctoral Thesis*, University of Duisburg, 1994.
21. Dilling, W. L.; Kroening, R. D.; Little, J. C. *J. Am. Chem. Soc.* **1970**, *92*, 928–948 and references cited therein.
22. (a) Reviews: Mizuno, K.; Otsuji, Y. *Top. Curr. Chem.* **1994**, *169* (Electron Transfer 1), 301–346, and refs. cited therein. (b) Mattay, J. *Synthesis* **1989**, 233–252. (c) Mattay, J. *Angew. Chem.* **1987**, *99*, 849–870; *Angew. Chem. Int. Ed. Engl.* **1987**, *26*, 825–845. (d) Gould, I. R.; Farid, S. *Acc. Chem. Res.* **1996**, *29*, 522–528.
23. (a) Mattes, S. L.; Farid, S. *J. Am. Chem. Soc.* **1982**, *104*, 1454–1456. (b) Murov, S. L.; Carmichael, F.; Hug, G. L. Handbook of Photochemistry, 2nd edition, M. Dekker, New York 1993, p. 278. (c) Arnold, D. R.; Maroulis, A. J. *J. Am. Chem. Soc.* **1976**, *98*, 5931–5937.
24. (a) Rehm, D.; Weller, A. *Ber. Bunsenges. Physik. Chem.* **1969**, *73*, 834–839. (b) Rehm, D.; Weller, A. *Israel J. Chem.* **1970**, *8*, 259–271. (c) Weller, A.; *Z. Phys. Chem. (N. F.)* **1982**, *133*, 93–98.
25. Bolton, J. R.; Archer, M. D. *Electron Transfer in Inorganic, Organic and Biological Systems*, Bolton, J. R.; Mataga, N.; McLendon, G.; Eds., *Advances in Chemistry Series Vol. 228*; ACS: Washington, 1991, p. 7–23, and refs. cited therein.
26. (a) Mattes, S. L.; Farid, S. *J. Am. Chem. Soc.* **1986**, *108*, 7356–7361, and refs. cited therein. (b) Kim, J. L.; Schuster, G. B. *J. Am. Chem. Soc.* **1992**, *114*, 9309–9317 and earlier papers.
27. (a) Griesbeck, A. G. *CRC-Handbook of Organic Photochemistry and Photobiology*, Horspool, W. M.; Song, P.-S., Eds., 1995, CRC-Press, Boca Raton, 522–535 and 550–559. (b) Mattay, J.; Conrads, R., Hoffmann, R. *Methoden der Organischen Chemie (Houben-Weyl) Vol. E 21c (Stereoselective Synthesis)*, Helmchen, G.; Hoffmann, R. W.; Mulzer, J.; Schaumann, E., Eds.; Thieme: Stuttgart **1995**, 3133–3178. (c) Porco, J. A.; Schreiber, S. L. *Comprehensive Organic Synthesis* Vol. 5, Trost, B. M. Ed., Pergamon Press: Oxford 1991, 151–192. (d) Jones II., G. *Organic Photochemistry*, Padwa, A., Ed., Vol. 5, Dekker: New York, 1981, 1–123.
28. Bach, T. *Angew. Chem.* **1996**, *108*, 976–977; *Angew. Chem. Int. Ed. Engl.* **1996**, *35*, 885–886.
29. Sengupta, D.; Chandra, A. K.; Nguyen, M. T. *J. Org. Chem.* **1997**, *62*, 6404–6406.
30. Döpp, D.; Fischer, M.-A. *Rec. Trav. Chim. Pays-Bas* **1995**, *114*, 498–503.
31. Döpp, D.; Memarian, H. R.; Fischer, M.-A.; van Eijk, A. M. J.; Varma, C. A. G. O. *Chem. Ber.* **1992**, *125*, 983–984.
32. Memarian, H.-R.; Henkel, G.; Döpp, D. *in preparation*.

33. Griesbeck, A. G.; Mauder, H.; Stadtmüller, S. *Acc. Chem. Res.* **1994**, *27*, 70–75 and references cited therein.
34. Griesbeck, A. G.; Buhr, S.; Fiege, M.; Schmickler, H.; Lex, J. *J. Org. Chem.*, **1998**, *63*, 3847–3854.
35. van Wolven, C., Döpp, D.; Fischer, M. A. *J. Inf. Recording*, in press.
36. Vasudevan, S.; Brock, C. P.; Watt, D. S.; Morita, H. *J. Org. Chem.* **1994**, *59*, 4677–4679.
37. Döpp, D.; Hassan, A. A.; Henkel, G. *Liebigs Ann.* **1996**, 697–700.
38. Bauschlicher, T. and Döpp, D. *unpublished.* Taken from planned doctoral thesis of Bauschlicher, T.; University of Duisburg.
39. (a) Hastings, D. J.; Weedon, A. C. *Tetrahedron Lett.* **1991**, *32*, 4107–4110 and references cited therein. (b) Hastings, D. J.; Weedon, A. C. *J. Org. Chem.* **1991**, *56*, 6326–6331, and references cited therein.
40. Hammond, G. S.; Stout, C. A.; Lamola, A. A. *J. Am. Chem. Soc.* **1964**, *86*, 3103–3106.
41. Krauch, C. H.; Farid, S.; Schenck, G. O. *Chem. Ber.* **1966**, *99*, 625–633.
42. Klaus, C. P.; Thiemann, C.; Kopf, J.; Margaretha, P. *Helv. Chim. Acta* **1995**, *78*, 1079–1082.
43. Margaretha, P. *Molecular and Supramolecular Photochemistry: Organic Photochemistry*, Ramamurthy, V.; Schanze, K. S., Eds.; Marcel Dekker: New York 1997, p. 85–110.
44. Blecking, A. *Doctoral Thesis*, Duisburg University, 1997.
45. (a) Döpp, D.; Neubauer, S. *unpublished.* (b) Neubauer, S. *Planned doctoral thesis*, University of Duisburg.
46. Suginome, H.; Kobayashi, K. *Bull. Chem. Soc. Jpn.* **1988**, *61*, 3782–3784.
47. Kirpichenok, M. A.; Mel'nikova, L. M.; Yufit, D. S.; Struchkov, Y. T.; Grandberg, I. I.; Denisov, L. K. *Khim. Geterotsikl. Soedin.* **1988**, *6*, 1176–1184.
48. Kirpichenok, M. A.; Mel'nikova, L. M.; Denisov, L. K.; Grandberg, I. I. *Khim. Geterotsikl. Soedin.* **1988**, *6*, 1169–1175.
49. Kirpichenok, M. A.; Mel'nikova, L. M.; Denisov, L. K.; Grandberg, I. I. *Khim. Geterotsikl. Soedin.* **1990**, *8*, 1022–1027.
50. Koziar, J. C.; Cowan, D. O. *Acc. Chem. Res.* **1978**, *11*, 334–341 and references cited therein.
51. (a) Meinwald, J.; Samuelson, G. E.; Ikeda, M. *J. Am. Chem. Soc.* **1970**, *92*, 7604–7606. (b) Hartmann, W.; Heine, H. G. *Angew. Chem.* **1971**, *83*, 291; *Angew. Chem. Int. Ed. Engl.* **1971**, *10*, 272.
52. Plummer, B. F.; Hall, R. A. *Chem. Commun.* **1970**, 44–45.
53. (a) Plummer, B. F.; Chihal, D. M. *J. Am. Chem. Soc.* **1971**, *93*, 2071–2072. (b) Feree, Jr., W. I.; Plummer, B. F. *J. Am. Chem. Soc.* **1973**, *95*, 6709–6717. (c) Feree, Jr., W. I.; Plummer, B. F.; Schloman, W. W. *J. Am. Chem. Soc.* **1974**, *96*, 7741–7746.
54. Plummer, B. F.; Songster, M. *J. Org. Chem.* **1990**, *55*, 1368–1372.
55. Weber, J.; Döpp, D. *J. Inf. Rec. Mats.* **1994**, *21*, 557–558.
56. Weber, J. *Doctoral Thesis*, University of Duisburg, 1995.
57. Döpp, D.; Pies, M.; *J. Chem. Soc., Chem. Commun.* **1987**, 1734–1735.

58. McCullough, J. J. *Chem. Rev.* **1987**, *87*, 811–860.
59. See Sections *"Photochemistry of Aromatic Compounds,"* Specialist Periodical Reports: Photochemistry, Vol. 1 (1970)–18 (1987), Bryce-Smith, D.; Ed., Vol. 19 (1988)–25(1994) Bryce-Smith, D.; Gilbert, A. Eds., Vol. 26 (1995)–29 (1999), Gilbert, A.; Ed., The Royal Society of Chemistry: London.
60. Gilbert, A.; Heath, P.; Kashoulis-Koupparis, A.; Ellis-Davies, G. C. R.; Firth, S. M. *J. Chem. Soc., Perkin Trans. 1*, **1988**, 31–36.
61. Bowman, R. M.; Chamberlain, R. T.; Huang, C.-W.; McCullough, J. J. *J. Am. Chem. Soc.* **1974**, *96*, 692–700.
62. Döpp, D.; Erian, A. W.; Henkel, G. *Chem. Ber.* **1993**, *126*, 239–242.
63. (a) Pac, C.; Sugioka, T.; Mizuno, K.; Sakurai, H. *Bull. Chem. Soc., Jpn.* **1973**, *46*, 238–243. (b) Mizuno, K.; Pac, C.; Sakurai, H. *J. Chem. Soc. Chem. Commun.* **1974**, 648–649. (c) Pac, C.; Mizuno, K.; Sakurai, H. *Nippon Kagaku Kaishi* **1984**, 110–118.
64. Al-Jalal, N. A. *Gazz. Chim. Ital.* **1994**, *124*, 205–207. Related investigations by the same author: Al-Jalal, N. A., Pritchard, R. G.; McAuliffe, C. A. *J. Chem. Res. (S)* **1994**, 452; (M), 2619–2638.-Al-Jalal, N. A. *J. Chem. Res. (S)* **1995**, *44*; (M) 0412–0423.
65. Miyamoto, T.; Mori, T.; Odeira, Y. *Chem. Commun.* **1970**, 1598.
66. (a) Creed, D.; Caldwell, R. A. *J. Am. Chem. Soc.* **1974**, *96*, 7369–7371. (b) Caldwell, R. A.; *J. Am. Chem. Soc.* **1973**, *95*, 1690–1692. (c) Farid, S.; Doty, J. C.; Williams, J. L. R. *J. Chem. Soc. Chem. Commun.* **1972**, 711. (d) Farid, S.; Hartmann, S. E.; Doty, J. C.; Williams, J. L. R. *J. Am. Chem. Soc.* **1975**, *97*, 3697–3702.
67. Kaupp, G. *Angew. Chem.* **1973**, *85*, 766–768; *Angew. Chem. Int. Ed. Engl.* **1973**, *12*, 765–767.
68. Mai, J. C.; Cheng, J. Y.; Ho, T. T. *J. Photochem. Photobiol. A.: Chem.* **1992**, *66*, 53–60.
69. Caldwell, R. A.; Smith, L. *J. Am. Chem. Soc.* **1974**, *96*, 2994–2996.
70. (a) Mizuno, K.; Pac, C.; Sakurai, H. *Chem. Lett.* **1973**, 309–310. (b) Mizuno, K.; Pac, C.; Sakurai, H. *J. Am. Chem. Soc.* **1974**, *96*, 2993–2994.
71. Caldwell, R. A.; Ghali, N. L.; Chien, C.-K.; DeMarco, D.; Smith, L. *J. Am. Chem. Soc.* **1978**, *100*, 2857–2863.
72. Caldwell, R. A.; Creed, D. *Acc. Chem. Res.* **1980**, *13*, 45–50.
73. Lewis, F. D.; DeVoe, R. J. *Tetrahedron* **1982**, *38*, 1069–1077.
74. Mizuno, K.; Caldwell, R. A.; Tachibana, A.; Otsuji, Y. *Tetrahedron Lett.* **1992**, *33*, 5779–5782.
75. Lewis, F. D.; Barancyk, S. V.; Burch, E. L. *J. Am. Chem. Soc.* **1992**, *114*, 3866–3870, and refs. cited therein.
76. Itoh, H.; Takada, M.; Asano, F.; Senda, Y.; Sakuragi, H.; Tokumaru, K. *Bull. Chem. Soc. Jpn.* **1993**, *66*, 224–230.
77. Itoh, H.; Shibata, H.; Wagatsuma, M.; Nanjoh, H.; Senda, Y.; Sakuragi, H.; Tokumaru, K. *Bull. Chem. Soc. Jpn.* **1993**, *66*, 340–343.
78. Sakuragi, H; Tokumaru, K.; Itoh, H.; Terakawa, K.; Kikuchi, K.; Caldwell, R. A.; Hsu, C. C. *Bull Chem. Soc. Jpn.* **1990**, *63*, 1049–1057.

79. Tscherny, I. *Doctoral Thesis*, University of Duisburg, 1999.
80. Trofimov, B. A.; Mel'der, U. Kh.; Pikver, R. I.; Vyalyk, E. P. *Theor. Exp. Chem.* (Translation from Russian) **1975**, *11*, 129–135.
81. Becker, H. G. O., Ed. *Einführung in die Photochemie*, 2nd ed., Thieme: Stuttgart, 1983, p. 117.
82. Tscherny, I.; Görner, H.; Döpp, D. *to be published.*
83. Tscherny, I.; Döpp, D.; Henkel, G. *J. Inf. Recording* **1998**, *24*, 341–347.
84. Brown, J. K., Williams, W. G. *Chem. Commun.* **1966**, 495–496.
85. Singh, P.; Boocock, D. G. B., Ullman, E. F. *Tetrahedron Lett.* **1971**, 3935–3938.
86. Lok, C. M.; den Boer, M. E.; Cornelisse, J.; Havinga, E. *Tetrahedron* **1973**, *29*, 867–872.
87. Arnold, D. R.; Gillis, L. B.; Whipple, E. B. *Chem. Commun.* **1969**, 918–919.
88. Scharf, H.-D.; Leismann, H.; Erb, W.; Gaidetzka, H. W.; Arctz, J. *Pure Appl. Chem.* **1974**, *41*, 581–600.
89. Somich, C.; Mazzocchi, P. H.; Ammon, H. L. *J. Org. Chem.* **1987**, *52*, 3614–3619.
90. McCullough, J. J.; McMurry, T. B.; Work, D. N. *J. Chem. Soc. Perkin Trans. 1* **1991**, 461–464.
91. Zupanic, N.; Sket, B. *J. Chem. Soc. Perkin Trans. 1* **1992**, 179–180.
92. Ciganek, E.; Wuonola, M. A.; Harlow, R. L.; *J. Heterocycl. Chem.* **1994**, *31*, 1251–1257.
93. Kohmoto, S.; Kobayashi, T.; Nishio, T.; Iida, I.; Kishikawa, K.; Yamamoto, M.; Yamada, K. *J. Chem. Soc. Perkin Trans. 1* **1996**, 529–535.
94. Noh, T.; Kang, S.; Yu, H.; Lee, S. *Bull. Korean Chem. Soc.* **1999**, *20*, 168–172.
95. Noh, T.; Kim, Ch.; Kim, D. *Bull. Korean Chem. Soc.* **1997**, *18*, 781–783.
96. (a) Teitei, T.; Wells, D.; Sasse, W. H. *Aust. J. Chem.* **1976**, *8*, 1783–1790. (b) Collins, P. J.; Roberts, D. B.; Sugowdz, G.; Wells, D.; Sasse, W. H. F. *Tetrahedron Lett.* **1972**, 321–324.
97. Al-Jalal, N. A. *J. Chem. Res. (S)* **1995**, *44*; (M) 0412–0423.
98. (a) Kubo, Y.; Noguchi, T.; Inoue, T. *Chem. Lett.* **1992**, *10*, 2027–2030. (b) Kubo, Y.; Inoue, T.; Sakai, H. *J. Am. Chem. Soc.* **1992**, *114*, 7660–7663.
99. (a) Kaupp, G. *Chimia* **1971**, *25*, 230–234. (b) Kaupp, G. *Angew. Chem.* **1972**, *84*, 251–261; *Angew. Chem. Int. Ed. Engl.* **1972**, *11*, 313–314. (c) Kaupp, G. *Angew. Chem.* **1972**, *84*, 718–719; *Angew. Chem. Int. Ed. Engl.* **1972**, *11*, 718–719. (d) Kaupp, G.; Dyllick-Brenzinger, R. *Angew. Chem.* **1974**, *86*, 523–524; *Angew. Chem. Int. Ed. Engl.* **1974**, *13*, 478–479. (e) Kaupp, G.; Dyllick-Brenzinger, R.; Zimmerman, I. *Angew. Chem.* **1975**, *87*, 520–521; *Angew. Chem. Int. Ed. Engl.* **1975**, *14*, 491–492. (f) Kaupp, G. *Liebigs Ann. Chem.* **1977**, 254–275. (g) Kaupp, G.; Grüter, H.-W. *Angew. Chem.* **1979**, *91*, 943–944. *Angew. Chem. Int. Ed. Engl.* **1979**, *18*, 881–882. (h) Kaupp, G.; Grüter, H.-W. *Chem. Ber.* **1980**, *113*, 1458–1471. (i) Kaupp, G.; Schmitt, D. *Chem. Ber.* **1981**, *114*, 1567–1571.
100. Epiotis, N. D.; Yates, R. L. *J. Org. Chem.* **1974**, *21*, 3150–3153.
101. Döpp, D.; Krüger, C.; Memarian, H. R.; Tsay, Y.-H. *Angew. Chem.* **1985**, *97*, 1059–1060. *Angew. Chem. Int. Ed. Engl.* **1985**, *24*, 1048–1049.
102. Döpp, D.; Memarian H. R. *Substituent Effects in Radical Chemistry*; Viehe, H. G.; Janousek, Z.; Merényi, R., Eds., D. Reidel: Dordrecht 1986, p. 383–385.

103. Pies, M. *Doctoral Thesis*, University of Duisburg, 1989.
104. Döpp, D.; Memarian, H. R.; Krüger, C.; Raabe, E. *Chem. Ber.* **1989**, *122*, 585–588.
105. Döpp, D.; Memarian, H. R. *Chem. Ber.* **1990**, *123*, 315–319.
106. Mühlbacher, B. *Doctoral thesis*, University of Duisburg, 1991.
107. Döpp, D.; Mlinaric, B. *ACH-Models in Chemistry* **1994**, *131*, 377–381.
108. Döpp, D.; Mlinaric, B. *Bull. Soc. Chim. Belg.* **1994**, *103*, 449–452.
109. Döpp, D.; Bredehorn, J.; Memarian, H. R.; Mühlbacher, B.; Weber, J. *Photochemical Key Steps in Organic Synthesis*, Mattay, J.; Griesbeck, A. G.; Eds., VCH: Weinheim, 1994, p. 186–187.
110. (a) Ixkes, U.; *Doctoral Thesis*, University of Duisburg, 1995. (b) Döpp, D.; Ixkes, U., *to be published*.
111. (a) Kruse, C.; *planned doctoral thesis*, University of Duisburg. (b) Döpp, D.; Kruse, C., *to be published*.
112. (a) Zerwes, L. *planned doctoral thesis*, University of Duisburg. (b) Döpp, D.; Zerwes, L., *to be published*.
113. Döpp, D.; Erian, A. W.; Henkel, G., *in preparation*.
114. Murov, S. L.; Carmichael, I.; Hug, G. L.; eds.; *Handbook of Photochemistry*, 2nd ed., M. Dekker, New York, 1993, p. 70.
115. Döpp, D.; Memarian, H. R.; van Eijk, A. M. J.; Varma, C. A. G. O. *J. Photochem. Photobiol., A: Chem.* **1990**, *53*, 59–67.
116. Buschmann, H.; Scharf, H.-D.; Hoffmann, N.; Plath, M. W.; Runsink, J. *J. Am. Chem. Soc.* **1989**, *111*, 5367–5373.
117. Bunnenberg, E.; Djerassi, C.; Mislow, K.; Moscowitz, A. *J. Am. Chem. Soc.* **1962**, *84*, 2823–2826.
118. Legrand, M.; Rougier, M. J.; *Stereochemistry: Fundamentals and Methods*; Kagan, H.; Ed.; Vol. 2 (Dipole Moments, CD or ORD) Thieme: Stuttgart, 1977, p. 33–183.
119. Memarian, H. R.; Nasr-Esfahani, M.; Boese, R.; Döpp, D. *Liebigs Ann./Recueil* **1997**, 1023–1027.
120. (a) Döpp, D.; Henkel, G.; Schmidt, M.; *in preparation*. (b) Schmidt, M. *planned doctoral thesis*, University of Duisburg.
121. Chow, Y. L.; Lin, X.-Y.; Hu, S. *J. Chem. Soc., Chem. Commun.* **1988**, 1047. The pertinent information is contained in a footnote.
122. Lanfermann, H.; Döpp, D. *J. Inf. Rec. Mats.* **1994**, 545–547.
123. Döpp, D.; Zalesiak, R., *unpublished results*.
124. (a) Kugelberg, A., *planned doctoral thesis*, University of Duisburg. (b) Döpp, D.; Kugelberg, A., *to be published*.
125. (a) Zimmerman, H. E.; *Organic Photochemistry and Photobiology*; Horspool, W. M.; Song, Pill-Soon, Eds.; CRC Press: Boca Raton 1994, p. 184–203, and pertinent references cited therein. (b) See appropriate sections in Ref. 59. (c) Döpp, D.; Zimmerman, H. E.; *Methoden der Organischen Chemie* (Houben-Weyl), Müller, E.; Ed.; Vol. IV/5a; Thieme: Stuttgart 1975, p. 413–432.
126. Jung, A.; Döpp, D. *J. Inf. Rec. Mats.* **1994**, *21*, 543–544.
127. Jung, A. *doctoral thesis*, University of Duisburg 1995; *Deutsche Hochschulschriften* 1060, Hänsel-Hohenhausen: Egelsbach 1995.

128. Tomlinson, W. J.; Chandross, E. A.; Fork, R. L.; Pryde, C. A.; Lamola, A. A. *Applied Optics* **1972**, *11*, 533–548.

129. Leppin, E. *Methoden der Organischen Chemie* (Houben-Weyl); Müller, E., Ed.; Vol. IV/5a; Thieme: Stuttgart 1975, p. 476–483, and references cited therein.

130. R. S. Givens; W. F. Oettle *J. Am. Chem. Soc.* **1971**, *93*, 3963–3968.

131. (a) Neumann, U., *planned doctoral thesis*, University of Duisburg. (b) Neumann, U.; Weber, J.; Döpp, D. *J. Inf. Recording, in press.*

132. Wagner, P. J.; Sakamoto, M. *J. Am. Chem. Soc.* **1989**, *111*, 9254–9256.

4

Photo- and Electroactive Fulleropyrrolidines

Michele Maggini
Università di Padova, Padova, Italy

Dirk M. Guldi
University of Notre Dame, Notre Dame, Indiana

I. INTRODUCTION

The covalent functionalization of fullerenes has over the last few years developed to such an extent that now C_{60}, by far the most abundant and most well-studied fullerene, can be considered a versatile building block in organic chemistry [1]. In contrast to graphite and diamond, C_{60} and the higher fullerenes display a considerable chemical reactivity that has stimulated a systematic modification of their structure [1,2]. The extensive investigation of the addition reactions to C_{60} revealed that the basic principles of its enhanced reactivity stem from the release of strain associated with the bent geometry of the C–C double bonds. Nonchemists might wonder why anybody would break the aesthetically pleasing geometry and perfect symmetry of the fullerenes. One simple answer is because it is challenging. A major problem, connected with the addition reactions to the fullerenes, is the high number of potential products. The second answer is more practical. Functionalized fullerenes are more soluble than their pristine counterparts in solvents of common use and less prone to aggregate. Decreased aggregation is a fundamental prerequisite to exploit and combine the unusual electronic and redox properties of the fullerenes with those of the appended functional groups for applications in materials science [3,4] and medicinal chemistry [5,6]. Thus, deter-

149

mination of the factors that govern the regiochemistry and regioselectivity of these addition reactions is a crucial aspect of this research [1].

Among the most striking properties of C_{60} is its ability to reversibly accept up to six electrons in solution [7] and its low energy of the first singlet excited state compared to other electron acceptors [8]. These properties caused many to believe that C_{60} could be a good partner in photo-induced redox processes. To this end, several research groups have prepared a wide variety of elaborate molecular systems in which C_{60}, acting as an electron acceptor unit, is covalently linked to electron donors such as aromatics [9–18], porphyrins [8,19–28], phthalocyanine [29,30], ferrocene [31–36], tetrathiafulvalene [31,37–40], carotene [41–43], rotaxanes [44,45], and ruthenium(II) polypyridine complexes [46–51]. The study of these systems is concerned with the fundamental understanding of their photophysical properties and of those factors that govern energy and electron transfer processes in relation to natural photosynthesis [52] and to practical applications such as photovoltaic devices for solar energy conversion [53–57]. The C_{60} chromophore is a particularly interesting electron acceptor not only for the above-mentioned reasons, but also because its three-dimensional structure, and its larger size than that of conventional planar acceptors, such as aromatics, quinones, or pyromellitic imides, accelerates the forward electron transfer while retarding the back electron transfer [58]. The distribution of the negative charge over 60 carbon atoms reduces the energy contribution to solvent reorganization. Also the contribution to the internal reorganization energy is probably small due to the rigid nature of the C_{60} framework [58].

II. SYNTHESIS

A. Electro- and Photoactive Fulleropyrrolidines via Azomethine Ylide Cycloaddition

Among the many reactions that have been successfully developed, the 1,3-dipolar cycloaddition of azomethine ylides provides a valuable synthetic procedure for C_{60} functionalization [59–62]. Azomethine ylides can be generated in several ways from readily accessible starting materials following well-established protocols [59]. One of the easiest approaches involves decarboxylation of immonium salts derived from condensation of α-amino acids with aldehydes (ketones can also be used). This affords the 1,3-dipolar intermediate in situ, which then adds to the fullerene yielding a stable fulleropyrrolidine derivative. The C_{2v} symmetrical N-methylfulleropyrrolidine 1, for instance, can be prepared in a one-pot synthesis in 41% yield by refluxing a solution of paraformaldehyde, N-methylglycine, and C_{60} by refluxing a toluene solution (Scheme 1) [59].

This simple cycloaddition is currently one of the most used ways of functionalizing C_{60} and, to a minor extent, C_{70}. The highly symmetrical C_{60} is monofunctionalized, affording a single fulleropyrrolidine, whereas C_{70}, under similar

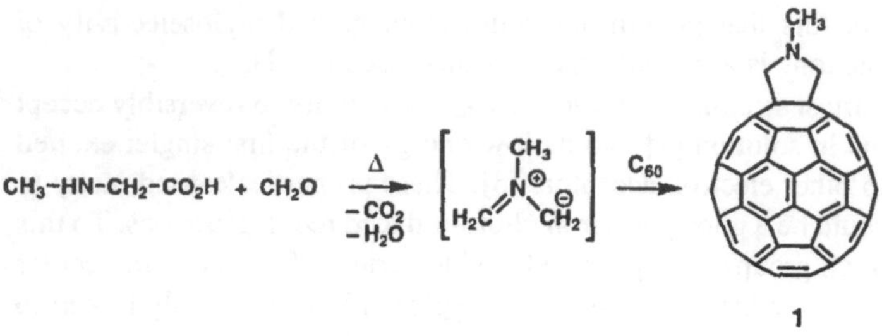

Scheme 1

conditions, gives rise to a mixture of regioisomers [63]. For this reason, current fulleropyrrolidine syntheses are predominantly carried out using C_{60}.

Multiple functionalization of C_{60} via azomethine ylide cycloadditions has brought about some very interesting results. Schick and co-workers [64] have reported the elegant syntheses and x-ray structures of the T_h hexaadduct **2** (Chart 1) and of the first D_3 hexa-adduct, which was isolated following the reaction of C_{60} with an excess acetone and α-aminoisobutyric acid. The unusual luminescence of the T_h hexaadduct **2** has been exploited in the fabrication of the first single layer, white light, C_{60}-based organic LED [65].

Five isomeric bisfulleropyrrolidines **3a–e** (*trans-1*, *trans-2*, *trans-3*, *trans-4*, and *equatorial*) have also been isolated and characterized by the UCLA group during their investigation of hexa-addition reactions of azomethine ylides [64]. The same sequence of bisadducts (**4a–e**, **5a–e**) was isolated upon treatment of C_{60} with an excess of *N*-methyl-, [66,67] or *N*-propyltriethoxysilylglycine [68], and paraformaldehyde respectively, as shown in Chart 2.

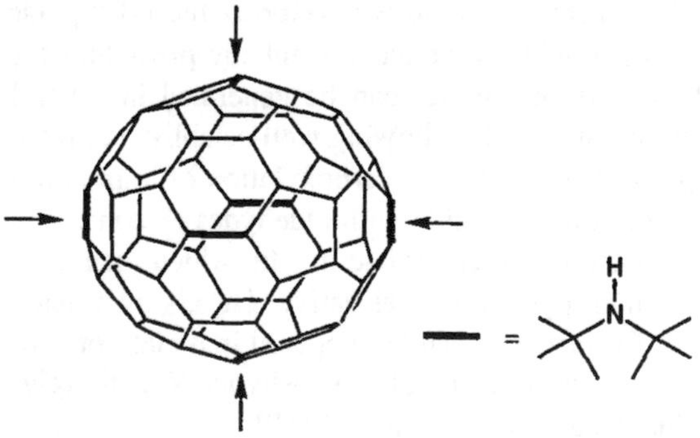

2

Chart 1

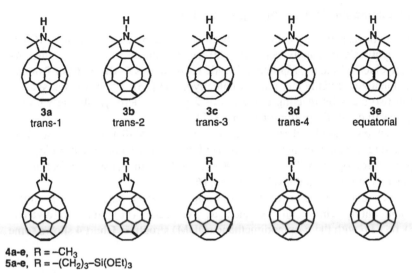

3a
trans-1

3b
trans-2

3c
trans-3

3d
trans-4

3e
equatorial

4a-e, R = –CH₃
5a-e, R = –(CH₂)₃–Si(OEt)₃

Chart 2

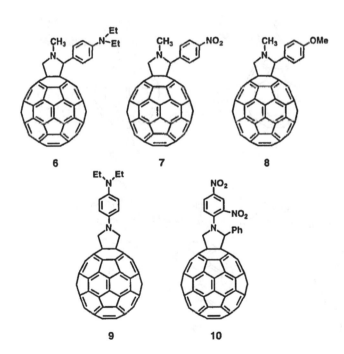

6

7

8

9

10

Chart 3

B. Electroactive Donor–Acceptor Systems

Simple donor–acceptor systems have been reported by several authors (Chart 3). The synthesis is usually very straightforward, starting from C_{60}, either N-methylglycine and the appropriate aldehyde (compounds **6–8**) [9,16] or from an N-functionalized glycine and paraformaldehyde (**9**) [15]. Dyad **10** was synthesized in two steps by reacting the anion of 2-phenylfulleropyrrolidine with the highly reactive 2,4-dinitrofluorobenzene [17].

In an early study, derivative **6** was used by Williams and co-workers for assessing whether or not functionalized fullerenes were suitable electron-accepting chromophores [9]. Dyads **7–8** were utilized to study the effect of solvent polarity on their photoexcited states [16].

Other elaborate electroactive fulleropyrrolidines bear more powerful electron-donating chromophores such as ferrocene (Fc), tetrathiafulvalene (TTF), or TTF with the p-quinodimethane (QDM) structure. Chart 4 shows some

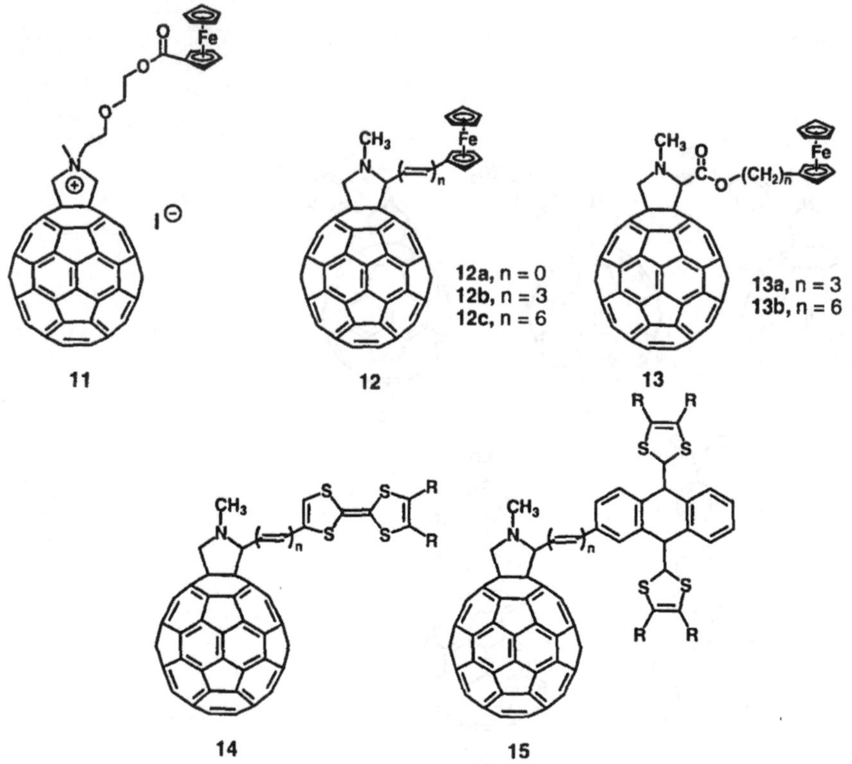

Chart 4

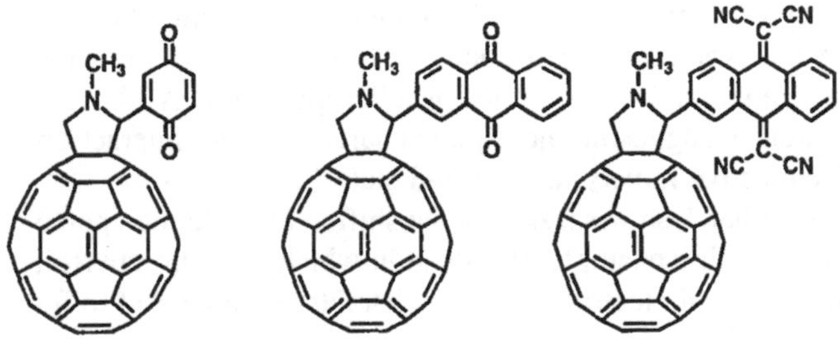

16   17   18

Chart 5

representative examples of dyads that have been prepared via the azomethine ylide decarboxylation route, from *N*-methylglycine and the appropriate Fc, [31,33,36], TTF [31,37,40], or TTF–QDM [39,40] carboxaldehyde. Derivatives **13a** and **13b** were prepared from the 3-ferrocenyl-*n*-propyl ester, and 6-ferrocenyl-*n*-hexyl ester of *N*-methylglycine and paraformaldehyde, respectively [31,33].

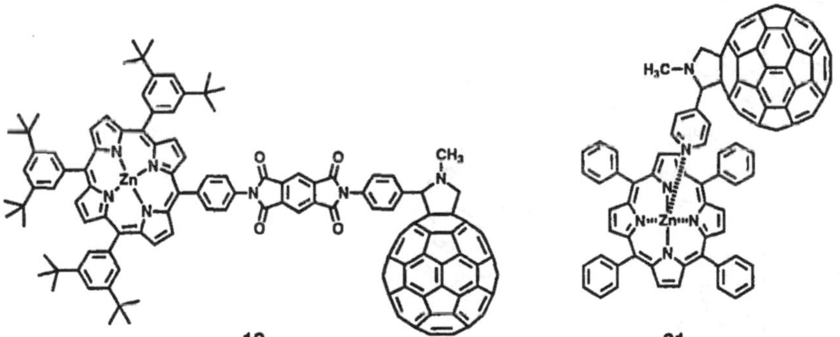

19   21

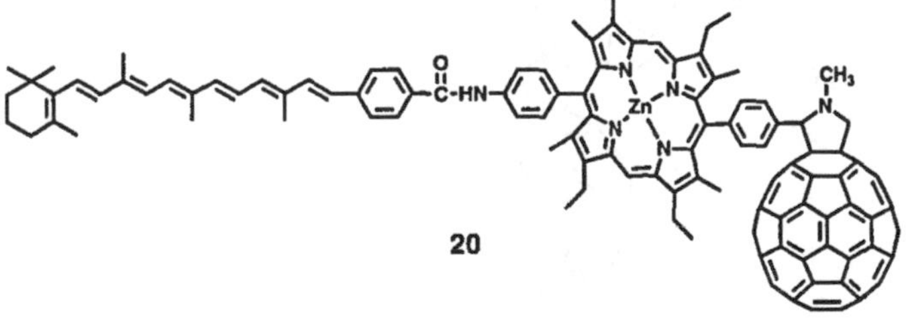

20

Chart 6

Cyclic voltammetry was employed to probe a possible interaction between donor and acceptor in the synthesized compounds. Only small shifts to more negative potentials were found in the $E_{1/2}$ values of fulleropyrrolidines **12–15** as compared to those of parent **1**, suggesting negligible charge–transfer interactions between the added electroactive moiety and ground state C_{60}.

Dyad **11**, on the other hand, shows a significant shift in the reduction potential and is therefore an interesting example of a dyad in which the fulleropyrrolidinium moiety exhibits enhanced electron acceptor character with respect to fulleropyrrolidine **1** and C_{60}[36].

The C_{60}-based electron acceptors, prepared via the azomethine ylide cycloaddition (Chart 5), exhibits interesting electrochemical properties, making them promising candidates for the preparation of charge transfer complexes with strong electron donor molecules [40,69].

C. Photoactive Donor–Acceptor Systems

The efficient conversion of solar energy into useful chemical or electrical energy by means of artificial molecular assemblies represents a long-standing multidisci-

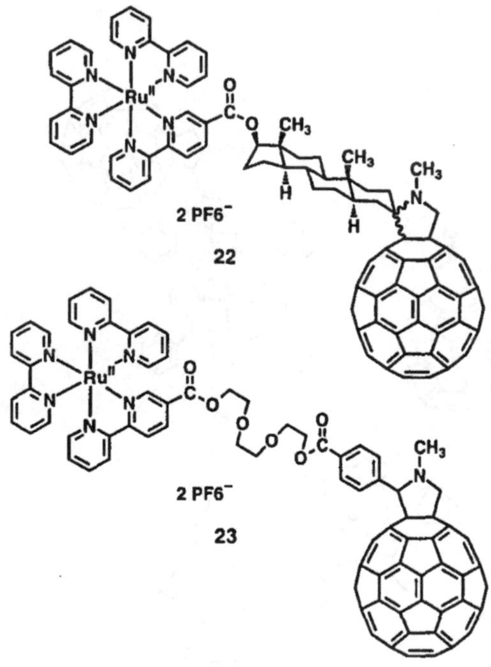

Chart 7

plinary objective of current photochemical research that encompasses chemistry, physics, and biology [52,53]. Inspired by the early events of the photosynthetic process in living organisms, many simple synthetic analogs of the photosynthetic reaction center have been produced with the aim of determining how to control the photo-induced electron transfer which plays a crucial role in the utilization of photon energy.

To this end C_{60} and, more recently, C_{70}, [28] have been used as acceptors in a variety of systems containing photoactive donor units such as porphyrins, carotenes, phtalocyanine, or ruthenium(II) polypyridine complexes. Porphyrin-based fullerenes are among the early fullerene derivatives synthesized and characterized in terms of photo-induced charge separation [8]. In this context, the 1,3-dipolar cycloaddition of azomethine ylides was profitably used to prepare both simple donor–bridge–acceptor dyads and more elaborate multicomponent redox assemblies such as the porphyrin–pyromellitimide–fulleropyrrolidine triad 19 [20], the fulleropyrrolidine–porphyrin–carotene triad 20 [42], and the noncovalently attached porphyrin–fulleropyrrolidine 21 (Chart 6) [70]. For more detail, the reader is referred to recent reviews and papers that have recently appeared in the literature on this subject [8,19,52,53,71].

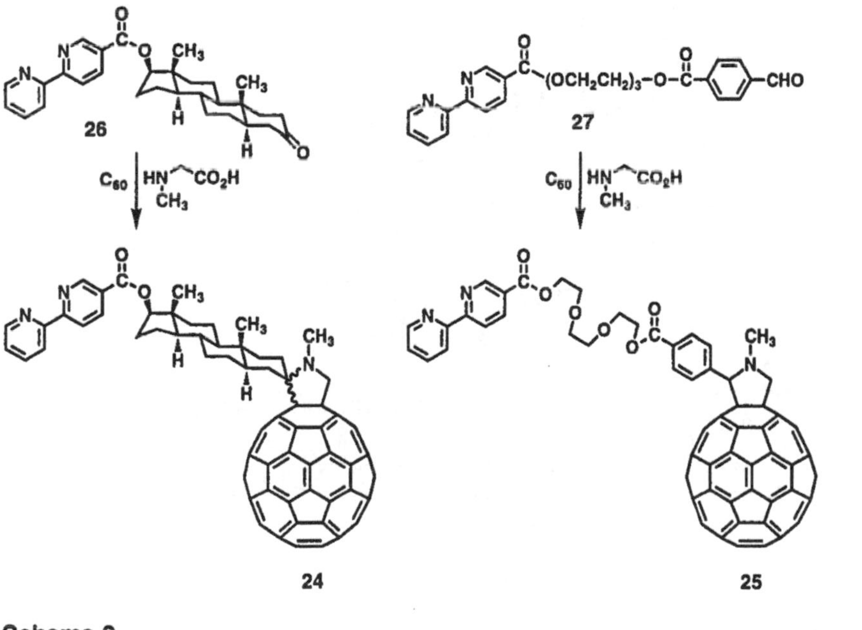

Scheme 2

Ruthenium(II)–polypyridine complexes compose another class of well-known and studied chromophores whose photophysical and redox properties are well understood [72]. In particular, the long lifetime and high-lying energy of the redox-active ruthenium(II) MLCT excited state, along with its prominent emission above 610 nm, makes the combination with C_{60} a promising approach toward a long-lived photo-induced charge-separated state. This led to the synthesis of a series of fulleropyrrolidine ligands, having a bidentate 2,2′-bipyridine (bpy) covalently linked to the C_{60} by means of a spacer, from which the corresponding ruthenium complexes were prepared [46–49]. Chart 7 illustrates the structure of ruthenium(II) trisbipyridine complexes 22–23 obtained using azomethine ylide cycloadditions for ligand preparation, followed by coordination of the ruthenium, using [Ru(bpy)$_2$]Cl$_2$.

C_{60}-bpy ligands 24 [48] and 25 [46] were synthesized starting from the ester-ketone 26 and 4-substituted benzaldehyde 27 with sarcosine in the presence of C_{60} respectively (Scheme 2).

Electrochemical characterization of dyads 22 and 23 was performed in or-

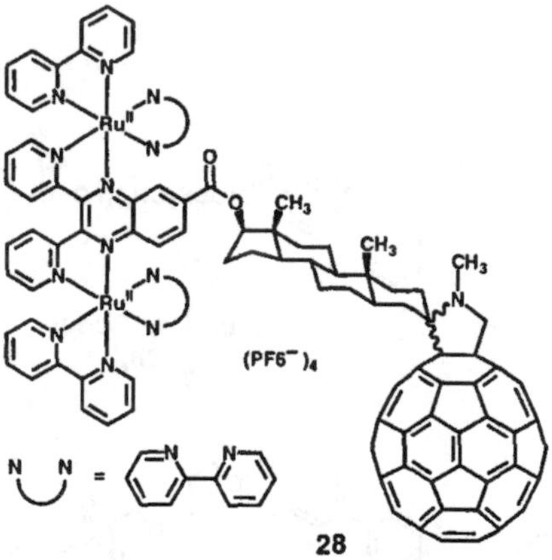

28

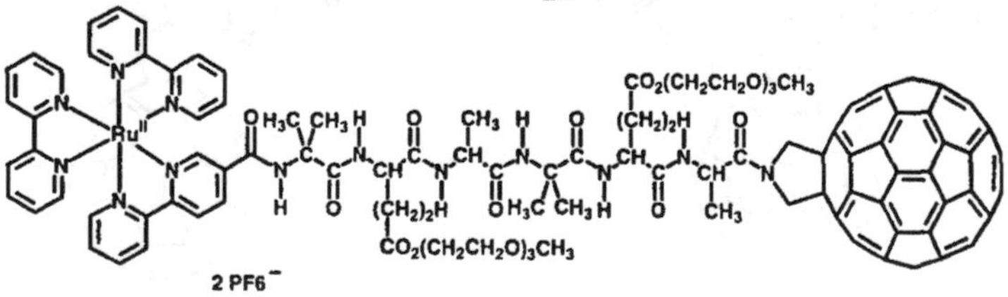

29

Chart 8

der to obtain information on mutual interactions between ruthenium and C_{60} redox sites in the ground state. It has been found that no substantial interaction between the two electroactive groups occurs through the different spacers. These findings are corroborated by the UV-Vis absorption spectrum that, for both dyads, display strong absorptions due to superimposition of those of the ruthenium complex and the C_{60} moiety (vide infra).

The C_{60}–bpy work has been extended to the synthesis of more elaborated systems, whose structures are shown in Chart 8.

Dyad **28** has the same rigid spacer as **22** but the portion responsible of complexation can coordinate two metallic centers [73]. This fulleropyrrolidine was prepared to shed light on the influence of two light-harvesting units on the overall photo-induced process (vide infra). The synthesis of ligand **33** was accomplished using the quinoxaline derivative **30** and androstane–alcohol **31** under

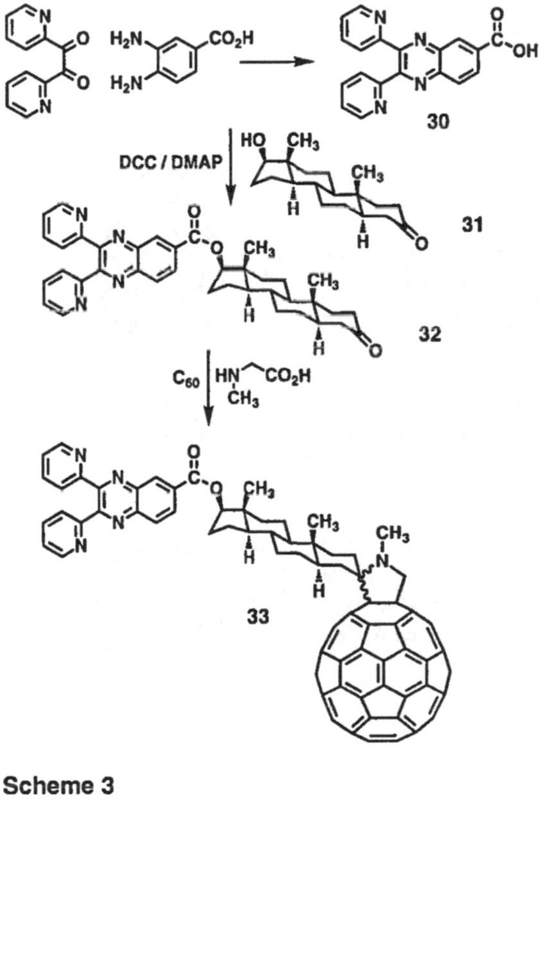

Scheme 3

standard coupling conditions (DCC/DMAP). Tetradentate ligand **33** was then obtained following a similar protocol to that used for preparing ligand **24** (Scheme 3). Coordination to ruthenium, using [Ru(bpy)$_2$]Cl$_2$, afforded the green bimetallic complex **28**.

The peptide-linked [Ru(bpy)$_3$]-C$_{60}$ dyad **29** [49] was prepared to study the influence of the solvent-dependent conformational changes of the peptide bridge on the electron transfer process that occurs upon photoexcitation.

Scheme 4 illustrates the synthetic route that afforded ligand **34**, the precursor to dyad **29**, through coordination of the ruthenium(II) center.

The solution conformation of the hexapeptide backbone was investigated by using a combination of Fourier transform infrared (FTIR) absorption and NMR techniques. In chlorinated hydrocarbons, the hexapeptide adopts a 3$_{10}$-helical structure. However, addition of a protic solvent promotes a conformational transition of the peptide backbone whose effect on the intramolecular electron transfer from ruthenium to C$_{60}$ were assessed by using flash photolytic experiments (vide infra).

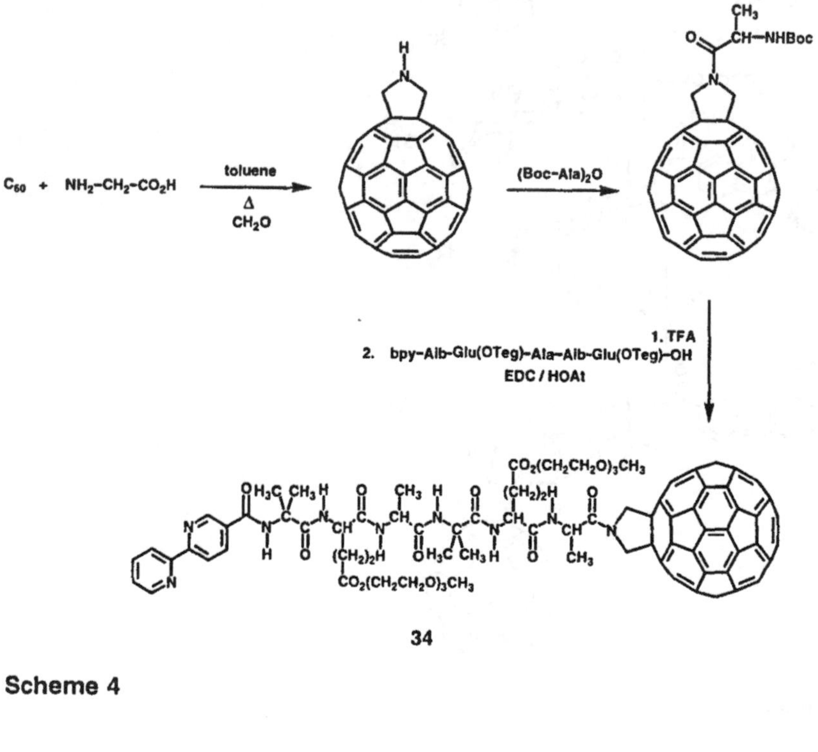

34

Scheme 4

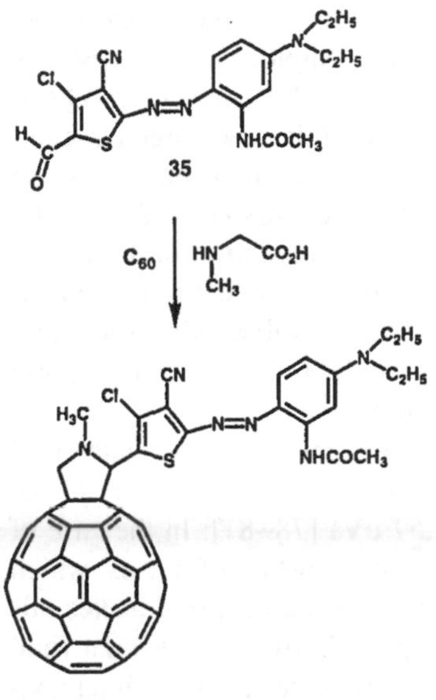

Scheme 5

A fulleropyrrolidine dyad (**36**) in which the donor unit is an azothiophene dye has also been prepared starting from C_{60}, sarcosine, and thiophenylazobenzeneamine **35** (Scheme 5).

Dyad **36** could be selectively excited in the visible region where the dye has an absorption maximum at 567 nm, giving rise to interesting photo-induced energy and electron transfer processes.

III. PHOTOPHYSICAL PROPERTIES AND ELECTRON TRANSFER REACTIONS

The scope of the first section is to summarize the intriguing electronic properties of the third modification of carbon, namely, C_{60} and its higher analogs in the ground state and in the excited states. We then follow the chemical functionalization of fullerenes to discuss in the following sections several novel fullerene materials, which bear electro- or photoactive moieties, covalently attached to the fullerene core.

A. Pristine Fullerene

Toluene solutions of C_{60} exhibit their strongest ground-state absorption in the UV region, with a series of bands at 220, 260, and 330 nm [74,75]. In contrast to the rather high extinction coefficients ($\sim 10^5$ M^{-1} cm^{-1}) of these UV transitions, C_{60} absorbs only weakly in the visible region. For example, the strongest Vis transition at 536 nm has an extinction coefficient of 710 M^{-1} cm^{-1} (Fig. 1) [76].

Excitation of C_{60} results in the population of the fullerene singlet excited state. Despite the constrained carbon network, the singlet excited state gives rise to a surprisingly low fluorescence quantum yield ($\Phi_{\text{FLUORESCENCE}}$) of $\sim 2.0 \times 10^{-4}$ [77]. The low Φ value is a direct consequence of a combination of 1) a short lifetime, 2) a quantitative intersystem crossing, and 3) a symmetry-forbidden nature of the lowest energy transition. In terms of transient absorption spectrum, the singlet excited state of C_{60} exhibits a singlet–singlet ($S_1^* \rightarrow S_n^*$) absorption in the near-infrared (near-IR) region around 920 nm.

Once formed, the lowest vibrational state of the singlet excited state (1.99 eV) undergoes a rapid and quantitative intersystem crossing ($\Phi_{\text{ISC}} = 1$) to the energetically lower lying triplet excited state (1.57 eV) [78–81]. In the case of C_{60}, this intersystem crossing takes place with a time constant of 1.8 ns, which is governed by a large spin-orbit coupling [78]. This strong coupling makes this process much faster than those reported for comparable two-dimensional rigid hydrocarbons. The triplet–triplet ($T_1^* \rightarrow T_n^*$) absorption spectrum reveals a maximum in the near-IR region (i.e., around 750 nm) which is similar to the ($S_1^* \rightarrow S_n^*$) features [82]. In the absence of alternative deactivation processes, such as triplet–triplet annihilation or ground-state quenching, the triplet lifetime is in the order of 100 μs [83]. The product of the spin-forbidden deactivation is the singlet ground state [74]. This triplet lifetime is, again, much shorter than the triplet lifetime of similar planar hydrocarbons. In this context, it is interesting to note that the highly constrained C_{60} carbon network prohibits any vibrational motion, C–C bond elongation, or even changes of the dipole moment. It has been proposed that an efficient spin-orbit coupling in C_{60}, as was seen for the singlet lifetime, is responsible for the short triplet lifetime.

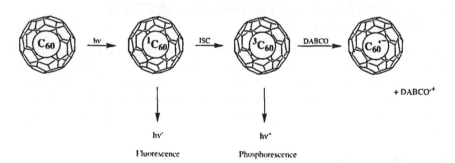

Figure 1 Photophysical processes in fullerene and functionalized fullerene derivatives. (Adapted from Ref. 91.)

B. Fulleropyrrolidines

C_{60} functionalization gives rise to a cancellation of the fivefold degeneracy of the highest occupied molecular orbital (HOMO) and the threefold degeneracy of the lowest occupied molecular orbital (LUMO) [7]. This stems, in a first-order approximation, from the perturbation of the fullerene π-electron system caused by saturation of one, or more, C=C double bonds [1]. In particular, fullerene functionalization raises the LUMO energy, producing a negative shift of the re-duction potentials of the derivative relative to pristine C_{60}. This shift typically amounts to a maximum of 100 mV in condensed media [7].

A particular sensitive set of parameters for studying the electronic proper-ties of fullerenes are the optical absorption and emission spectra [78–90]. The present review on mono- and multiply functionalized fullerene derivatives will demonstrate the impact of single and multiple addends on the photophysical and redox properties of the fullerene's ground and photoexcited states, by means of employing time-resolved and steady-state techniques. Since fulleropyrrolidines represent a general building block for the super- and supramolecular donor–acceptor assemblies discussed throughout this contribution, their excited-state properties are described in more detail.

In the excited state, fulleropyrrolidine 1 shows a well-resolved fluorescence spectrum with peaks at 703, 716, 724 (sh), 739, 754, 783, and 796 nm (sh) [91]. Interestingly, the fluorescence quantum yield ($\Phi_{FLUORESCENCE}$) of 1 (6.0×10^{-4})[91] is noticeably increased relative to that of pristine C_{60} ($\Phi_{FLUORESCENCE} \sim 2.0 \times 10^{-4}$) [77]. The emission spectrum is furthermore in excellent agreement with the mir-ror imaged UV-Vis absorption features [91]. A good mirror image (i.e., between the absorption band of the longest wavelength and the corresponding emission band of the shortest wavelength) is a valuable aid for an unambiguous assignment of the 0 → 0 transition bands. The good overlap of the 0 → 0 transitions (e.g., absorption and emission) can be ascribed to a small adjustment of the fullerene singlet excited state to the new solvent environment. The lack of any significant Stokes shifts suggests a comparable size and similar nuclear coordinates of the singlet ground and singlet excited state (Table 1).

Probing bisadducts 4a–e in emission experiments under similar conditions led to a number of interesting observations. One of them is that changing the symmetry of the functionalized fullerene derivative from 1 to the bisadducts 4a, 4b, 4c, 4d, and 4e resulted in successive blue shifts of the main emission (*0 → 0) band. There is no doubt that these spectra still originate from the fullerene emission since their excitation spectra resemble the spectral structure of the corre-sponding absorption spectra [62].

Acceleration of the spin-forbidden transformation from a given singlet ex-cited state to the corresponding triplet excited state can be accomplished by

Table 1 Photophysical Data of Pristine C_{60}, Derivative **1** and Fulleropyrrolidines Bisadducts **4b–e** in Toluene Solutions

Compound	$S_1^* \rightarrow S_n^*$ (nm)	$T_1^* \rightarrow T_n^*$ (nm)	$^1{*}E_{0-0}$ (eV)	$^3{*}E_{0-0}$ (eV)	$C_{60}^{\cdot-}$ (nm)[a]
C_{60}	920	747	1.99	1.57	1080
1	886	705	1.79	1.50	1010
4b	870	650	1.72	1.50	920
4c	895	680	1.74	1.50	990
4d	880	685	1.75	1.50	995
4e	880	700	1.81	1.50	1060

[a] Maximum of fullerene π-radical anion absorption.

applying an external heavy atom effect [86]. In fact, the fluorescence of fulleropyrrolidines **1, 4a–e** were completely abolished upon addition of ethyliodide to a methylcyclohexane solution of **1, 4a–e** where iodide is the heavy-atom provider. Instead, a new band appeared, which for fulleropyrrolidine **1** is located at 826 nm. This band has been assigned to the emissive deactivation of the fullerene triplet excited state (e.g., phosphorescence). This finding implies a notable lowering of the triplet excited-state energy (1.5 eV) relative to pristine C_{60} (1.57 eV) [74]. Despite the increase of the fluorescence quantum yields, the phosphorescence of the investigated fulleropyrrolidines (**1, 4a–e**) is still quite low ($\Phi_{\text{PHOSPHORESCENCE}} \sim 10^{-6}$).

In order to study the singlet excited-state behavior of functionalized fulleropyrrolidines in more detail, we probed solutions of **1, 4a–e** by means of transient absorption spectroscopy on the picosecond time scale (e.g., between 0–4000 ps). For example, differential absorption spectra of the fulleropyrrolidine monoadduct **1** in an oxygen-free toluene solution reveal the grow-in of an absorption maximum centered around 880 nm [33]. The grow-in kinetics are within the response time of our laser apparatus (~40 ps). Principally, this observation is in close agreement with that noted for pristine C_{60}, exhibiting a singlet–singlet ($S_1^* \rightarrow S_n^*$) absorption at 920 nm [78,79]. In the case of monoadduct **1**, this transient absorption is short-lived ($\tau = 1.54$ ns) and decreases via a clean first-order decay. In parallel to this deactivation, the formation of a new absorption develops around 705 nm. The close resemblance of the decay and grow-in kinetics lead to the conclusion that the underlying process is a direct transformation of the singlet excited state (1.76 eV) into the energetically lower lying triplet excited state (1.50 eV) via intersystem crossing (ISC) [33,91]:

$$\text{Fullerene} \xrightarrow{h\nu} {}^1{*}(\text{fullerene}) \xrightarrow{\text{ISC}} {}^3{*}(\text{fullerene}) \tag{1}$$

The differential absorption changes found for fulleropyrrolidines bisadducts **4a–e** are virtually identical to those described in detail for **1**. However, an overall blue shift of the singlet–singlet ($S_1^* \to S_n^*$) absorption, ranging from 920 nm, for pristine C_{60}, to 880 nm, for bisadduct **4d**, is noted [62].

In a similar manner, the triplet excited states of **1, 4a–d** were produced by nanosecond laser flash photolysis in toluene solutions and their optical absorption spectra were monitored in the visible range by kinetic spectrophotometry. The differential spectrum recorded about 50 ns after the laser pulse of **1** exhibits a set of maxima at 360 and 705 nm ($\varepsilon_{705\ nm} = 16,000\ M^{-1}\ cm^{-1}$) and the spectral features resemble those evolving from the intersystem crossing process (i.e., picosecond experiments). Under oxygen-free conditions, a determination of the triplet lifetime of around 30 μs was deduced from the kinetic traces.

It is noted that the maximum observed (705 nm) is blue-shifted by nearly 40 nm relative to the triplet maximum of pristine C_{60} (i.e., 750 nm). Further blue shifts, up to 650 nm, were found for the ($T_1^* \to T_n^*$) absorption of **4a–d** (Fig. 2). The only exception, within the investigated series of isomers, is the equatorial isomer **4e** (700 nm). In general, these hypsochromic shifts can best be rationalized

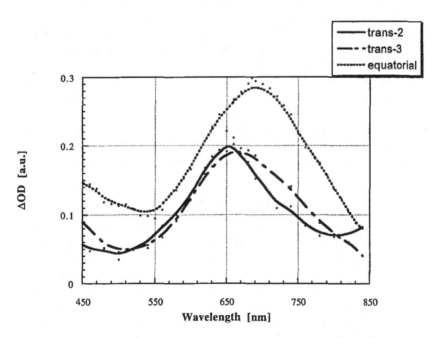

Figure 2 Differential absorption spectra obtained upon flash photolysis (337 nm) of $\sim 10^{-5}$ M solutions of *eq*-$C_{60}(C_3H_7N)_2$ (**4e**) (---), *trans*-3-$C_{60}(C_3H_7N)_2$ (**4c**) (— · —) and *trans*-2-$C_{60}(C_3H_7N)_2$ (**4b**) (—) in nitrogen saturated toluene.

in terms of a reduction of the resonance stabilization (see below) within the conjugated π system. The lowest energy, found for the equatorial bisadduct (**4e**), indicates, however, that this product is probably the best stabilized isomer among the various bisadducts.

Pulse radiolysis is another important tool for fast kinetic spectroscopic studies. This technique allows us to generate and characterize triplet excited states of fullerenes [89,92–95]. For instance, radiolysis of a deoxygenated toluene solution containing 0.02 M biphenyl affords the triplet excited state of biphenyl (i.e., ^{3}(biphenyl)) [89,92–95]. The high-lying triplet excited state of biphenyl ($E_{TRIPLET}$ = 2.80 eV) should, in principle, allow probing the intermolecular energy transfer to the energetically lower lying triplet excited state of fullerene derivatives **1**, **4a–d** ($E_{TRIPLET}$ = 1.5–1.7 eV) [91], summarized in the following equation:

$$^{3*}(\text{biphenyl}) + \text{fullerene} \xrightarrow{\ k_{ENT}\ } \text{biphenyl} + {}^{3*}(\text{fullerene}) \qquad (2)$$

The triplet excited state of biphenyl, which gives rise to a characteristic absorption maximum around 360 nm, suffers the expected accelerated decay upon addition of various concentrations of **1**, **4a–e**. Simultaneous to the ^{3*}(biphenyl) decay, formation of, for example, ^{3*}(**1**) was confirmed via monitoring the fullerene triplet maximum around 700 nm, which is in close agreement with the one noted in the photolytic measurements (see above). Also, the triplet–triplet absorption maxima for **4a** (710 nm), **4b** (705 nm), and **4c** (690 nm) parallel the flash photolytic findings and, thus, confirm the blue-shifted triplet–triplet (T_1^* → T_n^*) absorption of the investigated derivatives.

The second-order decay kinetics of the fullerene triplets, due to triplet–triplet annihilation, changes to pseudo-first order dynamics upon addition of a sacrificial electron donor, such as diazabicyclooctane (DABCO), ferrocene, etc. [33,82,92]. With respect to the dependence of the triplet energies upon fullerene functionalization (e.g., mono- and bisadducts), a significant change in the redox potentials were expected to dominate the quenching rates of the triplet excited states.

$$^{3*}(\text{fullerene}) + \text{DABCO} \xrightarrow{\ k_{ET}\ } (\text{fullerene})^{\cdot-} + (\text{DABCO})^{\cdot+} \qquad (3)$$

It should be noted that the use of DABCO in reductive electron transfer reactions to the fullerene triplet excited state bears the fundamental advantage that the molecule neither in its ground state nor in its oxidized form adds to the fullerene core [33,82,92]. It is noteworthy that addition reactions, by means of a covalent bond formation with the fullerene core, are reported to occur quite commonly with primary and secondary amines [1,2]. Mixing various concentrations of DABCO with **1** in toluene indeed resulted in the expected decay of the fullerene triplet excited state of **1**. More importantly, the observed first-order rate constants (k_{obs}) were linearly dependent on the DABCO concentration. This leads to the hypothesis that an intermolecular quenching reaction, as a rate-determining step, governs the fate of the fullerene triplet excited state. The bimolecular rate constant of $7.7 \times 10^7 \, M^{-1} \, s^{-1}$ [96] is, however, markedly slower than that described for the corresponding quenching of the triplet excited state of pristine C_{60} ($2.5 \times 10^9 \, M^{-1} \, s^{-1}$) [82]. This observation is again attributed to the perturbation of the electronic resonance structure of the functionalized fullerene core and corroborates the assumption that the quenching rates are impacted by the fullerene functionalization.

The electron transfer was then unequivocally confirmed in solvents of higher polarity, namely, n-butanol or benzonitrile [82]. These solvents permit a stabilization of the charge-separated radical pair via embedding the oppositely charged radicals into the dipoles of the solvent molecules. Simultaneous with the quenching of the fullerene triplet excited state, observable around 700 nm, the formation of a new product with a characteristic near-IR absorption pattern was observed. For example, the π-radical anion absorption of **1** is maximized at 1010 nm [33]. The resulting differential absorption spectrum resembles that obtained upon the radiolytically induced reduction of **1** in homogeneous systems and suggests that the monitored transient near-IR absorption can be attributed to the formation of the fullerene π-radical anion (i.e., $C_{60}{}^{\bullet-}$). Interestingly, the derived quenching rate constant of $9 \times 10^6 \, M^{-1} \, s^{-1}$ is subject to a further slow-down, relative to that in a nonpolar toluene solution [96]. It should be noted that in polar solvents decreasing quenching rates with DABCO and other sacrificial electron donors have also been reported for pristine C_{60} [82]. The time profile of this radical anion absorption indicates that this species decays over a few 100 µs. Since the fullerene π-radical anion (i.e., $C_{60}{}^{\bullet-}$) has been shown to be stable when produced in γ- or pulse-irradiated solutions [97], or by electrochemical reduction in polar and nonpolar solvents [7], this decay is attributable to the highly exothermic back electron transfer.

Among the various bisadducts (**4a–e**) the equatorial isomer (**4e**) exhibits the most effective triplet quenching by DABCO, well in line with the trend established for the triplet state energies [96]. It is interesting to note that the data of the two trans products seemingly contradict the trend in the triplet absorption. Possibly the optical properties reflect mainly the electronic features while the reaction kinetics are dominated by steric parameters. It is feasible that the latter play a bigger role for the bis- than for the monoadducts.

As an alternative to reductive triplet quenching, radical anions may be produced via radical-induced reduction of fullerenes, by means of pulse radiolysis [97]. The fullerene reduction was obtained in a solvent mixture containing toluene, 2-propanol, and acetone (8:1:1 v/v) [98]. Toluene was selected to guarantee a monomeric dissolution of the fullerene. On the other hand, acetone was chosen as an efficient electron scavenger to prevent a reaction between solvated electrons and toluene (i.e., generating excited states of toluene). Followed by a fast protonation a radical species, namely, $(CH_3)_2{}^{\bullet}COH$, with a reducing character is formed

[Eq. (4)]. In addition, the $(CH_3)_2\dot{C}OH$ species is identical to the main product of the radiolysis of the second cosolvent, 2-propanol [Eq. (5)].

$$(CH_3)_2CO + e_{sol}^- + H^- \rightarrow (CH_3)_2\dot{C}OH \tag{4}$$

$$(CH_3)_2CHOH \xrightarrow{radiolysis} (CH_3)_2\dot{C}OH \tag{5}$$

This reducing radical reacts quite rapidly with pristine fullerenes (e.g., C_{60}, C_{70}, C_{76}, C_{78}, and C_{84}) to yield the respective π-radical anions with nearly diffusion-controlled kinetics [98–100].

Accordingly, pulse irradiation of deoxygenated solutions containing about 2×10^{-5} M fullerenes derivatives 1, 4a–e resulted in the formation of distinct absorption patterns in the near-IR. For example, the differential absorption spectrum obtained upon pulse radiolysis of 1 exhibits a sharp maximum at 1010 nm [33,62], hypsochromicly shifted by about 70 nm relative to that of C_{60} ($\lambda_{max} =$ 1080 nm) [98]. Note that this maximum is in excellent agreement with the quenching studies (see above). The 1010-nm band and the corresponding maxima recorded for bisadducts 4a–d [62], which are all further shifted to the blue, are ascribed to the fullerene π-radical anions formed in the general reaction:

$$(CH_3)_2\dot{C}OH + fullerene \rightarrow (fullerene)^{\cdot-} + (CH_3)_2CO + H^+ \tag{6}$$

In contrast to this blue shift, one of the bisadduct π-radical anion absorptions, namely, that of the equatorial isomer (4e), is red-shifted compared to those of bisadducts 4a–d and that of the monoadduct 1 π-radical anions (Fig. 3).

Again, this clearly points to the interesting behavior of the equatorial isomer (4e), not only among the various bisadducts but also in relation to the monoadduct 1. Thus, we carried out a comparison of the HOMOs and LUMOs of the monoadduct 1 with those of pristine C_{60} and bisadducts 4a–e to gain further insight into the electronic properties of the fullerene reduced and excited states (Fig. 4a–c). The molecular orbitals of 1 show a significant electron deficit in the singlet ground state, especially in the area close to the equatorial position (Fig. 4b). Reduction, by means of a one-electron addition has been proposed to lead to an electron distribution with a notable localization particularly in this equatorial area. Thus, the substantial optical differences between the reduced-state and excited-state spectra of pristine C_{60} and monofunctionalized fullerene derivatives can be rationalized in terms of the electron distribution. Selective introduction of a second addend (e.g., bisadducts 4a–e), particularly into the equatorial area, has two major consequences (Fig. 4c). First, it clearly helps to intensify the electron distribution of the singlet ground state close to the equatorial position. Second, it homogenizes the corresponding LUMO level, resembling the one determined for C_{60} (Fig. 4a). It is interesting to note that this effect reaches its maximum for the equatorial bisadduct (4e). This emerges from the analysis of the corresponding electron distribution in the LUMO of the *trans*-3 (4c) and

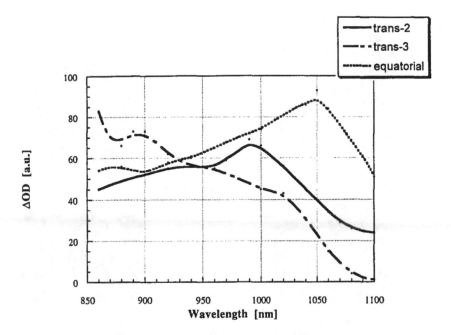

Figure 3 Differential absorption spectra in the near-IR region obtained upon pulse radiolysis of ~10^{-5} M solutions of eq-$C_{60}(C_3H_7N)_2$ (4e) (---), $trans$-3-$C_{60}(C_3H_7N)_2$ (4c) (-----) and $trans$-2-$C_{60}(C_3H_7N)_2$ (4b) (——) in nitrogen saturated solvent mixtures (toluene, 2-propanol, and acetone; 8:1:1 v/v).

$trans$-4 adducts (**4d**). Unmistakably, these finding imply a strong perturbation of the π-electron density, comparable to that shown for the corresponding monoadduct **1**.

C. Fulleropyrrolidinium Salts

1. Monopyrrolidinium Salts [101,102]

The insolubility of pristine fullerenes in polar solvents evokes a variety of interesting questions. For example: how is the fullerene redox behavior or reactivity toward a free radical attack affected in an aqueous environment? Despite the poor solubility, numerous biological implications have been explored since the advent of fullerenes [5,6]. A promising approach to overcome the water insolubility of pristine fullerenes encompasses functionalization of fullerenes via covalent attachment of hydrophilic addends [1–6]. In fact, functionalization with a hydrophilic addend, such as a quaternary ammonium cation, was thought to promote the water solubility of the fullerene core [36,101,102] (Chart 9).

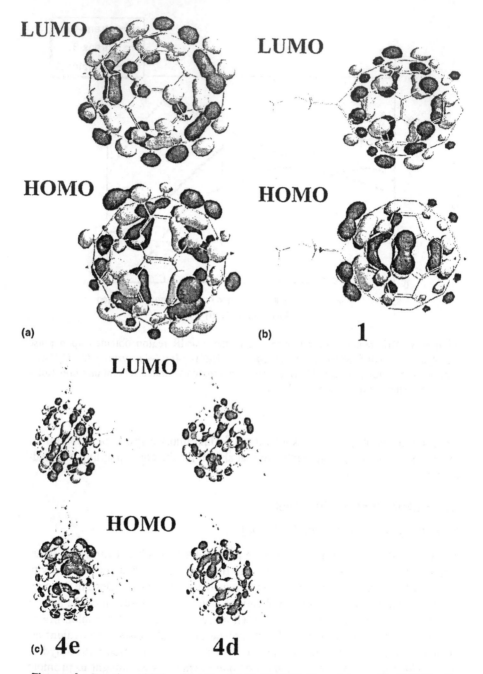

Figure 4 Frontier orbitals of (a) C$_{60}$. (b) C$_{60}$(C$_3$H$_7$N) (**1**) and (c) eq-C$_{60}$(C$_3$H$_7$N)$_2$ (**4e**) and $trans$-4-C$_{60}$(C$_3$H$_7$N)$_2$ (**4d**).

37

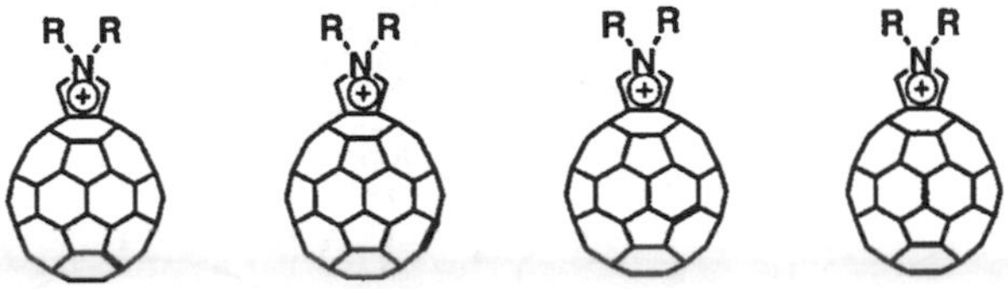

| **38a** | **38b** | **38c** | **38d** |
| trans-2 | trans-3 | trans-4 | equatorial |

38a-d, R = –CH$_3$

Chart 9

Despite the overall improved hydrophilic character of the fullerene core, dissolving the monopyrrolidinium salt **37** in aqueous solution led to the irreversible formation of colloidal fullerene clusters [101,102]. This was followed spectroscopically, e.g., the sharp absorption bands of monomeric fullerenes transform into broadly absorbing features. It should be noted that a similar cluster phenomenon has been proposed for pristine C$_{60}$, once dissolved in polar or vesicular media [104]. The data clearly demonstrate that a single hydrophilic addend is an insufficient means to prevent the strong hydrophobic interactions among the fullerene moieties and the resulting tendency to form aggregates [101–104]. As a consequence, formation of stable monomers of monofunctionalized fullerene derivatives in aqueous solution is precluded and instead the irreversible formation of clusters prevails.

$$n(\text{fullerene}) \;—aqueous\ solution\rightarrow\; \{\text{fullerene}\}_n \qquad\qquad (7)$$

Capping the surface of the water-soluble monopyrrolidinium salt with various surfactants, however, was found to be a successful approach to suppress the fullerene clustering [101,102]. A comparison of the surfactant-capped fullerene derivatives with fullerene clusters suggests that the fullerene monomers exists in a core (fullerene)–shell (surfactant) type of structure (Fig. 5).

$$\text{fullerene} \;—surfactant\rightarrow\; [\text{fullerene}]_{\text{surfactant}} \qquad\qquad (8)$$

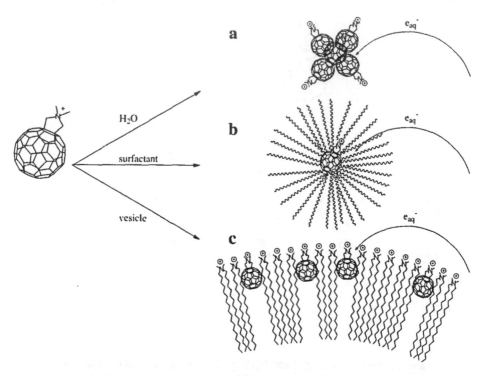

Figure 5 Reduction of a water-soluble monopyrrolidinium salt (**37**) in (a) a fullerene cluster, (b) a core (fullerene)–shell (surfactant) type of structure, and (c) a vesicle.

Flash photolytic (e.g., nanosecond- and picosecond-resolved) techniques were employed to generate excited and reduced states of fullerene monomers and fullerene clusters and to compare their spectral and kinetic properties. Differential absorption changes, as typically recorded upon picosecond excitation of the monopyrrolidinium salt, revealed spectral characteristics of the respective (S_i^* → S_n^*) absorption. These were found to be independent on the aggregation state, e.g., monomer vs. cluster [102]. Based on the similarity with other monofunctionalized fullerene derivatives, the spectral features (i.e., λ_{max} around 886 nm) indicate the immediate formation of the singlet excited state [62,91].

$$\{\text{fullerene}\}_n \xrightarrow{h\nu} {}^{1*}\{\text{fullerene}\}_n \xrightarrow{ISC} {}^{3*}\{\text{fullerene}\}_n \tag{9}$$

In these photoexcited fullerene clusters the dynamics of the spin-forbidden intersystem crossing rates surprisingly resemble those found for fullerene monomers in nonpolar solutions [62,91]. At first glance, this can be viewed as quite a remarkable observation, since the fullerene singlet excited state in a closely packed film is subject to rapid decay (e.g., lifetime on the order of ~2 ps) [105]. The density dependence of the relaxation process has been interpreted in terms of a fast depopulation of the singlet excited state via singlet–singlet annihilation. Since unambiguous evidence substantiates our cluster concept, the present data suggest weaker cohesive interaction between the fullerene moieties in aqueous solutions than in thin films. In conclusion, fullerene aggregation was found to have insignificant effects on photochemical properties, which are associated with the generation and fate of the fullerene singlet excited state (Table 2).

In contrast, the triplet excited-state properties give rise to a number of discrepancies between the monomeric and the colloidal derivatives [101–104]. First, noticeable differences were observed with respect to the absorption maxima of the corresponding ($T_1^* \rightarrow T_n^*$) absorption. In surfactant media, the maxima were generally blue-shifted relative to the analogous γ-CD complexes (e.g., 690 nm vs. 700 nm). The latter shifts parallel the blue shift noted for the fullerene clusters (680 nm) and in turn are ascribed to a close contact with the aqueous environment. Second, clustering caused drastic effects on the lifetime of the fullerene triplet excited states (~0.3 μs). Despite the fact that cluster formation leads to negligible effects on the singlet lifetimes, triplet–triplet annihilation leads to a marked acceleration of triplet decay. In particular, rate constants are typically two orders of magnitude higher than those for the corresponding fullerene monomers.

As demonstrated above, capping of the hydrophobic fullerene core with surfactants or incorporation into the cavity of γ-CD are important means to shield individual fullerene moieties from each other [106–109]. If the cluster-induced

Table 2 Photophysical Data of Monopyrrolidinium Salt 37 in Form of a Cluster and Embedded in Various Matrixes (e.g., γ-CD, Surfactant, and Vesicle) in Aqueous Solutions

Compound	$S_1^* \rightarrow S_n^*$ (nm)	$T_1^* \rightarrow T_n^*$ (nm)	Triplet lifetime (μs)	Rate constant k_{DABCO} (M^{-1} s^{-1})[a]	Rate constant $k_{ELECTRONS}$ (M^{-1} s^{-1})[b]	$C_{60}^{\cdot-}$ (nm)[c]
Cluster	920	680	0.3		0.36×10^{10}	1010
γ-CD	920	700	52	2.7×10^7	2.8×10^{10}	1015
Surfactant	910	690	56	7.8×10^6	3.5×10^{10}	1015
Vesicle		690	1.9		3.2×10^{10}	1015

[a] Rate constant for the reductive quenching of the fullerene triplet excited state with DABCO as a sacrificial electron donor.
[b] Rate constant for the one-electron reduction of the fullerene singlet ground state by radiolytically generated hydrated electrons.
[c] Maximum of fullerene π-radical anion absorption.

quenching truly applies, quenching of the triplet excited state should be suppressed. In line with this hypothesis, the monomeric pyrrolidinium salt 37 revealed a long-lived ($T_1^* \rightarrow T_n^*$) absorption with a lifetime exceeding even those observed in homogeneous solutions.

Reductive quenching of the triplet excited state of 37 with the sacrificial electron donor DABCO led to a twofold increase in the associated quenching rate constant (7.8×10^6 M^{-1} s^{-1}) in comparison to the analogous methanofullerenes (4.8×10^6 M^{-1} s^{-1}) [102]. It is notable that in γ-CD-incorporated complexes the rates for charge recombination were found to be on the order of $\sim 2 \times 10^4$ s^{-1}. This is indicative of a destabilization of the charge-separated radical pair by nearly two orders of magnitude, relative to the analogous surfactant-capped derivatives ($\sim 3 \times 10^2$ s^{-1}). A similar trend is also observed for pristine C_{60}. A reasonable hypothesis is that a more effective penetration of DABCO$^{\cdot+}$ molecules toward $C_{60}^{\cdot-}$ prevails when the fullerene is located within the cavity of the γ-CD host, relative to a penetration through the surfactant shell.

To probe the susceptibility of the surfactant-capped monopyrrolidinium salt 37 toward a reaction with reducing agents, complementary pulse radiolysis experiments were conducted [101]. This allowed for the differential absorption changes at 720 nm (i.e., absorption maximum of the hydrated electron) and throughout the near-IR to be explored in detail. The underlying decay rate (i.e., hydrated electron) was found to be dependent on the fullerene concentration (($0.2-4.0$) × 10^{-5} M) and, furthermore, in excellent agreement with the formation rate of the fullerene π-radical anion (1010 nm). This clearly documents the reduction of the fullerene complex by hydrated electrons.

It should be added that a different type of fullerene monomer [e.g., embedded within a vesicle matrix (DODAB)] and even fullerene clusters were also successfully reduced in pulse radiolysis experiments (Fig. 6).

Surprisingly, the rate constants for the reduction of the monopyrrolidinium salt 37 by hydrated electrons give rise to a marked enhancement over the negatively charged analogs (e.g., carboxylates) [110] and even over pristine C_{60} [102]. In a first-order approximation, this rate intensification can be ascribed to a less anodic reduction potential of the monopyrrolidinium salt relative to the carboxylate [36]. Another plausible rationale implies the coulombic forces which would decelerate a reaction between hydrated electrons and the negatively charged fullerenes while they would facilitate reduction of the positively charged fullerene. To rule out the last mentioned argument, electron transfer rates from uncharged $(CH_3)_2\dot{C}OH$ radicals, which should be less affected by the surface charge of the respective functionalized fullerene derivative, were studied. In fact, the associated rates substantiate the trend observed for reactions of the hydrated electron (i.e., a nearly twofold increase over the carboxylates). This suggests, in line with the quenching rates (e.g., with DABCO), an anodic shift of the reduction potential of the monopyrrolidinium salt 37 relative to those of the carboxylates.

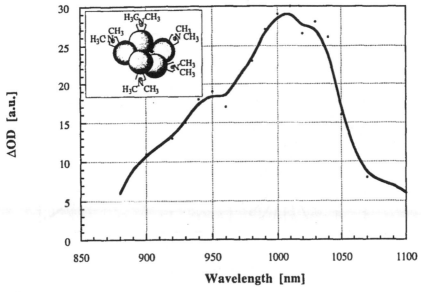

(a)

Figure 6 Differential absorption spectra in the near-IR region of (a) $\{(C_{60}^{\bullet-})(C_4H_{10}N^+)\}n$ obtained ~25 μs after pulse irradiation of 2×10^{-5} M $C_{60}(C_4H_{10}N^+)$ (37) in a nitrogen-saturated aqueous solution (10 vol % 2-propanol), (b) $(C_{60}^{\bullet-})(C_4H_{10}N^+)$/ surfactant obtained ~15 μs after pulse irradiation of 2×10^{-5} M $C_{60}(C_4H_{10}N^+)$/surfactant (37) in a nitrogen-saturated aqueous solution (10 vol % 2-propanol) and (c) $(C_{60}^{\bullet-})(C_4H_{10}N^+)$/DODAB obtained ~15 μs after pulse irradiation of 3.5×10^{-5} M $C_{60}(C_4H_{10}N^+)$ (37) and 5×10^{-5} M DODAB in a nitrogen-saturated aqueous solution (1 vol % methanol). (Adapted from Ref. 101.)

2. Bispyrrolidinium Salts [111]

In light of the fact that the monopyrrolidinium salt still exists in the form of clusters [102], introduction of a second pyrrolidinium addend should, under optimal conditions, lower the susceptibility of the resulting fullerene derivative to form clusters [111]. This argument is, of course, based on the fact that the resulting three-dimensional architectures display an enhanced surface coverage of the hydrophobic fullerene surface. In fact, these materials are quite soluble in aqueous solutions without revealing any evidence for the formation of fullerene clusters.

Photolysis experiments of **38a–d** in aqueous solutions give rise to singlet and triplet excited-state absorptions (Fig. 7) that closely resemble earlier observations for the pyrrolidine complexes, e.g., (**4a–d**), in deoxygenated toluene solu-

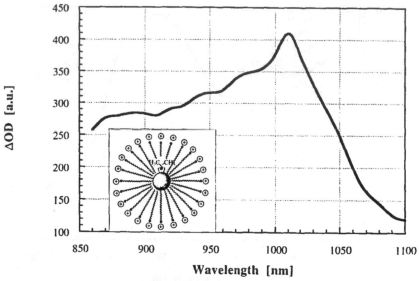

(b)

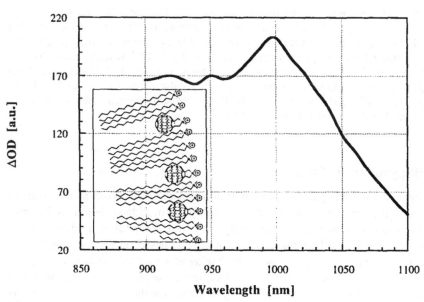

(c)

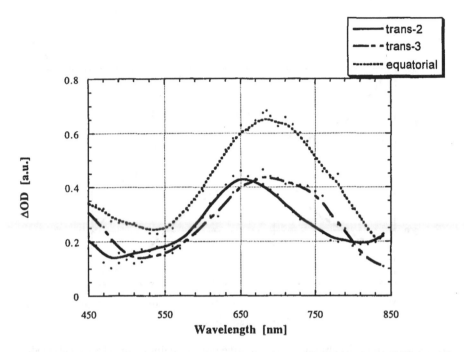

Figure 7 Differential absorption spectra obtained upon flash photolysis (337 nm) of
~10^{-5} M solutions of *eq*-C$_{60}$(C$_4$H$_{10}$N$^+$)$_2$ (**38d**) (---), *trans*-3-C$_{60}$(C$_4$H$_{10}$N$^+$)$_2$ (**38b**) (— · —),
and *trans*-2-C$_{60}$(C$_4$H$_{10}$N$^+$)$_2$ (**38a**) (—) in nitrogen saturated aqueous solutions (pH 9.3).

tions (Fig. 2). On the other hand, the reaction of the fullerene core with hydrated
electrons was directly monitored via the decay of the hydrated electron absorption
at 700 nm, which was found to parallel the grow-in of a new transient absorption,
with a distinct near-IR pattern (Fig. 8). In fact, the near-IR absorption is a characteristic fingerprint, indicative of the formation of fullerene π-radical anions.
Again, the similarity of the maxima observed of water-soluble bispyrrolidinium
salts **38a–d** with the corresponding water-insoluble pyrrolidine precursors (**4a–
d**) specifically suggests successful occurrence of the fullerene reduction.

Employing pyrrolidinium rather than pyrrolidine functionalities enables the
covalent linkage of two functionalizing addends (e.g., pyrrolidinium) to the fullerene core without constraining the susceptibility of the fullerene singlet ground
state to accept electrons [111]. This conclusion stems, for example, from the
measured rate constants for the radical-induced fullerene reduction, e.g., with
hydrated electrons and (CH$_3$)$_2$˙COH radicals. They are virtually identical to those
determined in earlier work for pristine C$_{60}$ [112]. Considering the reduction poten-

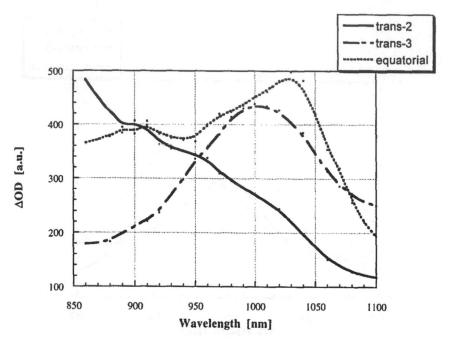

Figure 8 Differential absorption spectra in the near-IR region obtained upon pulse radiolysis of $\sim 10^{-5}$ M solutions of eq-$C_{60}(C_4H_{10}N^+)_2$ (**38d**) (---), $trans$-3-$C_{60}(C_4H_{10}N^-)_2$ (**38b**) (— · —) and $trans$-2-$C_{60}(C_4H_{10}N^+)_2$ (**38a**) (—) in nitrogen saturated aqueous solutions containing 10 vol % 2-propanol (pH 9.3).

tial of pristine C_{60} and that of the monopyrrolidinium salt -0.35 and -0.29 V vs. SCE, respectively, we gather that the reduction potentials of bispyrrolidinium salts **38a–d** are between -0.32 and -0.34 V vs. SCE [36]. This enhanced electronegativity can be reasonably interpreted in terms of inductive effects of the pyrrolidinium functionality. The latter affects the electron density within the fullerene π system (Table 3).

In the excited states, the lower triplet energies of the functionalized derivatives, relative to pristine C_{60}, impact the electron acceptor properties of the fullerene core in the bispyrrolidinium salts **38a–d** [111]. For instance, the rate constants for triplet quenching with DABCO reveal a 21-fold slow-down relative to C_{60}, which is, however, moderate compared to the difference between the biscarboxylates and C_{60} (e.g., a 108-fold decrease) [113]. As a general rule, the ground-state reduction potentials and the triplet excited-state energies of fullerenes are shifted to more negative values and lower energies, respectively, with increasing

Table 3 Photophysical Data of γ-CD Complexes of Pristine C_{60}, **1**, and **37** and Bispyrrolidinium Salts **38a–d** (Without a γ-CD Host) in Aqueous Solutions

Compound	$S_1^* \rightarrow S_n^*$ (nm)	$T_1^* \rightarrow T_n^*$ (nm)	Triplet lifetime (μs)	Rate constant k_{DABCO} $(M^{-1} s^{-1})^a$	Rate constant $k_{ELECTRONS}$ $(M^{-1} s^{-1})^b$	$C_{60}^{\cdot-}$ (nm)^c
C_{60}/γ-CD	920	747	100	1.6×10^8	1.8×10^{10}	1080
1/γ-CD	880	700	79	4.8×10^6	0.98×10^{10}	1010
37/γ-CD	920	700	52	7.8×10^6	2.8×10^{10}	1015
38a	880	660	36	4.7×10^6	2.2×10^{10}	900
38b	905	660	35	3.7×10^6	1.8×10^{10}	1000
38c	880	680	31	3.1×10^6	1.6×10^{10}	980
38d	890	690	31	2.5×10^6	0.88×10^{10}	1040

[a] Rate constant for the reductive quenching of the fullerene triplet excited state with DABCO as a sacrificial electron donor.
[b] Rate constant for the one-electron reduction of the fullerene singlet ground state by radiolytically generated hydrated electrons.
[c] Maximum of fullerene π-radical anion absorption.

degrees of functionalization [91]. In line with the hypothesis that the triplet energies in part compensate the effect of the ground-state reduction potential, the similarity of the values for **38a–d** with those of monofunctionalized fullerene derivatives should be noted. Also, among the different monofunctionalized derivatives the triplet excited state of the monopyrrolidinium salt is subject to the most efficient reductive quenching [102].

These observations make the bispyrrolidinium salts even better building blocks for three-dimensional donor–acceptor arrays. Furthermore, based on their water solubility, they are excellent probes for free-radical scavenging, including reactions with $O_2^{\cdot-}$, in aqueous media [5,6].

D. Donor–Acceptor Dyads

Fullerene-based donor–acceptor dyads encompass the linkage of a number of different sacrificial donor moieties, ranging from ferrocene (Fc) and aniline (An) to phenothiazine (PTZ) and tetrathiafulvalenes (TTF), to the fullerene core [4,8,40,114]. In these systems excitation takes place exclusively at the fullerene end, leaving the electron donor in its singlet ground state. The singlet excited-state energies of a variety of fullerene derivatives, such as methanofullerenes (1.796 eV) or fulleropyrrolidines (1.762 eV), enable a subsequent intramolecular electron transfer from the donor end to the fullerene moiety (Fig. 9) [91]. This

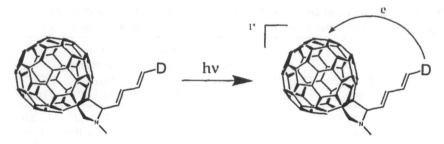

Figure 9 Schematic representation of a photo-induced electron transfer event from a sacrificial electron donor to the singlet excited state of a fullerene moiety.

generally yields the $C_{60}{}^{\bullet-}$–$D^{\bullet+}$ charge-separated state, in which $D^{\bullet+}$ denotes the oxidized donor moiety.

1. Fullerene–Ferrocene Dyads [33,34,115]

The absorption spectrum of dyad **13b** in hexane afforded no additional bands due to a potential charge transfer character or even perturbation of the fullerene π system. In turn, this indicates relatively little ground-state electronic interaction between the electron donor (ferrocene) and electron acceptor (fullerene) moieties. In dyads **12** and **13a**, however, the fullerene UV transition at 215 nm is subject to a 5- to 6-nm blue shift. It should be noted that addition of an equivalent amount of ferrocene to a hexane solution of **1** (e.g., fullerene reference compound) resulted in a similar blue shift from 215 nm to 207 nm. Therefore, we reach the conclusion that this absorption band evolves from a weak interaction between the ferrocene and the fullerene moieties, although differential solvation effects cannot be ruled out. In this context, it is important to consider that ground-state electronic coupling is known to occur typically across rigid bridges that contain up to six σ bonds [116,117].

Remarkable differences were observed with respect to the relative fluorescence yields of fullerene–ferrocene dyads **12a–c** and **13a–b** at different temperatures and in solvents of different polarity. Reference compound **1**, which, based on its structural similarity, serves as an internal standard, displaying a fluorescence quantum yield of $\Phi_{\text{FLUORESCENCE}} \sim 6.0 \times 10^{-4}$ (see above). The fullerene emission in all of the dyads investigated were extensively quenched. It is conceivable to attribute the decrease in fluorescence intensity on going from **1** to the ferrocene derivatives to an intramolecular quenching of the fullerene singlet excited state by the appended electron donor (e.g., ferrocene). It is interesting that the relative yields correlate well with the relative bridge length, which, on the other hand, is sought to determine the separation between the ferrocene and the fullerene moieties.

Picosecond-resolved photolysis of fullerene–ferrocene dyads **12** and **13** led to differential absorption changes that are distinctly different from that of the reference compound **1**. We still noted spectroscopic evidence for the involvement of the fullerene singlet excited state, by means of the characteristic $(S_1^* \rightarrow S_n^*)$ transition around 886 nm [91]. Its relative yield and, more importantly, its decay kinetics are, however, quite different from those of **1**. None of the ferrocene dyads reveal absorption features that are attributable to a triplet excited state. Instead, the fullerene singlet excited states of dyads **12** and **13** decay rapidly. Extending the relative length of the spacer enhances the lifetime of the singlet excited states, (e.g., 145 ps and 290 ps for dyad **13a** and **13b**, respectively). Furthermore, transient absorption spectroscopy of dyads **13a** and **13b** unravels that the initially quenched intermediate transforms into a broadly absorbing species. For example, in dyad **13a** the formation of this final product is completed around 2.9 ns after the pulse. Due to the flexible nature of the spacing alkyl chain for dyads **13a** and **13b**, the quenching of the fullerene singlet excited state is suggested to involve formation of an *intra*molecular excited state complex in which full charge separation is attained, but not before a later stage.

$$(C_{60})-(Fc) \xrightarrow{\;h\nu\;} {}^{1*}(C_{60})-(Fc) \rightarrow (C_{60}-Fc \text{ exciplex})$$
$$\xrightarrow{\text{electron transfer}} (C_{60}^{\cdot-})-(Fc^{\cdot+}) \quad (10)$$

To shed further light on the nature and lifetime of the product formed, nanosecond-resolved photolysis of dyads **13a** and **13b** were performed in different solvents. The resulting differential absorption spectra show sharp maxima in the near-IR, resembling the radiolytically induced reduction of **1**. Thus, *intramolecular* quenching of photoexcited fullerene–ferrocene dyads unambiguously involves electron transfer from the ferrocene to the triplet excited state of the fullerene moiety. In degassed benzonitrile solutions, for example, stabilization of the charge-separated radical pairs [τ = 2.6 µs (**13a**) and τ = 3.6 µs (**13b**)] were found. In polar solvents, the charge-separated radical pair is then regarded to diffuse semifreely, almost like two different entities in solution. The reported lifetimes are in clear support of this view. However, nonpolar solvents (e.g., toluene) lack the stabilization effect of the radical pair. Therefore, charge recombination, to yield the singlet ground-state or an excited-state product, is usually very fast.

2. Fullerene-Aniline Dyads [119,120]

Given the efficiency of electron transfer in fullerene–aniline dyads [9,15,118–121], a topographically controlled electron transfer process was the subject in a second generation of ortho- (**39**) and para-substituted (**40**) analogs [119] (Chart 10). Minimum energy conformations yielded a folded configuration for the ortho-substituted dyad **39**, while the para-substituted analog **40** adapted a stretched one (Fig. 10). An important aspect of this work is the quasi-flexibility of the chain

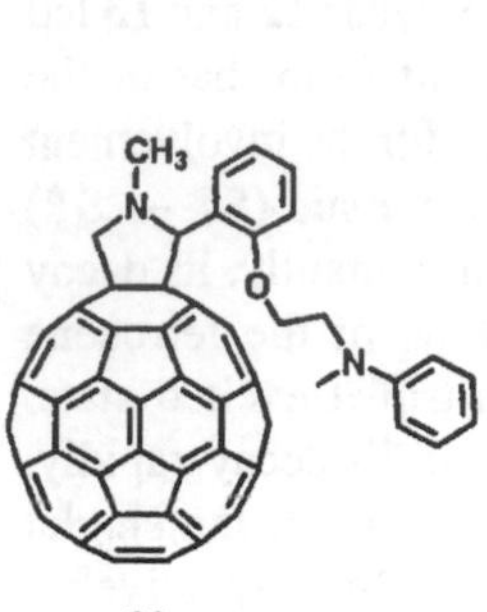

39 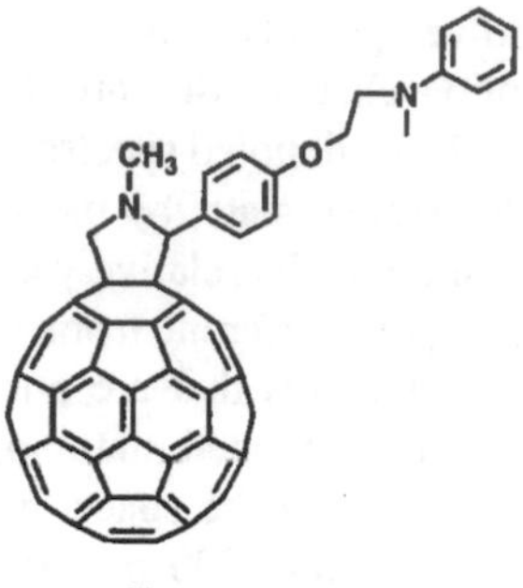   **40**

Chart 10

linking the phenyl spacer and the aniline donor leading to a degree of freedom for configurational rearrangements. The hereby implemented decrease in the spatial separation between the electron donor (aniline) and electron acceptor (fullerene) was beneficial in light of an efficient electron transfer. Experimental evidence, by means of singlet excited-state deactivation and quantum yield of charge separation, supports the computational view. Folding of the aniline group, into close proximity to the fullerene π system, enhances the efficiency of the electron transfer. To underline the importance of the orientation dependence on electron transfer reactions, it should be noted that the para-substituted dyad (**40**) gives rise to a very inefficient singlet excited-state quenching.

The controlled formation of fullerene clusters in polar media is another versatile approach to improve the performance of fullerene-containing dyads under the aspect of charge separation [120]. In particular, it is feasible that the electron delocalization within a fullerene cluster is distributed over all fullerene molecules. This, in turn, proved beneficial for the stabilization of charge-separated radical pairs. For example, the fullerene–aniline dyad, which lacks any electron transfer activity (i.e., para-substituted **40**) [119], forms a stable and optically transparent cluster in toluene with high acetonitrile contents. The fullerene cluster, in contrast to its monomer, gives rise to a strong quenching of the fullerene fluorescence and charge–transfer interactions between the photoexcited fullerene and aniline moieties are observed. The latter stems from the identification of the diagnostic absorption band of the fullerene π-radical anion in the near-IR region with lifetimes of several hundred nanoseconds.

3. Fullerene–[Ru(bpy)$_3$]$^{2-}$ Dyads [46–51,73,122,123] (Fig. 11)

Mononuclear Dyad [48,122]   The absorption spectrum of dyads **22** and **23** are superimpositions to that of the respective fullerene and ruthenium model compounds [46,48,122]. In particular, they reveal the fine-structured absorption

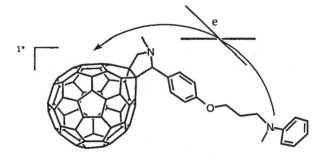

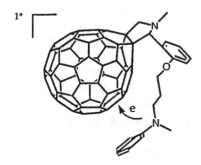

Figure 10 Schematic representation of a photo-induced electron transfer event in para-
and ortho-substituted fullerene–aniline dyads (**39,40**).

features of both chromophores (e.g., $[Ru(bpy)_3]^{2+}$ and fullerene). It is important
to note that the dyad spectrum lacks any evidence in form of additional transition
bands that may have suggested a possible charge transfer character in the ground
state (Fig. 12).

Quenching of the metal-to-ligand charge transfer (MLCT) excited states in
ruthenium-(II) complexes can be conveniently monitored by steady-state luminescence techniques. Upon excitation of the MLCT absorption at 460 nm, the
ruthenium model complex $[Ru(bpy)_3]^{2+}$ in CH_3CN is strongly luminescent with
a maximum at 640 nm [72]. Applying identical experimental conditions for recording the respective emission spectra of dyads **22** and **23** results, however, in
a noticeable quenching relative to the ruthenium model [48,122]. Since the intensity and frequency of the exciting light and the molar extinction coefficients were
kept constant, it is conceivable to attribute the decrease in luminescence intensity

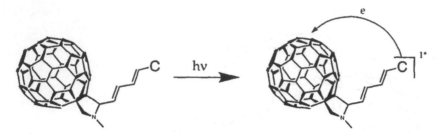

Figure 11 Schematic representation of a photo-induced electron transfer event from a photoexcited chromophore to the ground-state fullerene.

of the dyads to an intramolecular electron transfer process between the excited ruthenium MLCT state and the electron-accepting fullerene core.

Further support for this assumption emerges from the characteristic luminescence dependence on the solvent dielectric constant. Specifically, increasing the solvent polarity from a nonpolar solvent mixture (e.g., CH_2Cl_2/toluene; 1:1 v/v) ($\varepsilon = 5.73$) to CH_3CN ($\varepsilon = 37.5$) results in a substantial decrease of the $^{3*}[Ru(bpy)_3]^{2-}$ emission yield in dyad **22** [48]. The dielectric continuum model helps to quantify these solvent effects by handling the charge-separated radical pair as two spherical ions separated by a distance (R) and submerged into a solvent of a static dielectric constant (ε) [124,125]. In essence, the free-energy change ($-\Delta G^0$), associated with an intramolecular electron transfer, becomes more exothermic with increasing solvent polarity. Thus, the current observation corroborates a through-bond electron transfer mechanism mediated through the rigid σ-bond framework of the androstane skeleton in photoexcited dyad **22**.

$$[Ru(bpy)_3]^{2-}-C_{60} \xrightarrow{h\nu} {}^{3*}[Ru(bpy)_3]^{2-}-C_{60}$$
$$\xrightarrow{electron\ transfer} [Ru(bpy)_3]^{3-}-(C_{60}^{\cdot-}) \quad (11)$$

On the contrary, the data presented for dyad **23** reveals no significant acceleration of the quenching rate upon increasing the solvent dielectric constant from $\varepsilon = 5.73$ (CH_2Cl_2/toluene; 1:1 v/v) to 37.5 (CH_3CN) [122]. In addition, the relative length of the spacer and presumably the distance between the two reactive centers should be noted. Although the length increases in the following order **22** (12 σ bonds) < **23** (17 σ bonds), the MLCT state quenching, e.g., in CH_2Cl_2/toluene (1:1 v/v), is opposite to the one expected.

At this point the ground-state absorption of both dyads (e.g., **22** and **23**) should be considered again. They rule out a folded conformer in which the two moieties are located in close proximity to each other. The number of σ bonds (17) in dyad **23** disfavors a through-bond electron process. This leads to the

electron transfer

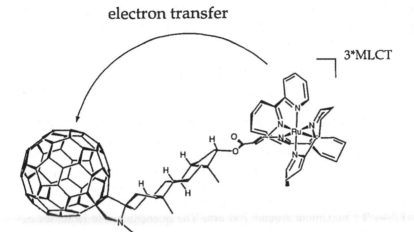

energy transfer

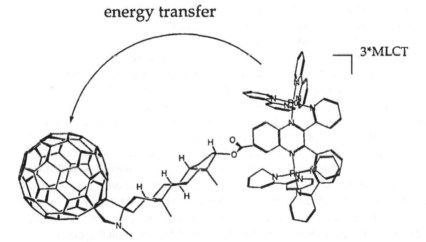

Figure 12 Schematic representation of photo-induced electron transfer and energy transfer events from ruthenium-excited MLCT states, i.e., [Ru(bpy)$_3$]$^{2+}$ (**22**) and [(bpy)$_2$ Ru(BL)Ru(bpy)$_2$]$^{4+}$ (**28**), respectively, to the ground-state fullerene. (Adapted from Ref. 123.)

assumption that intramolecular quenching in dyad **23** necessitates conformational rearrangement of the flexible spacer probably via formation of a transient excited-state complex (see fullerene–ferrocene dyads).

$$[Ru(bpy)_3]^{2+}-C_{60}—h\nu→^{3*}[Ru(bpy)_3]^{2+}-C_{60}→([Ru(bpy)_3]^{\delta+})-C_{60}^{\delta-})$$
$$—electron\ transfer→[Ru(bpy)_3]^{3+}-(C_{60}^{·-}) \quad (12)$$

Picosecond-resolved photolysis of dyads **22** and **23** in deaerated CH_2Cl_2 shows that the characteristics of the photoexcited MLCT state transform rapidly into a broadly absorbing product. It should be noted that the MLCT excited state is practically stable on this time scale for the $[Ru(bpy)_3]^{2+}$ reference compound with a lifetime of 150 ns [72]. The broad absorption features recorded 4000 ps after the pulse are distinctly different from the fullerene triplet excited state (e.g., a sharp $T_1^* → T_n^*$ maximum around 700 nm. The quenching rate increases as the dielectric constant of the solvating medium increases, with rates typically in the general range of 10^9 s^{-1}. It should be noted that the quenching rates closely follow the luminescence efficiencies. In particular, a minor solvent dependence characterizes the luminescence yields for flexible dyad **23**, while, on the other hand, drastic changes are noted for rigid dyad **22** [48,122].

Differential absorption changes in the UV-Vis range, recorded about 50 ns after laser excitation of dyads **22** and **23**, display the same series of maxima at 390, 500, 570, and 640 nm. Surprisingly, no direct spectral evidence for the reduced fullerene moiety ($C_{60}^{·-}$) [98] and the oxidized ruthenium chromophore ($[Ru(bpy)_3]^{3+}$) were found [126]. This necessitated the generation of the fullerene π-radical anion and the one-electron oxidized ruthenium (III) complex separately in a pulse radiolysis experiment. While reduction experiments in a toluene/2-propanol/acetone solvent mixture (8:1:1 v/v) [98] allowed us to form the π-radical anion of the fullerene reference, oxidation of $[Ru(byp)_3]^{2+}$ was followed by reaction with the oxidizing $Cl_2^{·-}$ radical in aqueous solutions [126]. By overlaying the differential absorption changes monitored for a $[Ru(bpy)_3]^{3+}$ complex and a $C_{60}^{·-}$ moiety with the photolytically generated spectrum we could show that the transient spectrum is virtually the sum of the individual spectra.

$$Cl_2^{·-} + [Ru(bpy)_3]^{2+} → 2Cl^- + [Ru(bpy)_3]^{3+} \quad (13)$$

Furthermore, a well-resolved near-IR absorption band with λ_{max} at 1040 nm, monitored upon nanosecond photolysis, resembles the radiolytically generated fullerene π-radical anion (e.g., $C_{60}^{·-}$) [98] and thus completes the spectral characterization of the radical pair. This clearly documents the existence of a long-lived charge-separated state, namely, $[Ru(bpy)_3]^{3+}—(C_{60}^{·-})$, evolving from reductive quenching of the ruthenium MLCT excited state and corroborates the above low luminescence yields.

In light of stabilizing the transient $[Ru(bpy)_3]^{3+}$—$(C_{60}^{\bullet-})$ radical pair, the flexible spacer in dyad 23 implies a similar scenario as was encountered earlier in the series of flexible spaced fullerene–ferrocene dyads. Accordingly, only solvents that possess a high dielectric constant should help to separate the radical pair, namely, $(C_{60}^{\bullet-})$ and $([Ru(byp)_3]^{3+})$ and, in turn, stabilize the charge-separated state. This is indeed the case: Photolysis in CH_3CN and CH_3OH yields the unmistakable signature of the fullerene π-radical anion, whereas the insufficient polarity of, for example, CH_2Cl_2/toluene (1:1 v/v) or CH_2Cl_2 results in a fast back electron transfer. Consequently, no fullerene π-radical anion absorption was observed. Instead, the characteristics of the fullerene triplet excited state with two maxima at 360 and 700 nm were noted [91]. It should be noted that the lifetimes of the charge-separated state are 742 ns and 178 ns in CH_3CN and CH_3OH solutions, respectively.

On the other hand, in dyad 22 the large spatial separation retards the back electron transfer and, in turn, leads to a charge-separated radical pair, which is quite stable even in CH_2Cl_2 solutions. In the latter solvent the radical pair decays with a lifetime of 304 ns, quantitatively yielding the singlet ground state.

Dinuclear Dyad [73] The ground-state absorption of the dinuclear $[(bpy)_2Ru(BL)Ru(bpy)_2]^{4+}$ chromophore in dyad 28 (BL = 2,3-bis(2-pyridyl)quinoxaline) reveals a strong red shift relative to the corresponding mononuclear dyad 22 [46,48,122]. In contrast to the long-wavelength absorption of the reference $[Ru(bpy)_3]^{2+}$ complex, which is centered around 460 nm (2.69 eV), the photoactive MLCT transition in $[(bpy)_2Ru(BL)Ru(bpy)_2]^{4+}$ is about 625 nm (1.98 eV). In general, these absorption features imply a significant advancement of the photosensitizing properties and are promising for light-driven energy transfer reactions to the covalently linked fullerene moiety.

In line with the ground-state absorption, the emission spectrum of the $[(bpy)_2Ru(BL)Ru(bpy)_2]^{4+}$ complex, upon excitation of the MLCT absorption, is subject to a parallel red shift to 710 nm. This corresponds to an energy of the photoexcited MLCT state of 1.74 eV, relative to the higher MLCT state energy of 1.97 eV for the $[Ru(bpy)_3]^{2+}$ complex. The luminescence behavior of the rigidly spaced $[(bpy)_2Ru(BL)Ru(bpy)_2]^{4+}$–$C_{60}$ donor–bridge–acceptor dyad (28) was complementarily studied in strongly polar CH_3CN and moderately polar CH_2Cl_2 solutions. Under identical experimental conditions, only a moderate loss of intensity (about 27%) was noted in both solvents, relative to the emission of $[(bpy)_2Ru(BL)Ru(bpy)_2]^{4+}$ model complex. It is interesting to note that the associated digression is in sharp contrast to the effective quenching processes observed for the mononuclear $[Ru(bpy)_3]^{2+}$ dyad 22. As stated above, the Born formula predicts that $-\Delta G^0$ for an intramolecular electron transfer becomes more exothermic with increasing solvent polarity. Consequently, the current observation contradicts a through-bond electron transfer mechanism for photoexcited dyad **28**.

A similar behavior was noted in picosecond-resolved photolytic experiments. In particular, the differential absorption changes, upon excitation of a dinuclear ruthenium model lacking the fullerene moiety, are dominated by bleaching of the ground-state MLCT transition between 550 and 750 nm. A kinetic analysis of the absorption time profiles documented, however, a fast deactivation of the photoexcited MLCT state. The derived lifetime of 30 ns is substantially shorter than that of $[Ru(bpy)_3]^{2+}$ (180 ns). Despite the short-lived nature, the MLCT state in dyad **28** is subject to a further decrease (i.e., 22 ns in CH_2Cl_2). The relative MLCT lifetime, noted in dyad **28** and in the corresponding model complex, is in good agreement with the luminescence intensities and, in turn, suggests the occurrence of an intramolecular deactivation process.

To shed light on the energetics associated with a possible intramolecular electron or energy transfer mechanism the energy levels of the excited state and the charge-separated states were estimated. Thermodynamic evaluation of an electron transfer from the photoexcited MLCT state to the electron-accepting fullerene in dyad **28** led to endothermic free-energy changes in CH_2Cl_2 and CH_3CN, respectively. In accordance with a thermodynamically unfavorable electron transfer process, no spectral evidence, in particular in the fullerene π-radical anion characteristic near-IR region, was monitored. Parallel to the near-IR region, spectral features in the UV-Vis range lack clear evidence for the formation of either an oxidized ruthenium chromophore (i.e., $[(bpy)_2Ru(BL)Ru(bpy)_2]^{5-}$) or a reduced fullerene transient (i.e., $C_{60}^{\bullet-}$).

Nanosecond-resolved kinetics of dyad **28** in deaerated CH_2Cl_2 shows that the photoexcited MLCT state slowly transformed into an absorbing product. The transient species displays a series of maxima at 360 and 700 nm and decays with a lifetime of 43 μs. The broad absorption features are in remarkably good agreement with the fullerene triplet excited state, with a sharp ($T_1^* \rightarrow T_n^*$) maximum around 700 nm [91]. In line with the characteristic triplet–triplet absorption (700 nm) and the triplet lifetime (43 μs) is the following intramolecular energy transfer reaction:

$$[(bpy)_2Ru(BL)Ru(bpy)_2]^{4+}-C_{60} \xrightarrow{h\nu} {}^{3*}[(bpy)_2Ru(BL)Ru(bpy)_2]^{4+}-C_{60}$$

$$\xrightarrow{ENT} [(bpy)_2Ru(BL)Ru(bpy)_2]^{4+}-({}^{3*}C_{60}) \quad (14)$$

In summary, the dinuclear dyad shows red-shifted and intensified ground-state absorption. In light of aspects that are concerned with solar energy conversion, this effect is of importance. The excited-state energy of the MLCT state (1.74 eV) is, however, substantially impacted and, in turn, formation of a charge-separated radical pair is thermodynamically unfavorable. Nevertheless, intramolecular energy transfer to the energetically lower lying triplet excited state of the fullerene moiety (1.50 eV) takes place. It should be noted that a decrease of the spatial distance between the fullerene and ruthenium centers may result in a reactivation of the electron transfer reaction, evolving from reductive quenching of the ruthenium MLCT excited state.

Peptide Spaced Mononuclear Dyad [49] Peptides have also been used as molecular spacers to separate, for example, a fulleropyrrolidine acceptor unit from a ruthenium(II) trisbipyridine complex. The investigated peptide-spaced dyad **29** [49] is a complementary extension of earlier work on a rigid (with an androstane spacer **22** [48]) and a flexible spaced (with a polyglycol spacer **23** [122]) fullerene–[Ru(bpy)₃]²⁺ analog. Spacers, such as the investigated hexapeptide which in chlorinated hydrocarbons adopts a helical structure, are prone to conformational changes of their secondary structure upon addition of protic solvents [127–129]. In chlorinated solvents the helical secondary structure of the peptide spacer places the two redox-active moieties into close proximity, which is favorable for their mutual electronic interaction. An edge-to-edge distance of about 12 Å provides the means for an intramolecular electron transfer from the [Ru(byp)₃]²⁺ MLCT excited state to the electron-accepting fullerene. In contrast, protic solvents disrupt the helical structure of the peptide backbone and the separation between the two chromophores, [Ru(bpy)₃]²⁺ and C₆₀, located at the N and C termini of the peptide chain, tends to increase to a point that disfavors their mutual electronic interactions. Despite the general flexibility of the peptide backbone, the encountered scenario does not support either a through-bond or an excited-state complex electron transfer mechanism.

After careful removal of the protic component from the solvent mixture the luminescence intensity of [Ru(bpy)₃]²⁺ chromophore became comparable to that recorded for the original nonprotic solution, prior to the addition of protic solvent. The reversible activation/deactivation of the ET mechanism was repeated successfully many times and, thus, is a sensitive probe for the secondary structure of peptides (Fig. 13).

4. Fullerene-Thiophenylazobenzeneamine Dyad [130]

The distance between the donor and acceptor moiety is one of the important parameters determining the free-energy changes of an intramolecular electron transfer. As stated above, the excited-state energy of the MLCT state in the dinuclear [(bpy)₂Ru(BL)Ru(bpy)₂]⁴⁺ complex takes a disadvantageous turn [73]. We suggested that a shorter donor–acceptor separation would, however, be beneficial to activate an intramolecular electron transfer event for dyad **28**. In this light we directly linked a thiophenylazobenzeneamine chromophore (**35**) to the pyrrolidine ring of monoadduct **1** [130].

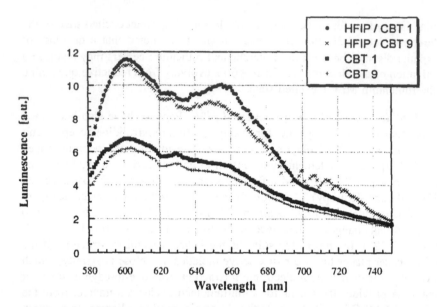

Figure 13 Emission spectra (excitation at 460 nm) of peptide-spaced fullerene [Ru(bpy)₃]²⁻ dyad **29** (initial cycle) in CBT (CBT = *n*-chlorobutane) (■) and in 1:1 CBT/HFIP (HFIP = hexafluoroisopropanol) (+) and (after nine cyles) in CBT (●) and 1:1 CBT/HFIP (×). As all samples were studied under identical conditions, the relative intensities represent relative emission quantum yields.

In fullerene–dye dyad **36** notable ground-state interactions prevail between the fullerene and dye units. In fact, the dye electronic absorption is subjected to a red shift from 554 nm (**35**) to 567 nm (**36**), corroborating the above hypothesis. Additional evidence for this hypothesis stems from the first fullerene-centered reduction and the cathodic shift of the dye-based reductions in dyad **36**. However, the electrochemical characterization shows no additional stabilization of the dye-centered radical cation.

Complementary emission and transient absorption measurements revealed a rapid deactivation of the dye singlet excited state. In view of the steady-state and time-resolved photophysical experiments, the quenching mechanism of the ^{1*}dye in dyad **36** (1.97 eV) involves both an energy transfer process leading to $^1*C_{60}$ state and an electron transfer, affording the (dye$^{•+}$)–($C_{60}^{•-}$) charge-separated state.

$$\text{dye-}C_{60} \xrightarrow{h\nu} {}^{1*}\text{dye-}C_{60} \xrightarrow{\text{ENT}} \text{dye-}{}^{1*}C_{60} \tag{15}$$

$$\text{dye-}C_{60} \xrightarrow{h\nu} {}^{1*}\text{dye-}C_{60} \xrightarrow{\text{electron transfer}} (\text{dye}^{•+})\text{-}(C_{60}^{•-}) \tag{16}$$

The energy difference between $^{1*}C_{60}$ (1.73 eV)[91] and the charge-separated state, namely, $(dye^{\bullet+})–(C_{60}{}^{\bullet-})$ (1.71 eV), is very small in CH_2Cl_2 leading to the suggestion that a rapid exchange between the two states occurs.

$$dye–^{1*}C_{60} \rightarrow (dye^{\bullet+})–(C_{60}{}^{\bullet-}) \tag{17}$$

With the support of pulse radiolysis, we demonstrated that the proposed $(dye^{\bullet+})–(C_{60}{}^{\bullet-})$ radical pair exhibits finely structured characteristics of the reduced fullerene (e.g., 1010 nm) [98] and the oxidized dye. For instance, radiolytic oxidation of dye **37** was carried out in dilute solutions of aerated CH_2Cl_2. The oxidation reaction involves radiolytically generated $CH_2ClO_2{}^{\bullet}$ and $CHCl_2O_2{}^{\bullet}$ radicals, which are formed during the radiolysis of the solvent [131]. In particular, the oxidized dye exhibits bleaching of the singlet ground state and formation of a new Vis band at 560 and 650 nm, respectively. For details on the fullerene reduction see Section III.B.

Finally, we have, to the best of our knowledge, reported the first example of a TiO_2 sensitization with a fullerene-based donor–acceptor dyad. However,

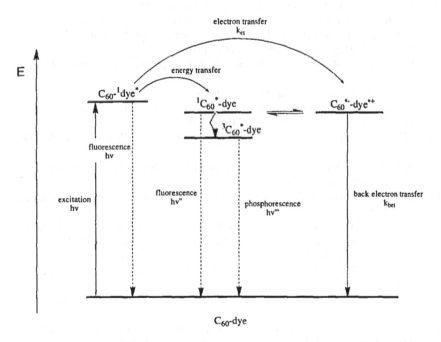

Figure 14 Schematic representation of photo-induced electron transfer and energy transfer events in fullerene–thiophenylazobenzeneamine dyad **36**. (Adapted from Ref. 130.)

observation of a relatively low sensitization suggests that the lifetime of the dye singlet excited state (i.e., ^{1*}dye) is too short. Furthermore, the rapid decay transforms the dye singlet excited state into a state which is thermodynamically unable to inject an electron into the semiconductor conduction band. Interestingly, the sensitized electrodes are very stable to daylight exposure, as compared to those coated with derivative **35**. This effect has been ascribed to UV filtering exerted by the fullerene moiety in **36** (Fig. 14).

ACKNOWLEDGMENTS

Part of this work has been supported by CNR (legge 95/95) and by MURST (contract No. 9803194198) and by the Office of Basic Energy Sciences of the Department of Energy. This is document NDRL-4171 from the Notre Dame Radiation Laboratory. D.G. and M.M. thank NATO for a travel grant (CRG 960099).

REFERENCES

1. Hirsch, A., Ed. *Fullerenes and Related Structures*, Series: Topics in Current Chemistry. Springer-Verlag: Berlin/Heidelberg, 1998, Vol. *199*.
2. Hirsch, A. *The Chemistry of the Fullerenes*; Thieme: Stuttgart, 1994.
3. Mirkin, C. A.; Caldwell, W. B. *Tetrahedron* **1996**, *52*, 5113–5130.
4. Prato, M. *J. Mater. Chem.* **1997**, *7*, 1097–1109.
5. Jensen, A. W.; Wilson, S. R.; Schuster, D. I. *Bioorg. Med. Chem.* **1996**, *4*, 767–779.
6. Da Ros, T.; Prato, M. *Chem. Commun.* **1999**, 663.
7. Echegoyen, L.; Echegoyen, L. E. *Acc. Chem. Res.* **1998**, *31*, 593–601.
8. Imahori, H.; Sakata, Y. *Adv. Mater.* **1997**, *9*, 537–546.
9. Williams, R. M.; Zwier, J. M.; Verhoeven, J. W. *J. Am. Chem. Soc.* **1995**, *117*, 4093–4099.
10. Nakamura, Y.; Minowa, T.; Tobita, S.; Shizuka, H.; Nishimura, J. *J. Chem. Soc., Perkin Trans. 2* **1995**, 2351–2357.
11. Khan, S. I.; Oliver, A. M.; Paddon-Row, M. N.; Rubin, Y. *J. Am. Chem. Soc.* **1993**, *115*, 4919–4920.
12. Lawson, J. M.; Oliver, A. M.; Rothenfluh, D. F.; An, Y.-Z.; Ellis, G. A.; Ranasinghe, M. G.; Khan, M. G.; Franz, A. G.; Ganapathi, P. S.; Shephard, M. J.; Paddon-Row, M. N.; Rubin, Y. *J. Org. Chem.* **1996**, *61*, 5032–5054.
13. Belik, P.; Gügel, A.; Kraus, A.; Walter, M.; Müllen, K. *J. Org. Chem.* **1995**, *60*, 3307–3310.
14. Matsubara, Y.; Tada, H.; Nagase, S.; Yoshida, Z. *J. Org. Chem.* **1995**, *60*, 5372–5373.
15. Williams, R. M.; Koeberg, M.; Lawson, J. M.; An, Y. -Z.; Rubin, Y.; Paddon-Row, M. N.; Verhoeven, J. W. *J. Org. Chem.* **1996**, *61*, 5055–5062.
16. Luo, C.; Fujitsuka, M.; Watanabe, A.; Ito, O.; Gan, L.; Huang, Y.; Huang, C.-H. *J. Chem. Soc., Faraday Trans.* **1998**, *94*, 527–532.

17. Deviprasad, G. R.; Rahman, M. S.; D'Souza, F. *Chem. Commun.* **1999**, 849–850.
18. Guldi, D. M.; Torres-Garcia, G.; Mattay, J. *J. Phys. Chem.* **1998**, *102*, 9679–9685.
19. Sun, Y.; Drovetskaja, T.; Bolskar, R. D.; Bau, R.; Boyd, P. D. W.; Reed, C. A. *J. Org. Chem.* **1997**, *62*, 3642–3649.
20. Imahori, H.; Yamada, K.; Hasegawa, M.; Taniguchi, S.; Okada, T.; Sakata, Y. *Angew. Chem. Int. Ed. Engl.* **1997**, *36*, 2626–2629.
21. Baran, P. S.; Monaco, R. R.; Khan, A. U.; Schuster, D. I.; Wilson, S. R. *J. Am. Chem. Soc.* **1997**, *119*, 8363–8364.
22. Kuciauskas, D.; Liddell, P. A.; Moore, A. L.; Moore, T. A.; Gust, D. *J. Am. Chem. Soc.* **1998**, *120*, 10880–10886.
23. Dietel, E.; Hirsch, A.; Zhou, J.; Rieker, A. *J. Chem. Soc., Perkin Trans. 2* **1998**, 1357–1363.
24. Nierengarten, J. -F.; Oswald, L.; Nicoud, J.- F. *Chem. Commun.* **1998**, 1545–1546.
25. Nierengarten, J. -F.; Schall, C.; Nicoud, J. -F. *Angew. Chem. Int. Ed.* **1998**, *37*, 1934–1936.
26. Bourgeois, J.-P.; Diederich, F.; Echegoyen, L.; Nierengarten, J.-F. *Helv. Chim. Acta.* **1998**, *81*, 1835–1844.
27. Cheng, P.; Wilson, S. R.; Scuster, D. I. *Chem. Commun.* **1999**, 89–90.
28. Tamaki, K.; Imahori, H.; Nishimura, Y.; Yamazaki, I.; Shimomura, A.; Okada, T.; Sakata, Y. *Chem. Lett.* **1999**, 227–228.
29. Linssen, T. G.; Dürr, K.; Hirsch, A.; Hanack, M. *J. Chem. Soc., Chem. Commun.* **1995**, 103–104.
30. Durr, K.; Fiedler, S.; Linssen, T.; Hirsch, A.; Hanack, M. *Chem. Ber.* **1997**, *130*, 1375.
31. Prato, M.; Maggini, M.; Giacometti, C.; Scorrano, G.; Sandonà, G.; Farnia, G. *Tetrahedron* **1996**, *52*, 5221–5234.
32. Maggini, M.; Karlsson, A.; Scorrano, G.; Sandonà, G.; Farnia, G.; Prato, M. *J. Chem. Soc., Chem. Commun.* **1994**, 589–590.
33. Guldi, D.; Maggini, M.; Scorrano, G.; Prato, M. *J. Am. Chem. Soc.* **1997**, *119*, 974–980.
34. Guldi, D. M.; Maggini, M.; Scorrano, G.; Prato, M. *Res. Chem. Intermed.* **1997**, *23*, 561–573.
35. Deschenaux, R.; Even, M.; Guillon, D. *Chem. Commun.* **1998**, 537–538.
36. Da Ros, T.; Prato, M.; Carano, M.; Ceroni, P.; Paolucci, F.; Roffia, S. *J. Am. Chem. Soc.* **1998**, *120*, 11645–11648.
37. Martin, N.; Sánchez, L.; Seoane, C.; Andreu, R.; Garín, J.; Orduna, J. *Tetrahedron Lett.* **1996**, *37*, 5979–5982.
38. Llaclay, J.; Mas, M.; Molins, E.; Veciana, J.; Powell, D.; Rovira, C. *Chem. Commun.* **1997**, 659.
39. Martin, N.; Perez, I.; Sánchez, L.; Seoane, C. *J. Org. Chem.* **1997**, *62*, 5690–5695.
40. Martin, N.; Sánchez, L.; Illescas, B.; Pérez, I. *Chem. Rev.* **1998**, *98*, 2527.
41. Imahori, H.; Cardoso, S.; Tatman, D.; Lin, S.; Noss, L.; Seely, G.; Sereno, L.; Chessa de Silber, J.; Moore, T. A.; Moore, A. L.; Gust, D. *Photochem. Photobiol.* **1995**, *62*, 1009–1014.
42. Liddell, P. A.; Kuciauskas, D.; Sumida, J. P.; Nash, B.; Nguyen, D.; Moore, A. L.; Moore, T. A.; Gust, D. *J. Am. Chem. Soc.* **1997**, *119*, 1400–1405.

43. Carbonera, D.; Di Valentin, M.; Corvaja, C.; Agostini, G.; Giacometti, G.; Liddell, P. A.; Kuciauskas, D.; Moore, A. L.; Moore, T. A.; Gust, D. *J. Am. Chem. Soc.* **1998**, *120*, 4398–4405.

44. Diederich, F.; Dietrich-Buchecker, C.; Nierengarten, J.-F.; Sauvage, J.-P. *J. Chem. Soc., Chem. Commun.* **1995**, 781–782.

45. Armaroli, N.; Diederich, F.; Dietrich-Buchecker, C.; Flamigni, L.; Marconi, G.; Nierengarten, J.-F.; Sauvage, J.-P. *Chem. Eur. J.* **1998**, *4*, 406.

46. Maggini, M.; Donò, A.; Scorrano, G.; Prato, M. *J. Chem. Soc., Chem. Commun.* **1995**, 845–846.

47. Sariciftci, N. S.; Wudl, F.; Heeger, A. J.; Maggini, M.; Scorrano, G.; Prato, M.; Bourassa, J.; Ford, P. C. *Chem. Phys. Lett.* **1995**, *247*, 210–214.

48. Maggini, M.; Guldi, D. M.; Mondini, S.; Scorrano, G.; Paolucci, F.; Ceroni, P.; Roffia, S. *Chem. Eur. J.* **1998**, *4*, 1992.

49. Polese, A.; Mondini, S.; Bianco, A.; Toniolo, C.; Scorrano, G.; Guldi, D. M.; Maggini, M. *J. Am. Chem. Soc.* **1999**, *121*, 3446–3452.

50. Armspach, D.; Constable, E. C.; Diederich, F.; Housecroft, C. E.; Nierengarten, J.-F. *Chem. Commun.* **1996**, 2009–2010.

51. Armspach, D.; Constable, E. C.; Diederich, F.; Housecroft, C. E.; Nierengarten, J.-F. *Chem. Eur. J.* **1998**, *4*, 723–733.

52. Gust, D.; Moore, T. A.; Moore, A. L. *Pure. Appl. Chem.* **1998**, *70*, 2189–2200.

53. Sakata, Y.; Imahori, Y.; Tsue, H.; Higashida, S.; Akiyama, T.; Yoshizawa, E.; Aoki, M.; Yamada, K.; Hagiwara, K.; Taniguchi, S.; Okada, K. *Pure Appl. Chem.* **1997**, *69*, 1951–1956.

54. Kraabel, B.; Hummelen, J. C.; Vacar, D.; Moses, D.; Sariciftci, N. S.; Heeger, A. J.; Wudl, F. *J. Chem. Phys.* **1996**, *104*, 4267–4273.

55. Nierengarten, J.-F.; Eckert, J.-F.; Nicoud, J.-F.; Ouali, L.; Krasnikov, V.; Hadziioannou, G. *Chem. Commun.* **1999**, 617–618.

56. Akiyama, T.; Imahori, H.; Ajawakom, A.; Sakata, Y. *Chem. Lett.* **1997**, 907–908.

57. Imahori, H.; Azuma, T.; Ozawa, S.; Yamada, H.; Ushida, K.; Ajavakom, A.; Norieda, H.; Sakata, Y. *Chem. Commun.* **1999**, 557–558.

58. Imahori, H.; Hagiwara, K.; Akiyama, T.; Aoki, M.; Taniguchi, S.; Okada, T.; Shirakawa, M.; Sakata, Y. *Chem. Phys. Lett.* **1996**, *263*, 545.

59. Maggini, M.; Scorrano, G.; Prato, M. *J. Am. Chem. Soc.* **1993**, *115*, 9798–9799.

60. Zhang, X.; Willems, M.; Foote, C. S. *Tetrahedron Lett.* **1993**, *34*, 8187–8188.

61. Prato, M.; Maggini, M. *Acc. Chem. Res.* **1998**, *31*, 519–526.

62. Guldi, D. M.; Maggini, M. *Gazz. Chim. It.* **1997**, *127*, 779–785.

63. Wilson, S. R.; Lu, Q. *J. Org. Chem.* **1995**, *60*, 6496–6498.

64. Schick, G.; Levitus, M.; Kvetko, L. D.; Johnson, B. A.; Lamparth, I.; Lunkwitz, R.; Ma, B.; Khan, S. I.; Garcia-Garibay, M. A.; Rubin, Y. *J. Am. Chem. Soc.* **1999**, *121*, 3246–3247.

65. Hutchison, K.; Gao, J.; Schick, G.; Rubin, Y.; Wudl, F. *J. Am. Chem. Soc.* **1999**, *121*, 5611–5612.

66. Lu, Q.; Schuster, D. I.; Wilson, S. R. *J. Org. Chem.* **1996**, *61*, 4764–4768.

67. Pasimeni, L.; Hirsch, A.; Lamparth, I.; Herzog, A.; Maggini, M.; Prato, M.; Corvaja, C.; Scorrano, G. *J. Am. Chem. Soc.* **1997**, *119*, 12896–12901.

68. Maggini, M.; Zanellato, S.; Guldi, D. M.; Scorrano, G., unpublished results.
69. Iyoda, M.; Sultana, F.; Kato, A.; Yoshida, M.; Kuwatani, Y.; Komatsu, M.; Nagase, S. *Chem. Lett.* **1998**, 63–64.
70. Da Ros, T.; Prato, M.; Guldi, D. M.; Alessio, E.; Ruzzi, M.; Pasimeni, L. *Chem. Commun.* **1999**, 635–636.
71. D'Souza, F.; Deviprasad, G. R.; Rahman, S.; Choi, J.-P. *Inorg. Chem.* **1999**, *38*, 2157–2160.
72. Juris, A.; Balzani, V.; Barigelletti, F.; Campagna, S.; Belser, P.; Zelewsky, A. *Coord. Chem. Rev.* **1988**, *84*, 85.
73. Maggini, M.; Mondini, S.; Scorrano, G.; Guldi, D. M.; Paolucci, F.; Ceroni, P., unpublished results.
74. Foote, C. S. *Top. Curr. Chem.* **1994**, *169*, 347–363.
75. Kamat, P. V.; Asmus, K.-D. *Interface* **1996**, *5*, 22–25.
76. Leach, S.; Vervloet, M.; Despres, A.; Brcheret, E.; Hare, P.; Dennis, T. J. S.; Kroto, H. W.; Taylor, R.; Walton, D. R. M. *Chem. Phys.* **1992**, *160*, 451–466.
77. Sun, Y.-P.; Wang, P.; Hamilton, N. B. *J. Am. Chem. Soc.* **1993**, *115*, 6378–6381.
78. Ebbesen, T. W.; Tanigaki, K.; Kuroshima, S. *Chem. Phys. Lett.* **1991**, *181*, 501–504.
79. Tanigaki, K.; Ebbesen, T. W.; Kuroshima, S. *Chem. Phys. Lett.* **1991**, *185*, 189–192.
80. Arbogast, J. W.; Darmanyan, A. P.; Foote, C. S.; Rubin, Y.; Diederich, F. N.; Alvarez, M. M.; Anz, S. J.; Whetten, R. L. *J. Phys. Chem.* **1991**, *95*, 11–12.
81. Arbogast J. S.; Foote, C. S. *J. Am. Chem. Soc.* **1991**, *113*, 8886–8889.
82. Guldi, D. M.; Huic, R. E.; Neta, P.; Hungerbühler, H.; Asmus, K.-D. *Chem. Phys. Lett.* **1994**, *223*, 511–516.
83. Ausman, K. D.; Benedetto, A. F.; Samuels, D. A.; Weisman, R. B. In *Recent Advances in the Chemistry of Fullerenes and Related Materials.*, Vol. 6, edited by Kadish, K. M.; Ruoff R. S. The Electrochemical Society Proceedings Series, Pennington, NJ, 1998.
84. Sension, R. J.; Phillips, C. M.; Szarka, A. Z.; Romanow, W. J.; McGhie, A. R.; McCauley, J. P. J.; Smith III, A. B.; Hochstrasser, R. M. *J. Phys. Chem.* **1991**, *95*, 6075–6078.
85. Palit, D. K.; Sapre, A. V.; Mittal, J. P.; Rao, C. N. R. *Chem. Phys. Lett.* **1992**, *195*, 1–6.
86. Biczok, L.; Linschitz, H.; Walter, R. I. *Chem. Phys. Lett.* **1992**, *195*, 339–346.
87. Etheridge, H. T.; Weisman, R. B. *J. Phys. Chem.* **1995**, *99*, 2782–2787.
88. Caspar, J. V.; Wang, L. S. *Chem. Phys. Lett.* **1994**, *218*, 221–228.
89. Dimitrijevic, N. M.; Kamat, P. V. *J. Phys. Chem.* **1992**, *96*, 4811–4814.
90. Levanon, H.; Meiklyar, V.; Michaeli, S.; Gamliel, D. *J. Am. Chem. Soc.* **1993**, *115*, 8722–8727.
91. Guldi, D. M.; Asmus, K.-D. *J. Phys. Chem. A* **1997**, *101*, 1472–1481.
92. Guldi, D. M.; Hungerbühler, H.; Asmus, K. D. *J. Phys. Chem.* **1995**, *99*, 9380–9385.
93. Bensasson, R. V.; Hill, T. J.; Lambert, C.; Land, E. J.; Leach, S.; Truscott, T. G. *Chem. Phys. Lett.* **1993**, *201*, 326–335.

94. Bensasson, R. V.; Hill, T. J.; Lambert, C.; Land, E. J.; Leach, S.; Truscott, T. G. *Chem. Phys. Lett.* **1993**, *206*, 197–202.
95. Priyadarsini, K. I.; Mohan, H.; Birkett, P. R.; Mittal, J. P. *J. Phys. Chem.* **1996**, *100*, 501–506.
96. Guldi, D. M.; Maggini, M., unpublished results.
97. Guldi, D. M. In *Radiation Chemistry: Present Status and Future Prospects*, Jonah, C. D.; Rao, B.S.M. Eds.. Elsevier: Amsterdam, **2000**, in press.
98. Guldi, D. M.; Hungerbühler, H.; Janata, E.; Asmus, K. D. *J. Chem. Soc., Chem. Commun* **1993**, 84–85.
99. Guldi, D. M.; Hungerbühler, H.; Wilhelm, M.; Asmus, K.-D. *J. Chem. Soc. Faraday Trans.* **1994**, *90*, 1391–1396.
100. Guldi, D. M.; Liu, D.; Kamat, P. V. *J. Phys. Chem. A* **1997**, *101*, 6195–6201.
101. Guldi, D. M.; Hungerbühler, H.; K.-D. Asmus, *J. Phys. Chem. A* **1997**, *101*, 1783–1786.
102. Guldi, D. M. *J. Phys. Chem. A* **1997**, *101*, 3895–3900.
103. Guldi, D. M.; Hungerbühler, H.; Asmus, K. -D. *J. Phys. Chem.* **1995**, *99*, 13487–13493.
104. Guldi, D. M. *Res. Chem. Intermed.* **1997**, *23*, 653–673.
105. Thomas, T. N.; Taylor, R. A.; Ryan, J. F.; Mihailovic, D.; Zamboni, R. In *Electronic Properties of Fullerenes*; Kuzmany, H.; Fink, J.; Mehring, M.; Roth, S., Eds.; Springer-Verlag: Berlin, **1993**, pp. 292–296.
106. Yamakoshi, Y. N.; Yagami, T.; Fukuhara, K.; Sueyoshi, S.; Miyata, N. *J. Chem. Soc., Chem. Commun.* **1994**, 517–518.
107. Beeby, A.; Eastoe, J.; Heenan, R. K. *J. Chem. Soc., Chem. Commun.* **1994**, 173–175.
108. Beeby, A.; Eastoe, J.; Crooks, E. R. *J. Chem. Soc., Chem. Commun.* **1996**, 901–902.
109. Andersson, T.; Nilsson, K.; Sundahl, M.; Westman, G.; Wennerström, O. *J. Chem. Soc., Chem. Commun.* **1992**, 604–605.
110. Guldi, D. M.; Hungerbühler, H.; Asmus, K.-D. *J. Phys. Chem.* **1995**, *99*, 13487–13493.
111. Guldi, D. M. *J. Phys. Chem. B* **2000**, *104*, 1483–1489.
112. Priyadarsini, K. I.; Mohan, H.; Mittal, J. P.; Guldi, D. M.; Asmus, K.-D. *J. Phys. Chem.* **1994**, *98*, 9565–9569.
113. Guldi, D. M.; Hungerbühler, H.; Asmus, K.-D. *J. Phys. Chem. B* **1999**, *103*, 1444–1453.
114. Balch A. L.; Olmstead, M. M. *Chem. Rev.* **1998**, *98*, 2123–2165.
115. Guldi, D. M.; Maggini, M.; Scorrano, G.; Prato, M.; Bianco A, Toniolo, C. *J. Inf. Recording* **1998**, *24*, 33–39.
116. Paddon-Row, M. N. *Acc. Chem. Res.* **1994**, *27*, 18.
117. Kroon, J.; Verhoeven, J. W.; Paddon-Row, M. N.; Oliver, A. M. *Angew. Chem. Int. Ed. Engl.* **1990**, *30*, 1358.
118. Thomas, K. G.; Biju, V.; George, M. V.; Guldi; D. M.; Kamat, P. V. *J. Phys. Chem. A* **1998**, *102*, 5341–5348.
119. Thomas, K. G.; Biju, V.; Guldi, D. M.; Kamat P. V. George, M. V. *J. Phys. Chem. B* **1999**, *103*, 8864–8869.

120. Thomas, K. G.; Biju, V.; Guldi, D. M.; Kamat P. V. George, M. V. *J. Phys. Chem. A* **1999**, *103*, 10755–10763.
121. Luo, C.; Fujitsuka, M.; Huang, C.-H.; Ito, O. *J. Phys. Chem. A* **1998**, *102*, 8716–8721.
122. Maggini, M.; Mondini, S.; Scorrano, G.; Guldi, D. M.; Paolucci, F.; Ceroni, P. Unpublished results.
123. Guldi, D. M.; Maggini, M.; Martin, N.; Prato, M. *Carbon* **2000**, in press.
124. Weller, A. *Z. Physik. Chem.* **1982**, *132*, 93.
125. Gaines, G. L. I.; O'Neil, M. P.; Svec, W. A.; Niemczyk M. P.; Wasielewski, M. R. *J. Am. Chem. Soc.* **1991**, *113*, 719.
126. Mulazzani, Q. G.; Venturi, M.; Bolletta, F.; Balzani, V. *Inorg. Chim. Acta* **1986**, *113*, L1.
127. Toniolo, C.; Bonora, G. M.; Barone, V.; Bavoso, A.; Benedetti, E.; Di Blasio, B.; Grimaldi, P.; Lelj, F.; Pavone V.; Pavone, C. *Marcomolecules* **1985**, *18*, 895.
128. Toniolo C.; Benedetti, E. *Trends Biochem. Sci.* **1991**, *16*, 350.
129. Hanson, P.; Millhauser, G.; Formaggio, F.; Crisma M.; Toniolo, C. *J. Am. Chem. Soc.* **1996**, *118*, 7612.
130. Cattarin, S.; Ceroni, P.; Paolucci, F.; Roffia, S.; Guldi, D. M.; Maggini, M.; Menna, E.; Scorrano, G. *J. Mater. Chem.* **1999**, *9*, 2743–2750.
131. Shank, N. E.; Dorfman, L. M. *J. Chem. Phys.* **1970**, *52*, 4441–4417.

5

Applications of Time-Resolved EPR in Studies of Photochemical Reactions

Hans van Willigen
University of Massachusetts, Boston, Massachusetts

I. INTRODUCTION

The last two decades has seen a significant broadening of the field of applications of electron paramagnetic resonance (EPR) spectroscopy. For the most part, this can be attributed to major developments in the area of instrumentation. Whereas for a long time the majority of EPR studies were carried out with cw (continuous wave) X-band (~9 GHz) spectrometers operating at field strengths around 0.3 tesla, currently measurements are performed at frequencies ranging from 1.5 GHz to 670 GHz [1–4]. In addition, pulsed EPR instruments have developed to the point where measurements similar to those performed with pulsed nuclear magnetic resonance are becoming routine [5–7] Accounts of recent innovations in instrumentation and the implications for research carried out with EPR can be found in a number of recent publications [8–10].

The present chapter is concerned with applications of time-resolved EPR (TREPR*) techniques in the study of molecular photochemistry and photophysics, a field of research whose scope has grown significantly in recent years. The increase in research activity is due to a number of factors. First, there is interest

---

* The label TREPR will be used to denote time-resolved techniques in general. Labels to distinguish different methodologies will be introduced later in the text.

in acquiring a better understanding of photochemical processes because of applications in such diverse areas as chemical synthesis, photodegradation of pollutants, solar energy conversion and storage, and molecular scale optoelectronic devices. Since photochemical reactions in many cases involve transient paramagnetic molecules, it is evident that EPR can be an important source of information on processes that are of current interest. Second, with modern EPR instrumentation combined with pulsed lasers, the evolution of photogenerated paramagnetic species can be monitored with nanosecond time resolution combined with high spectral resolution and sensitivity. Third, theories for the quantitative interpretation of time profiles of resonance signals given by transient free radicals have been formulated and make it possible to use TREPR techniques to get detailed information on reaction mechanisms and dynamics.

In terms of time resolution, TREPR is similar to nanosecond transient optical (UV, Vis, IR) absorption spectroscopy. However, because of a number of unique features data from TREPR measurements constitute a valuable complement to those provided by optical spectroscopy. For instance, evidently only paramagnetic species contribute to the EPR spectra. This, together with the (typically) small linewidths of resonance peaks, generally leads to well-resolved, relatively simple spectra. Information on g values and hyperfine coupling constants (hfcc) derived from the spectra in most cases serves to identify the paramagnetic molecules unequivocally. The spectral parameters (hfcc's, linewidths) also report on intermolecular interactions which can serve, for instance, to get an insight into the spatial location of paramagnetic species in a microheterogeneous environment. That signal contributions from different species are resolved and readily identified facilitates the extraction of kinetic data from the time profiles of signal intensities.

Probably the most striking and valuable characteristic of TREPR spectra is that the time profiles not only reflect the (chemical) kinetics of radical formation and decay, but also are affected strongly by what is known as chemically induced dynamic electron polarization (CIDEP) [11–14]. CIDEP effects arise because the spin system of paramagnetic molecules formed in a chemical reaction initially will not be at thermal equilibrium. In pulsed-laser-initiated reactions, radical formation steps typically occur over a time period that is shorter than, or of the same order as, the spin-lattice relaxation time (T_1) of the electron spins. (T_1's of organic free radicals in liquid solution can range from tens of nanoseconds to 10 μs or more.) Consequently, a TREPR spectrum acquired at time τ_d following the laser pulse will display anomalous signal intensities reflecting the non-Boltzmann spin polarization if τ_d is of the order of T_1.

As will be discussed in more detail further on, CIDEP can give rise to enhanced absorption and/or stimulated emission peaks in TREPR spectra. The effects originate in the spin selectivity of chemical and physical processes involved in free-radical formation and decay, as well as in the spin state evolution in transient paramagnetic precursors of the species monitored with TREPR. For this reason, CIDEP constitutes a unique probe of the mechanistic details of photochemical reactions.

A. Instrumental Aspects

It was recognized early on that EPR could be a valuable source of information on transient free radicals formed in chemical reactions. Even so, the work reviewed here calls (at a minimum) for modifications of the data acquisition circuitry of conventional cw EPR spectrometers. For this reason, TREPR studies in the nanosecond and microsecond time regimes are of relatively recent vintage.

Commercial cw EPR spectrometers use the field modulation/phase-sensitive detection technique to provide optimum sensitivity [10]. As a consequence, the lower limit of the time response of the instruments is about 0.1 ms. By contrast, spin-lattice relaxation times of organic free radicals lie in the nanosecond to microsecond range, and this is also the time domain in which many photochemical reactions occur. Therefore, operating in the normal fashion the spectrometers do not have the time resolution required for detailed studies of the chemical and spin dynamics associated with photochemical reactions. However, the instruments can be adapted to these requirements relatively easily. For instance, by using light modulation combined with phase-sensitive detection, kinetic data can be obtained from an analysis of the dependence of signal intensity and phase on light modulation frequency [15].

Alternatively, in what is called the direct-detection mode, or cw TREPR, modulation/phase-sensitive detection (the limiting factor with respect to the response time of cw EPR spectrometers) is done away with altogether. Instead, the signal from the microwave detector/amplifier is directly routed to a data acquisition system with nanosecond time resolution such as a boxcar integrator or digital oscilloscope [16–19]. The spectrum of paramagnetic molecules present at a selected time, τ_d, after laser excitation of the sample is obtained by sweeping the magnetic field while recording the signal present at τ_d. Alternatively, the time profile of the intensity of a resonance peak can be recorded at a fixed field by varying τ_d. The loss in signal-to-noise caused by operating the instrument in the direction detection mode is made up, in part, by the use of ensemble averaging. As a result, spectra from thermalized paramagnetic systems can be recorded under favorable conditions [20,21]. Even so, in many cases cw TREPR studies are possible only by virtue of the strong signal amplification given by CIDEP, and signals from free radicals at Boltzmann equilibrium may remain buried in the noise.

Time-resolved measurements with higher sensitivity became possible with the introduction of pulsed EPR instruments. In recent years, pulsed EPR techniques have proven to be ideally suited for the study of the dynamics of photochemical reactions. The simplest method is to obtain the spectrum of free radicals present at some selected time τ_d after pulsed laser excitation of the sample by delivering a $\pi/2$ microwave pulse at that time and acquiring the ensuing free induction decay (FID) signal given by the transverse magnetization. The Fourier transform of the FID then gives the frequency domain spectrum of the paramagnetic species present at τ_d [6,22–24].

The FT-EPR technique offers higher sensitivity and spectral resolution than the cw TREPR method. In addition, the analysis of the time evolution of signal intensities is simplified by the fact that it is not affected by the continuous presence of a microwave field as is the case in cw TREPR measurements. However, FT-ESR has its own limitations. First, the bandwidth covered by the $\pi/2$ microwave pulse (~100 MHz) in many cases is substantially less than the spectral width. This limitation can be overcome by performing measurements at a series of fixed fields, so that the complete spectrum can be assembled from the discrete frequency ranges covered at these field settings. Second, typically FID signal acquisition cannot be initiated for 100–200 ns following the microwave pulse. Evidently if the signal decays into the baseline during this dead time, because of chemical decay of the free radicals or short relaxation times, the radicals cannot be detected. In practice this means that systems that give rise to broad resonance peaks (typically more than 2 gauss or so) cannot be studied with FT-EPR. The cw TREPR technique does not suffer from this limitation.

In some cases it may be advantageous to use the electron spin-echo (ESE) technique in investigations of photogenerated paramagnetic species [25]. The spectrum of a transient free radical can be obtained by measuring the echo signal at fixed τ_d as a function of magnetic field. Measurements of the echo signal at fixed field for a range of τ_d settings give information on reaction and spin dynamics. ESE measurements can be useful in cases where broad resonance peaks, not caused by short radical lifetime, preclude the application of FT-EPR. In the case of narrow-line spectra, application of the ESE pulse sequence followed by FID acquisition [14] circumvents, at the expense of some loss of signal intensity, the missing-data-points problem associated with FT-EPR.

B. Scope of Review

In this chapter an overview will be given of recent TREPR studies. Applications discussed were selected to give the reader an idea of the areas of photochemistry where the technique has been used and the insights the spectroscopic data can provide. As noted earlier, a unique aspect of TREPR measurements is that they provide information on the role played by electron and nuclear spins in chemical reactions [26–28]. The theory of the various mechanisms that produce spin polarization (CIDEP) can be found in a number of previous reviews [11–14]. Here only a brief summary of CIDEP mechanisms will be given. For early work and discussions of instrumental techniques, the reader is also referred to the literature [13,24,25]. Reviews of advances in selected areas of TREPR research can be found in recent publications [29,30]. In particular, for a review of the extensive work carried out in the field of covalently linked radical pairs the reader is referred to reference [30]. Finally, it is noted that TREPR studies continue to make important contributions to the understanding of the mechanistic details of photosynthesis [12,31,32]. The present chapter will not cover this important area of research.

II. CIDEP MECHANISMS

In this section, the various mechanisms that can give rise to spin polarization in photochemically generated paramagnetic molecules will be discussed. For an understanding of CIDEP effects and the insights they provide into reaction mechanisms; a rigorous quantum mechanical description of spin state evolution generally is not required and the discussion here will be purely phenomenological. It is important to note that spin polarization can originate in processes not directly related to the mechanism of radical formation. The identification of signal contributions from such processes is essential for TREPR studies to be useful. For this reason, these "non-CIDEP" mechanisms are included in the discussion.

A. Triplet Mechanism

Triplet mechanism (TM) CIDEP can be exhibited by free radicals produced in a reaction involving a reactant molecule in the photoexcited triplet state [33,34]. Taking photo-induced electron transfer as example, the sequence of steps that can produce TM spin polarization in the (doublet spin state) redox products is shown in the following reaction scheme.

$$D \xrightarrow{h\nu} {}^1D^* \xrightarrow{isc} {}^3D^*_{SP} \tag{1}$$

$$ {}^3D^*_{SP} \xrightarrow{T_1} {}^3D^* \tag{2}$$

$$ {}^3D^*_{SP} + A \xrightarrow{k_f} D^-_{SP} + A^-_{SP} \tag{3}$$

$$ {}^3D^* + A \xrightarrow{k_f} D^+ + A^- \tag{4}$$

Photoexcitation of the donor (D) to a singlet excited state followed by intersystem crossing (*isc*) produces triplet excited-state molecules. In many cases, spin-polarized triplets (${}^3D^*_{SP}$) are produced because of the spin selectivity of *isc*. If the rate of radical formation (k_f) can compete with spin-lattice relaxation (T_1^{-1}), giving triplets at thermal equilibrium (${}^3D^*$), the triplet spin polarization will be transferred to free-radical products. If the $\beta(1)\beta(2)$ ($M_s = -1$) triplet state is preferentially populated by *isc*, electron transfer quenching of the triplets can produce doublet radicals with spin polarization, $P = (N_\beta - N_\alpha)/(N_\beta + N_\alpha)$, larger than the polarization at Boltzmann equilibrium (P_B). Then the TREPR spectra of the redox products will display *enhanced absorption* (A) signals. On the other hand, if the $\alpha(1)\alpha(2)$ ($M_s = +1$) triplet state is preferentially populated, free radicals generated by electron transfer may carry negative spin polarization, in which case TREPR spectra can show *stimulated emission* (E) signals.

The magnitude of spin polarization created in rotating triplet state molecules in a magnetic field is given by [34]

$$P_T = \frac{4}{15}\{D\hat{K} + 3E\hat{I}\}\left\{\frac{\omega_0}{\omega_0^2 + \tau_r^{-2}} + \frac{4\omega_0}{4\omega_0^2 + \tau_r^{-2}}\right\} \tag{5}$$

with $\hat{K} = \frac{1}{2}(p_x + p_y) - p_z$ and $\hat{I} = \frac{1}{2}(p_y - p_x)$.

In the equation, D and E denote the zero-field splitting (zfs) parameters of the triplet, ω_0 stands for the Larmor frequency, and τ_r for the rotational correlation time. The values of p_i give the probabilities of *isc* into the zero-field states ($|x\rangle$, $|y\rangle$, $|z\rangle$) of the triplet manifold. The fraction of the polarization with which the triplets are "born" ultimately captured by doublet radical products is given by $k_f T_1/(1 + k_f T_1)$. Triplet spin-lattice relaxation times depend on the magnitudes of the zfs parameters, τ_r, and the Zeeman splitting [35]. They can range from less than 1 ns to more than 100 ns. For most triplet state molecules in low-viscosity solvents, however, T_1 values fall in the 1- to 100-ns time domain. For this reason, rates of radical formation in bimolecular reactions, such as photo-induced electron transfer, must be near the diffusion-controlled limit for triplet spin polarization to be transferred to free-radical products. Furthermore, because relaxation to thermal equilibrium normally occurs in a time that is much shorter than the triplet lifetime, signal contributions from triplet spin polarization transfer develop over a time span that is short compared to the time it takes for the radical formation reaction to run to completion. Monomolecular reactions, such as photo-induced bond homolysis and photoionization, can occur on a time scale short compared to T_1, so that free radical products can exhibit strong TM CIDEP.

TM CIDEP is characterized by a rapidly developing* net absorptive or emissive signal contribution in the TREPR spectrum of a photogenerated free radical. It provides evidence that the radical formation step involves a triplet excited state. The selectivity of *isc* (i.e., the values of p_x, p_y, p_z) is determined by the symmetry character of the excited states involved. Hence, TM spin polarization gives an insight into the electronic state of the triplet precursor. For in-

---

* Since the triplet T_1 in most cases is less than the instrument response time (10–100 ns), a TM signal contribution typically will show an instrument-controlled rise time.

stance, in studies of the photooxidation of triphenylporphyrins (TPP) by quinones, it is found that for ZnTPP the TREPR spectra of the quinone anion radicals exhibit absorptive TM CIDEP [23,36,37]. With photoexcited MgTPP [36] or H$_2$TPP [38] as electron donor, the TM signal contribution is emissive. The switch from absorption to emission reflects the effect of metal ion binding on the character of the triplet excited state of these porphyrins. Similarly,one can expect solvent effects on the ordering of the lowest $^3n\pi^*$ and $^3\pi\pi^*$ in aromatic ketones to affect TM CIDEP found for free radicals produced via these triplets.

Data on the values of the zfs parameters D and E as well as p_x, p_y, and p_z can be derived, with the aid of simulations, from rigid matrix TREPR spectra of aromatics in the photoexcited triplet state [39].

In most cases, photoexcitation of a molecule is followed by relaxation to the thermalized first excited singlet state from which *isc* to the triplet excited state occurs. Hence, TM CIDEP generally will be independent of excitation wavelength. However, if *isc* occurs before internal conversion and vibrational relaxation has generated the thermalized singlet excited state, the spin polarization may depend on the wavelength of excitation. This may account for wavelength-dependent CIDEP observed in FT-EPR spectra of radicals produced in photoinduced reactions of xanthone with alcohols [40] and bond homolysis in a [Ru(alkyl)(α-diimine)] complex [41].

A two-laser, two-color study [42] of the rigid matrix TREPR spectra given by halogen-substituted anthracenes in the photoexcited triplet state suggests that $T_n \rightarrow S_1$ reverse *isc* (*risc*) can affect the observed TM spin polarization as well. In the experiment, excitation with a UV laser pulse generated the anthracene triplets. With a subsequent pulse of visible light these triplets were excited to a higher triplet state. In the case of 9-bromoanthracene and 9,10-dibromoanthracene it was found that excitation with the second laser caused a change in the polarization pattern of the rigid matrix EPR spectra. The effect was attributed to $T_n \xrightarrow{\text{risc}} S_n$, which, after internal conversion and vibrational deactivation of the remaining triplets, leaves the T_1 state with a spin population distribution modified by spin-selective *risc*. The finding is relevant for TREPR studies of photochemical reactions involving photoexcited triplets because of the fact that the characteristic time of singlet–triplet *isc* in many cases is less than the duration of the laser pulse (10–20 ns). It follows that even with a single laser pulse, $T_1 \rightarrow T_n$ excitation may occur.* In systems where *risc* can compete with $T_n \rightarrow T_1$ relaxation, TM CIDEP observed in the spectra of doublet radical products can, in part, be determined by this process.

---

* The pulse widths of lasers used in TREPR (and flash photolysis) measurements makes it likely that two-photon processes are a common occurrence. To what extent this affects the results has remained largely unexplored and is an interesting area of future study.

Where radical formation occurs from triplets with the spin system at Boltzmann equilibrium ($T_1^{-1} \gg$ rate of radical formation), products will be formed with $4/3xP_B$ because of the fact that the Zeeman splitting in the triplets is twice as large as that in the doublet spin systems. It follows that TREPR spectra of these products initially will exhibit enhanced absorption signals [43].

B. Radical Pair Mechanism

Radical pair mechanism (RPM) CIDEP is generated by the spin-state evolution in the transient radical pair that is formed in a photochemical reaction that leads to free-radical formation [44–46]. For instance, the oxidative quenching of $^3D^*$ by acceptor A:

$$^3D^* + A \rightarrow {}^3[D^+ \cdots A^-] \rightarrow D^+ + A^- \tag{6}$$

generates the $^3[D^+ \cdots A^-]$ radical pair intermediate. The spin state of this transient species evolves in time because of the difference in precession frequencies of the two unpaired electrons caused by differences in g values and/or hfcc's. Normally the mixing terms that cause $T \leftrightarrow S$ interconversion are small compared to the Zeeman splitting of the radical pair spin states, so that the spin-state evolution is governed by $T_0 \leftrightarrow S$ $[\alpha(1)\beta(2) + \beta(1)\alpha(2)] \leftrightarrow [\alpha(1)\beta(2) - \beta(1)\alpha(2)]$ mixing. Even then the exchange interaction J between the unpaired electrons in a contact ion pair causes a splitting between the T_0 and S levels that is large compared to mixing terms. As a consequence, in *low-viscosity* solvents spin-state evolution will be negligible during the contact ion pair lifetime. To account for the spin-state mixing that is reflected in RPM CIDEP, a three-step process is invoked [45]. In the first step the contact radical pair is generated. In the second, the radicals diffuse apart so that $J \sim 0$ and $T_0 \leftrightarrow S$ interconversion can take place. In the third, the radicals reencounter and the effect of the ST_0 mixing is expressed in the form of excess α-electron spin for one radical and excess β-electron spin for the other.

The theoretical expression for ST_0 RPM spin polarization in radical 1 with nuclear spin state a in contact with radical 2 with nuclear spin state b is given by [46]:

$$P_1^{ab} = C[Q_{ab}^{1/2} - \gamma Q_{ab}] \tag{7}$$

with

$$Q_{ab} = \frac{1}{2}\left\{\mu_B B_0 \Delta g + \sum_m a_{1m}m_{1m}^a - \sum_n a_{2n}m_{2n}^b\right\} \tag{8}$$

Q_{ab} represents half the difference in resonance frequencies of radicals 1 and 2 determined by $\Delta g = g_1 - g_2$ and the difference in hyperfine interactions, represented by the last two terms in (8). Hfcc's in radicals 1 and 2 are given by a_{1m} and

a_{2n}, respectively. The spin quantum number of the mth (nth) nucleus of radical 1 (2) in overall nuclear spin state a (b) is given by m_{1m}^a (m_{2n}^b). A calculation of the RPM spectrum of radical 1 involves a summation of contributions from all nuclear spin states of radical 2 weighed by the spin state degeneracy d_b:

$$P_1^a = \sum_b d_b P_1^{ab} \tag{9}$$

The dependence of P_1^{ab} on $Q_{ab}^{1/2}$ [Eq. (7)] stems from the separation–reencounter mechanism described above. In media where radical pair dissociation is inhibited, i.e., high-viscosity solvents, or in systems where the ST_0 mixing terms are large, RPM spin polarization can be generated directly [44]. The contribution of this direct process is represented by the last term in Eq (7). This term contains the weighing factor γ, in most cases $\gamma \approx 0$.

The absolute magnitude of C depends on the characteristics of the radical pair, whereas its sign depends on the sign of J and whether radical pair formation involves a singlet or triplet excited state. It can be shown that all resonance peaks from radical 1 positioned on the low-field side of the center of the spectrum from radical 2 will be in absorption (A) or emission (E), while the peaks on the high-field side will be in emission or absorption. Whether an AE or EA polarization pattern is observed in the spectrum from a given radical depends on 1) the spin state with which the radical pair is "born" and 2) the sign of the exchange interaction J. In most systems, $J < 0$ (S radical pair state lies below T_0). In that case, radical formation via a triplet excited state will give rise to an EA RPM CIDEP pattern. For $J > 0$ and triplet state precursor, the pattern changes to AE. An AE pattern will be found as well if radical formation occurs via a singlet excited state and $J < 0$. ST_0RPM CIDEP does not produce net spin polarization. Thus, if it is the single source of signal intensity, integration of the TREPR spectrum will yield zero overall signal intensity.

RPM spin polarization not only originates in the radical formation step but is also generated in radical–radical reactions producing diamagnetic products because of the spin selectivity of the radical termination step:

$$R_1 + R_2 \Leftrightarrow {}^{1,3}[R_1 \cdots R_2] \to D \tag{10}$$

Radical pairs formed with the singlet state will react to give D. Triplet radical pairs left behind engage in the separation–reencounter steps that generate spin polarization. CIDEP stemming from the radical formation step is labeled *geminate pair* spin polarization, that generated in the termination step is labeled *F-pair* spin polarization. The polarization pattern produced by F pairs is identical to that given by radical formation via a triplet excited-state precursor.

Since the energy gap between the singlet radical pair state and the T_{-1} (or T_{-1}) state generally is large compared to the interactions that can give rise to ST_{-1} (or ST_{+1}) mixing, this process rarely contributes to spin polarization. However, if hyperfine interactions are very large or if J and the Zeeman interaction are of similar magnitude, so that the ST_{-1} (or ST_{+1}) energy gap becomes very small, mixing can produce an RPM contribution. Under the usual condition that $J < 0$, the effect of ST_{-1} mixing must be considered. Theory predicts that this mixing gives rise to an emission spectrum if the radical pair is formed via a triplet state precursor, whereas an absorption spectrum is produced by a singlet state precursor. The mixing terms depend on the magnitude of the hyperfine fields experienced by the electron spins, so that the spin polarization produced by the ST_{-1} mechanism is nuclear spin–dependent. The procedure of calculating relative intensities of hyperfine components given by ST_{-1} RPM CIDEP is described in Ref. 13.

A number of features distinguish TM from RPM CIDEP. First, spin polarization transfer from a precursor photoexcited triplet state leaves the spectra of *both* doublet radicals in absorption or emission while *relative intensities of hyperfine components in each of the spectra are identical to those found at Boltzmann equilibrium.* In ST_0 RPM CIDEP, by contrast, *no net polarization is found*, the spectra will exhibit an *EA* or *AE* pattern, and *relative intensities of hyperfine components in the spectrum of one radical depend strongly on their position relative to the center of the spectrum of the other radical.* The ST_{-1} mechanism does produce net polarization, but here as well relative intensities of hyperfine components deviate from those found at thermal equilibrium. Second, the rise time of TM CIDEP is controlled by the spin-lattice relaxation time of the triplet precursor, which normally is one or two orders shorter than the characteristic time of the doublet radical formation step. On the other hand, *(geminate)* RPM spin polarization is generated throughout the radical formation step and the time profile of RPM buildup can serve to determine the reaction rate. The time profile of F-pair polarization is linked to the kinetics of the radical termination step.

C. Spin-Correlated Radical Pairs

The RPM provides indirect information on radical pairs through spin polarization observed in TREPR spectra of cage escape products. In a number of cases, geminate radical pairs have been observed directly with TREPR. Direct detection requires a lifetime (more than 100 ns) that exceeds that normally found for radical pairs in low-viscosity solvents and a relatively small exchange interaction. (If the doublet radicals are not linked by a covalent bond so that considerable variation in radical pair structure is likely, a value of J well below 10 MHz may be required to produce detectable signals.)

The spectra of the radicals that make up these spin-correlated radical pairs (SCRPs) are characterized by the fact that all peaks have a derivative-like line shape. This special feature can be accounted for qualitatively by treating the

$[R_1^{\bullet} \cdots R_2^{\bullet}]$ radical pair as an AX spin system [47]. That is, the difference in resonance frequencies of the two unpaired electrons ($\Delta\omega = \omega_1 - \omega_2$) is assumed to be large compared to the exchange interaction J. In that case, each hyperfine component in the spectrum of radical 1 will split into a doublet because of the interaction with the electron spin of radical 2. For each nuclear spin state the two spin transitions are $\beta_1\alpha_2 \leftrightarrow \alpha_1\alpha_2$ and $\beta_1\beta_2 \leftrightarrow \alpha_1\beta_2$. Assuming radical pair formation via a triplet excited-state precursor, the relative populations of the four spin states $\beta_1\beta_2$, $\alpha_1\beta_2$, $\beta_1\alpha_2$, and $\alpha_1\alpha_2$ in first approximation will be $1:0.5:0.5:1$. Hence, of the two resonance transitions in radical 1, the first will be in emission and the second in absorption. With a singlet excited-state precursor, the $\alpha_1\beta_2$ and $\beta_1\alpha_2$ spin states will carry excess population and the polarization pattern will turn from EA to AE. These two cases are illustrated in Fig. 1. Typically, the separation between the peaks ($2J$) is of the order of the linewidth because of the distribution in J values. As a result, the two peaks merge to give a single, derivative-like, resonance peak. Figure 2 gives examples of SCRP spectra. One spectrum is from a covalently linked carotenoid–porphyrin–diquinone tetrad in which the $C^{\bullet-}$-P-Q_A-$Q_B^{\bullet-}$ SCRP is formed via the porphyrin singlet excited state [48]. The resonance peaks of $Q^{\bullet-}$ show AE polarization. The other spectrum is from a (not covalently linked) porphyrin/quinone system captured in a micelle [49].

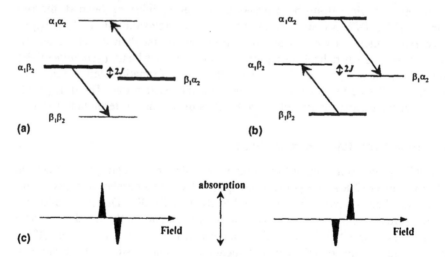

Figure 1 Energy level diagram of the spin states of an SCRP. Line thickness is an indication of relative populations given by formation via singlet and triplet state precursors. Transitions induced in radical 1 are indicated by arrows. Polarization patterns found in the spectrum of radical 1 are sketched at the bottom of the figure. Note that hyperfine splittings have not been taken into account.

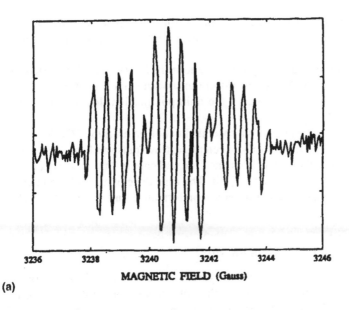

MAGNETIC FIELD (Gauss)

(a)

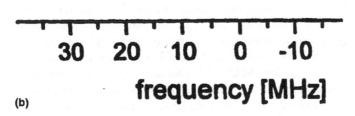

(b)

Figure 2 (a) Continuous wave TREPR spectrum from a covalently linked carotenoid–porphyrin–diquinone tetrad. The $C^{\cdot+}$-P-Q_A-$Q_B^{\cdot-}$ SCRP is formed via the porphyrin singlet excited state and all hyperfine components of the quinone anion radical show an AE polarization pattern. (Reprinted with permission from Ref. 48.) (b) FT-EPR spectrum from a (not covalently linked) porphyrin/quinone system captured in a micelle. The hyperfine components in the $Q^{\cdot-}$ spectrum exhibit EA polarization, reflecting the fact that electron transfer takes place from the porphyrin triplet state. (Adapted with permission from Ref. 49.)

Here electron transfer takes place from the porphyrin triplet state, which is reflected in the fact that the hyperfine components in the $Q^{\cdot-}$ specrum exhibit *EA* polarization.

It is noteworthy that the spin polarization in SCRPs can far exceed that produced by TM and RPM mechanisms in the separated radicals. Because of the strong signal amplification caused by the spin polarization, low concentrations of SCRPs can be detected.

In pulsed EPR measurements a special situation arises if the SCRPs dissociate in the time interval (dead time) between microwave pulse and data acquisition. It can be shown that the signal from the free radicals in that case acquires an out-of-phase (dispersive) component [50]. The time development of the intensity of this dispersive signal carries information on formation and decay of the SCRPs [51].

D. Radical-Triplet Pair Mechanism

It has been found that the electron spin system of stable free radicals in solution can become polarized by interaction with molecules in the photoexcited triplet state [52–54]. Polarization patterns observed can be accounted for quantitatively on the basis of a theory [55] similar to that developed for RPM CIDEP. In the case of doublet radical–triplet encounters, polarization is generated by transitions between quartet and reactive (leading to triplet quenching) doublet spin states of the pair. Quartet–doublet transitions are induced by electron spin–spin and hyperfine interactions. Data from fluorescence and photoacoustic measurements support this radical–triplet pair mechanism (RTPM) [56].

RTPM CIDEP generally gives rise to emissively polarized doublet radical spectra showing a small dependence of polarization on nuclear spin state [55,57]. Because it is generated by encounters of free radicals with triplets, RTPM signals develop at near-diffusion-controlled rates (under the conditions of most experiments over a time period of 0.1–1 μs). The relatively slow growth distinguishes the RTPM from spin polarization transfer from photoexcited triplets to free radicals. The rise time of the latter process is determined by the triplet spin-lattice relaxation time, as is the case for TM CIDEP. That direct spin polarization transfer can give a significant signal contribution has been shown in a TREPR study of porphyrin/nitroxide radical systems [58]. As illustrated in Fig. 3, in the case of Zn octaethylporphyrin (ZnOEP)/2,2,6,6-tetramethyl-1-piperidinyloxyl (TEMPO) in toluene, the TREPR spectrum of TEMPO, following photoexcitation of the porphyrin, initially (0.1–0.3 μs) shows an enhanced absorption signal. At later times (2.0–2.2 μs), the spectrum is in emission. The initial *A* signal reflects spin polarization of the ZnOEP triplets transferred directly to TEMPO. The *E* polarization stems from the RTPM.

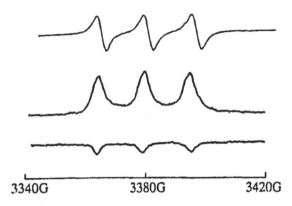

3340G   3380G   3420G

Figure 3   (top) Conventional EPR spectrum of the nitroxide radical TEMPO in toluene. (middle) Continuous wave TREPR spectrum given by ZnOEP/TEMPO in toluene. Detection in the time interval 0.1–0.3 μs following excitation of the porphyrin. (bottom) Continuous wave TREPR spectrum detection time 2.0–2.2 μs. TREPR spectra give the *change* in signal intensity caused by photoexcitation. Peaks pointing up denote enhanced absorption, peaks pointing down denote emission. (Adapted with permission from Ref. 58.)

Quartet–doublet transitions can also generate spin polarization in encounters of doublet radicals with molecules in the singlet excited state [59]. In this case the mechanism gives rise to enhanced absorption signals. Evidently, to play a role the lifetime of the singlet excited state must be long enough to allow for radical-excited singlet encounters.

These processes are not directly linked to the chemistry involved in free-radical formation. However, it is important to realize that they can make significant signal contributions. In the use of CIDEP as a mechanistic probe, their effect has to be taken into consideration.

E. Reversed Triplet Mechanism

As noted in Section II.A, TM CIDEP originates in the spin selectivity of the S → T intersystem crossing (*isc*) process. A magnetic field effect on radical ion yield from donor–acceptor triplet exciplexes, where the donor contained heavy-atom substituents, was interpreted in terms of the reverse process [60]. Here the cage escape of redox ions competes with deactivation of the charge–transfer complex, giving ground-state donor and acceptor molecules. The magnetic field effect establishes that this T → S_0 process is spin-selective, and the heavy-atom effect points to spin orbit coupling (SOC) as the interaction responsible for reverse *isc*.

Evidence for spin polarization in photochemically generated free radicals stemming from this *risc* process has been found in TREPR studies of oxidative quenching of triplet xanthene dyes by quinones [61,62] and reductive quenching of triplet duroquinone by halogen-substituted *N,N*-dimethylanilines [63,64]. The role played by SOC is evident from the fact that net spin polarization is observed, with the polarization increasing with increasing atomic number of the halogen substituent on the xanthene and aniline molecules. Concomitant with this increase, radical yields show a significant decrease consistent with enhanced *risc* due to the heavy-atom effect. SOC is a short-range interaction and will play a role only if photochemical free-radical formation involves a strongly coupled, contact radical pair. The buildup rate of the polarization is determined by the encounter rate of donor and acceptor molecules, and therefore is the same as that of RPM polarization.

F. An Overview of Spin Polarization Contributions

In conclusion of this section it is useful to review the processes that can contribute to spin polarization in photogenerated free radicals by tracing the steps involved in free-radical formation. The scheme shown in Fig. 4, which represents the

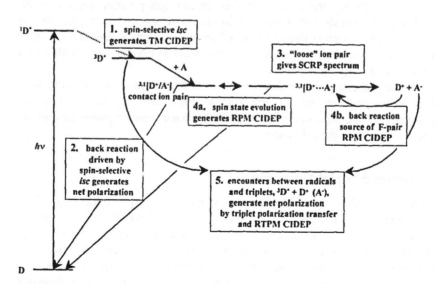

Figure 4 Reaction scheme for the oxidative quenching of triplet state molecules (^{3}D*) by an acceptor A giving the paramagnetic redox products D$^+$ and A$^-$ displaying the processes that give rise to electron spin polarization.

mechanism of oxidative quenching of triplet state molecules ($^3D^*$) giving the paramagnetic redox products D^+ and A^-, serves as an example:

1. *TM CIDEP*. Photoexcitation of D can produce spin-polarized triplets ($^3D^*_{SP}$) and this polarization can be carried over to D^+ and A^-. The mechanism produces net polarization (*E* or *A*) in both radicals. The buildup rate of the TM contribution to the TREPR signal depends on the T_1 of the triplets.

2. *SOC-generated spin polarization*. $^3D^*$–A encounters generate contact radical pairs, $[D^+/A^-]$. SOC-induced *risc* can produce net spin polarization (*E* or *A*) in free-radical cage escape products. The buildup rate is controlled by the rate of electron transfer. The process apparently plays a significant role only if one or more heavy-atom centers are present in the reactant molecules.

3. *SCRP spectrum*. Given a sufficiently long lifetime, the weakly coupled $[D^+ \cdots A^-]$ spin-correlated radical pair can be detected. The characteristic spectral signature is that each hyperfine component displays a "derivative-like" *EA* (triplet precursor) or *AE* (singlet precursor) lineshape.

4. *RPM CIDEP*. (a) In the weakly coupled geminate radical pair ST_0 mixing gives rise to nuclear spin state–dependent (multiplet) spin polarization. Generally, for pairs generated via a triplet excited state, the polarization pattern is low field emission, high field absorption. The pattern reverses if a singlet excited state is involved. The process does not give rise to net polarization. ST_{-1} (or ST_{+1}) mixing can play a role if the radical pair lifetime is long and/or hfcc's are large. It produces nuclear spin state–dependent net polarization (*E* or *A*). The buildup rate of RPM CIDEP reflects the reaction rate.

   (b) After cage escape, reencounters of D^+ and A^- can generate F-pair CIDEP. This process produces multiplet polarization identical to that produced by geminate pairs "born" in the triplet state. The buildup rate reflects the rate of the radical–radical termination reactions.

5. *Radical–Triplet Pair Mechanism*. Encounters of D^+ or A^- with $^3D^*$ can produce net polarization (*E*) with a minor multiplet contribution. The buildup rate depends on the rate of radical–triplet encounters. The rate of generation and relative magnitude of this polarization contribution is a sensitive function of laser power. Encounters of D^+ or A^- with $^3D^*_{SP}$ can lead to direct polarization transfer. The buildup rate of this polarization contribution reflects the triplet T_1.

Notwithstanding the complexity of this scheme, CIDEP patterns have been interpreted in terms of TM and ST_0 RPM contributions only in the majority of cases [13,14,24,25].

III. APPLICATIONS

A. Photoionization

Photoionization of aromatics in homogeneous fluid solution and solubilized in micellar solution has been the subject of a number of TREPR investigations. Since the pathway of this process is expected to be linked to a characteristic CIDEP signature, TREPR spectra, in principle, should be a valuable source of mechanistic information. In fact, the interpretation of the spectroscopic data in many cases turns out not to be a straightforward matter and is the subject of ongoing discussions.

1. Phenols

The photoionization of phenols and phenolates continues to be the subject of studies that try to shed light onto the photophysics and photochemistry involved. The interest stems in part from the fact that phenols are an important constituent of many biochemical systems and that photoionization and/or OH–bond dissociation of the aromatic amino acid tyrosine may play an important role in protein photodegradation [65]. Issues addressed in the studies are whether the process is mono- or biphotonic and the characterization of the dissociative excited state.

A series of papers dealing with steady-state and flash photolysis studies of phenols and phenolates in aqueous solution and nonpolar solvents illustrate the complexity of the problem. In alkaline aqueous solution, photoexcitation of phenolates can lead to electron ejection [66–69]. The main channel of the process is monophotonic, but at high photon flux (i.e., pulsed laser excitation) a biphotonic path involving phenolate triplets may contribute. Electron ejection may occur from an "electron-donating" or [PhO$^{\bullet}$ $\cdots$ e^{-}] radical pair state that can be reached from the S$_1$ as well as the S$_2$ state in a thermally activated step [69].

In neutral aqueous solution, phenoxyl radicals (PhO$^{\bullet}$) can be formed in two reactions [68]. One involves electron ejection,

$$PhOH \xrightarrow{h\nu} PhO^{\bullet} + H^{-} + e_{aq}^{-} \tag{11}$$

the other OH–bond cleavage,

$$PhOH \xrightarrow{h\nu} PhO^{\bullet} + H^{\bullet} \tag{12}$$

The quantum yields of H$^{\bullet}$ and e$_{aq}^{-}$ formation show a pronounced increase with increasing excitation energy. It is assumed that this is due to the presence of two parallel reaction paths: 1) photochemistry from the S$_2$ state, in competition with relaxation to the fluorescent state, and 2) photochemistry involving S$_1$. At relatively low excitation intensities the processes are monophotonic but at the highlight fluxes used in pulsed laser flash photolysis, a consecutive two-photon process involving the phenol triplet state can make the dominant contribution to free-radical formation.

Finally, in nonpolar solvents OH–bond cleavage is the main photochemical process. The process is monophotonic and its quantum yield increases as excitation shifts from the S_1 absorption band to the S_2 band [70,71].

Whereas distinct monophotonic reaction paths involving S_1 and S_2 have been identified under a variety of conditions, the spin multiplicity of the ultimate dissociative state has not been established unequivocally with the optical studies. A cw TREPR study of photoionization of phenol and some substituted phenols in aqueous solutions at pH 11 by Jeevarajan and Fessenden [72] addressed this question. The cw TREPR spectra given by basic solutions of phenol, p-cresol, and tyrosine upon pulsed laser excitation show the resonances of the phenoxyl radicals on the low-field side and the resonance peak of e_{aq}^- on the high-field side ($g_{PhO^.} > g_e$). The spectra of the phenoxyl radicals are in emission while the resonance from e_{aq}^- is in enhanced absorption. The EA polarization is characteristic for the spin-sorting process occurring in a radical pair, [PhO$^.$ $\cdots$ e_{aq}^-], prior to cage escape of free radicals (RPM CIDEP). If the exchange interaction between the unpaired electrons in the radical pair J is less than 0, the EA pattern signifies that the radical pair is "born" in the triplet state; alternatively, with $J > 0$ the radical pair state is a singlet. Jeevarajan and Fessenden assume that radical formation occurs via the singlet excited states of the phenolates,

$$\text{PhO}^- \xrightarrow{h\nu} {}^1\text{PhO}^-* \rightarrow {}^1[\text{PhO}^. \cdots e_{aq}^-] \rightarrow \text{PhO}^. + e_{aq}^- \tag{13}$$

and conclude that $J > 0$. Generally, in cases where the multiplicity of the precursor excited state is well established, the sign of J in radical pairs is negative (singlet radical pair state below triplet state), so that this is an unusual result.

FT-EPR measurements carried out in the laboratory of the author [73] confirm that photoionization of phenol and substituted phenols in basic (pH 11) aqueous solution yields the phenoxyl radicals and hydrated electron. Representative spectra given by p-cresol, 3,4-methylenedioxyphenol (sesamol, SEOH), and tyrosine are displayed in Fig. 5. The spectra show the low-field (high-frequency) E/high-field (low-frequency) A polarization pattern reported by Jeevarajan and Fessenden [72].

The question of whether electron ejection takes place from a singlet or triplet excited state may be resolved by checking for a net absorption or emission signal component. As pointed out in Section II.B, ST_0 RPM CIDEP does not produce net spin polarization. Therefore, integration of the spectrum from radicals produced by singlet excited-state ionization should show that there is no net signal (the integral over the entire spectrum is zero). A process involving the triplet excited state is expected to involve TM spin polarization giving a net A

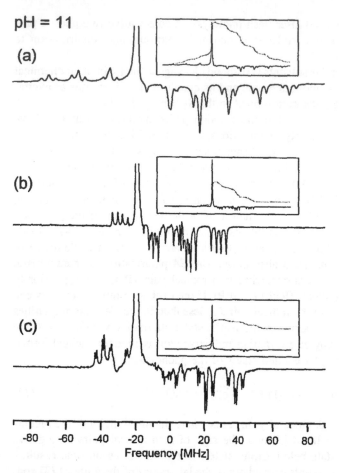

pH = 11

Figure 5 FT-EPR spectra given by aqueous solutions (pH 11) of (a) *p*-cresol, (b) sesa-mol, and (c) tyrosine for a delay time between laser (308 nm) excitation and $\pi/2$ micro-wave pulse of 100 ns [73]. Insets show the relative amplitudes of the resonance from the hydrated electron (strong narrow absorption peak) and phenoxyl radicals (displaying *EA* polarization patterns, note that increasing frequency corresponds to decreasing magnetic field). The integration of the spectra is shown by the dotted lines.

or *E* signal contribution in addition to the *EA* RPM signal component, so that the integral is expected to be nonzero.

Integration of the spectra shown in Fig. 5 (shown by the dotted lines) re-veals that all three have a net *A* signal contribution. In the case of the cresol and sesamol systems the net signal contribution is small (~10%) compared to the

RPM signal contribution. However, the net A signal intensity for tyrosine is of similar magnitude as the EA component. The results suggest that photoionization must take place from the triplet excited state of these phenols.

Clancy and Forbes [74] made a cw TREPR study of the photoionization of the tyrosine anion in alkaline aqueous solution. The authors confirm the finding [72] that the free radicals formed give rise to an EA CIDEP pattern. It was established that addition of 2,4-hexadienoic acid (sorbic acid) to the solution leads to quenching of the TREPR signal. The effect is attributed to quenching of the tyrosine triplets by sorbic acid, and it is concluded that the dominant photoionization route must be

$$TyO^- \xrightarrow{h\nu} {}^1TyO^-* \xrightarrow{isc} {}^3TyO^-* \rightarrow {}^3[TyO^\bullet \cdots e_{aq}^-] \rightarrow TyO^\bullet + e_{aq}^- \quad (14)$$

In that case, the radical pair is "born" in the triplet state and, with $J < 0$, the ST_0 RPM accounts for the EA polarization pattern. A measurement of signal intensity vs. excitation light intensity established that the monophotonic process delineated in Eq. (14) dominates at low light intensity, whereas excitation of ${}^3TyO^-*$,

$$^3TyO^-* \xrightarrow{h\nu} {}^3[TyO^\bullet \cdots e_{aq}^-] \rightarrow TyO^\bullet + e_{aq}^- \quad (15)$$

contributes at high laser pulse power.

It is noteworthy that the triplet photoionization channel apparently can be shut off by 0.5 M sorbic acid [74]. Assuming a diffusion-controlled triplet energy transfer rate constant of the order of 10^{10} M^{-1} s^{-1}, this implies that the lifetime of ${}^3TyO^-*$ in the absence of triplet quencher must be more than 0.2 ns, so that electron ejection must occur from the thermalized triplet excited state.

Clancy and Forbes [74] also find evidence for a monophotonic singlet ionization channel [Eq (13)]. This channel was isolated by adding sorbic acid and 2-bromo-2-methylpropionic acid (BMPA) to the tyrosinate solution. Sorbic acid serves as tyrosine triplet quencher while BMPA is a good electron acceptor that debrominates upon capture of e_{aq}^- to give the 2-methylpropionate (MPA$^\bullet$) free radical:

$$TyO^- \xrightarrow{h\nu} {}^1TyO^-* \rightarrow {}^1[TyO^\bullet \cdots e_{aq}^-]$$

$$\rightarrow TyO^\bullet + e_{aq}^- \xrightarrow{BMPA} TyO^\bullet + MPA^\bullet \quad (16)$$

The TREPR spectrum of the MPA$^\bullet$ radical shows a net emissive signal contribution that is attributed to capture of spin polarization from e_{aq}^- in its reaction with the spin trap. The switch from triplet-channel e_{aq}^- formation to singlet-channel formation must be accompanied by a change in spin polarization of the hydrated electron from A to E, so that the results are consistent with $J < 0$ in the radical pair.

That the isolation of the singlet photoionization channel requires the addition of BMPA is puzzling. It is argued that the formation of e_{aq}^- is established by the net polarization signal contribution in the multiline spectrum of MPA$^\cdot$ formed according to Eq. (16). Then the single narrow resonance peak of e_{aq}^- should be readily detectable in the absence of BMPA since its intensity, at a minimum, would correspond to the integrated intensity of the MPA$^\cdot$ spectrum. The absence of this emissive signal from e_{aq}^- in the TREPR spectrum given by a tyrosinate solution upon addition of sorbic acid (figure 1B of reference 74) remains unexplained.

Forbes and co-workers applied similar methods to address the question of the mechanism of photoionization of N, N, N', N'-tetramethylphenylenediamine (TMPD) in alcohol [75]. TREPR studies of this system have been reported as well by the research groups of Murai [76] and Trifunac [77]. Photoionization of aromatic ketones has been investigated by Murai et al. [78] and Beckert et al. [79,80].

The photochemistry of SEOH in aqueous solution shows a remarkable pH dependence [73]. FT-EPR spectra from SEO$^\cdot$ (identified with the aid of published hfcc's [81]) obtained upon excitation of solutions at pH 11 and 6 are given in Fig. 6. Two features distinguish the spectrum at pH 11 from that obtained at pH 6:1) The resonance due to e_{aq}^- is absent and 2) the AE polarization pattern shows the transition from A to E close to the center of the SEO$^\cdot$ spectrum rather than near the high-field (low-frequency) wing of the spectrum. From this it can be concluded that the radical pairs that generate the AE polarization must comprise two radicals with nearly identical g values. Another distinguishing characteristic is that the SEO$^\cdot$ spectrum develops over a period of 1–2 μs whereas it displays an instrument-controlled rise time at pH 11 (cf. Fig. 7). The rate of formation increases with SEOH concentration but is independent of laser intensity. Finally, the spectrum shows a complex pattern of additional resonance peaks in the wings of the spectrum (extending well outside the spectral range displayed in Fig. 6). The rate of formation of the radicals responsible for these signals is slightly less than that of SEO$^\cdot$ formation.

The main features of the signals found in the wings of the spectrum can be accounted for by assuming that photoexcitation of the phenol leads to hydrogen atom transfer producing SEO$^\cdot$ and cyclohexadienyl-type radicals. The concentration-dependent formation rate is consistent with a reactive encounter between ^{3}SEOH* and ground-state SEOH in which hydrogen-transfer from the OH group to the phenyl ring takes place. That photoexcitation of SEOH produces a triplet state in which OH–bond fission is facile is also evident from the fact that with 2-propanol as solvent, photoexcitation yields the 2-hydroxypropan-2-yl radical in addition to the hydrogen-addition products and SEO$^\cdot$ [73].

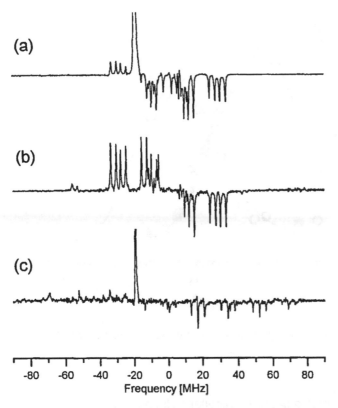

Figure 6 FT-EPR spectra given by aqueous solutions of (a) sesamol (pH 11), (b) sesamol (pH 6), and (c) *p*-cresol (pH 6). Delay time between laser (308 nm) excitation and $\pi/2$ microwave pulse of 100 ns [73]. The spectrum given by the SEOH solution at pH 6 extends well beyond the frequency range shown in the figure.

Flash photolysis measurements on phenols in neutral aqueous solution suggest that H atoms, as products of the OH–bond cleavage reaction, may be detectable with TREPR. However, no signals from H atoms were observed in the FT-EPR study of SEOH. It is also noted that the results obtained with SEOH in neutral aqueous solution apparently reflect the specific characteristics of its excited states. As Fig. 6c shows, with *p*-cresol in aqueous solution at pH 6, the FT-EPR spectrum obtained following laser excitation is similar to that given by a solution at pH 11. The only difference is a significant reduction in the signal-to-noise ratio. The reduction is consistent with the reported drop in quantum yield of the electron ejection reaction as the pH is lowered [68].

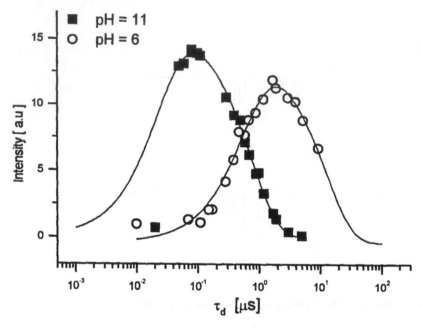

Figure 7 Time profiles of the signal intensities of the resonances from SEO˙ at pH 11 and 6 [73].

2. Photoionization of Aromatics in Micellar Solution

Alkaitis et al. [82] have shown that the quantum yield of photoionization of aromatics, as well as the lifetimes of the cation and solvated electron that are produced, can be increased by solubilization in aqueous micellar solutions. TREPR investigations have been carried out with the aim of shedding light on the mechanism of e_{aq}^- formation and decay in these microheterogeneous systems.

A study by Ishiwata et al. [78] shows that photoionization of aromatic ketones in sodium dodecyl sulfate (SDS) micelles gives rise to emissively polarized cw TREPR spectra of paramagnetic products. The spin polarization was attributed to TM CIDEP and it was concluded that $S \rightarrow T$ *isc* precedes the electron ejection process. The decay of the e_{aq}^- signal was found to be first order and was assumed to be due to relaxation of the spin system to thermal equilibrium.

TREPR studies of the photoionization of phenothiazine (PTH) in micellar solutions carried out by a number of research groups have resulted in conflicting interpretations. An FT-EPR study of PTH in SDS by Tero-Kubota and co-workers [83] shows that the hydrated electron is formed with emissive spin polarization and that the signal form e_{aq}^- decays exponentially. These findings are in accord with those reported by Turro et al. [84]. However, the first group notes that signal intensity is linearly dependent on light intensity and concludes that photoionization is a monophotonic process. They propose that electron ejection takes place from the singlet excited state, with spin polarization generated as a result of the evolution of the $^1[\mathrm{PTH}^+ \cdots e^-]$ radical pair state (RPM CIDEP). This interpretation implies that the radical pair lifetime is long enough (of the order of a nanosecond or so) to cause the spin dephasing required for generation of spin polarization. Since time-resolved optical absorption measurements [85] indicate that escape of the photoelectron into the aqueous phase occurs on a femtosecond to picosecond time scale, it can be questioned as to whether the RPM mechanism invoked by Tero-Kubota et al. [83] can play a role.

The Turro group, on the other hand, reports that the e_{aq}^- signal intensity is not linearly dependent on laser power [84]. Also, the cw TREPR spectrum given by PTH in SDS shows emissive signals from PTH^+ as well as c_{aq}^- rather than the AE polarization expected for RPM CIDEP generated in $^1[\mathrm{PTH}^+ \cdots e^-]$. The authors propose that the spectrum is dominated by the contribution from a two-photon process:

$$\mathrm{PTH} \xrightarrow{\;h\nu\;} {}^1\mathrm{PTH}^* \xrightarrow{\;isc\;} {}^3\mathrm{PTH}^* \xrightarrow{\;h\nu\;} \mathrm{PTH}^+ + e^-. \qquad (17)$$

The presence of this channel in homogeneous solution has been established with flash photolysis measurements [86]. Even though its contribution to free-radical formation may be minor compared to singlet excited-state photoionization, the contribution to TREPR signal intensity can dominate because of amplification caused by triplet spin polarization created in the isc step.

The interpretation of the exponential decay of the emissive signal from the hydrated electron also is a point of dispute. Tero-Kubota et al. [83] attribute it to spin-lattice relaxation. However, a number of experimental findings argue against this interpretation. 1) The decay time is appreciably shorter than the T_1 reported for e_{aq}^- [72]. 2) A change from SDS (anionic head group) to Triton X-100 (neutral head group) is found to lead to a substantial increase in the decay rate [87]. In a micellar solution made up of surfactant molecules with a cationic head group, no e_{aq}^- resonance can be detected. 3) Signal decay depends on micelle concentration, [Mic], as well as [PTH] [88]. These findings point to chemical decay as the cause of the exponential decay of the signal from e_{aq}^-. The effect of surfactant head group charge suggests that electron capture by micellized PTH plays a major role [87,88]. By reducing the electrostatic barrier encountered by e_{aq}^-, a change from anionic to neutral head group facilitates capture by PTH. The dependence of decay rate, k_d on [Mic] and [PTH] also fits a mechanism in which the electron is captured by PTH [88]. The value of k_d is determined by the rate of "penetration" of a micelle by e_{aq}^- and the probability that the micelle contains at least one PTH molecule. The concentration of micelles containing one or more PTH molecules, [Mic_{PTH}], is given by the Poisson distribution law. For a fixed [PTH], [Mic_{PTH}] increases as the surfactant concentration is increased. This accounts for the observed increase in k_d with increased surfactant concentration [88]. At constant surfactant concentration, an increase in [PTH] leads to an increase in [Mic_{PTH}] until the level of saturation is reached. Finally, that e_{aq}^- signal decay is due to scavenging by micellized PTH is consistent with the finding that the rate of decay can be varied by introducing an alkyl substituent on the nitrogen of PTH, giving PTH(CH_2)$_n$$CH_3$. It is found [88] that k_d remains constant for $n \le 15$ but then increases as n is increased. The effect is attributed to the fact that the PTH moiety gets pushed toward to micelle–water interface as the alkyl chain length increases, so that it becomes more accessible to the hydrated electron.

3. Photoionization of Aromatic Anions

Photoexcitation of aromatic anions produced by alkali metal reduction in ethereal solvents can lead to photoionization. The mechanism of the reaction and primary products formed depend in part on aromatic anion–alkali cation ion pair characteristics. Levanon and co-workers made a detailed FT-EPR study of the photoionization of the pyrene dianion (Py^{2-}) in THF solution to explore the effect of the alkali metal counterion on the electron ejection process as reflected in CIDEP effects displayed in the spectra of the paramagnetic species produced [89].

Photoexcitation (532 nm) of the dianion generates the anion radical Py^-, as evident from its characteristic EPR spectrum, and a species that gives rise to a single resonance peak at $g = 2.0023$. Under the conditions of the measurements, the dianion is assumed to be associated with one or two cations and excitation can give the anion radical via a singlet or triplet channel reaction:

$$[nM^+, Py^{2-}] \xrightarrow{h\nu} [nM^-, {}^1Py^{2-}{}^*] \rightarrow [M^+ \cdots e^-] + Py^- \qquad (18)$$

$$[nM^+, Py^{2-}] \xrightarrow{h\nu} [nM^-, {}^1Py^{2-}{}^*]$$

$$\xrightarrow{isc} [nM^-, {}^3Py^{2-}{}^*] \rightarrow [M^+ \cdots e^-] + Py^- \qquad (19)$$

Here $n = 1$ or 2 and [$M^+ \cdots e^-$] denotes an alkali cation–electron pair of undefined structure that gives rise to the resonance at $g = 2.0023$. From CIDEP patterns of the Py^- spectrum it is concluded that with Li^-, Na^+, Rb^+, or Cs^- as counterion, electron transfer involves the triplet excited-state ${}^3Py^{2-}{}^*$. It is proposed that with K^- as counterion the singlet excited-state reaction path is active. The remarkable counterion effect on the reaction path is attributed to a change in the relative magnitudes of electron transfer and isc rates caused by an alkali ion–induced change in ΔG^0 of the electron transfer reaction. Specifically, it is argued that in the case of $[nK^+, {}^1Py^{2-}*]$ electron transfer via Eq. (18) is an activationless process, i.e., the reorganization energy $\lambda_s \approx -\Delta G^0$, so that $k_{et} > k_{isc}$.

Generally, lasers (excimer and Nd-YAG) used in TREPR measurements have a pulse width of the order of 10–30 ns. As a consequence, two-photon processes can play an important role in many studies. Since signal intensity in TREPR measurements is dependent on the product of spin polarization and free-radical concentration, even a minor two-photon process can give rise to a dominant signal contribution if the associated CIDEP is large [84]. Apart from the possibility that a minor biphotonic reaction channel controls the spin polarization pattern in the TREPR spectrum of a primary reaction product, it also can give rise to secondary (paramagnetic) products. An example of such an effect apparently is provided by a FT-EPR study of the reductive quenching of anthraquinone disulfonate (AQDS) triplets, formed by pulsed laser (308-nm) excitation of the quinone, by methionine (MET) in aqueous solution [90]:

$$AQDS \xrightarrow{h\nu} {}^1AQDS* \xrightarrow{isc} {}^3AQDS* \xrightarrow{MET} AQDS^- + MET^+ \qquad (20)$$

At low AQDS concentration, the FT-EPR spectrum initially ($\tau_d < 500$ ns) shows the emissively polarized resonance peaks from the anthraquinone anion radical and the hydrated electron. The resonance peak from e_{aq}^- cannot be observed at longer delay times and an increase in AQDS concentration causes an increase in its decay rate with a concomitant reduction in signal amplitude. Formation of e_{aq}^- is attributed to the follow-up reaction:

$$AQDS^- \xrightarrow{h\nu} AQDS^{-*} \rightarrow AQDS + e_{aq}^- \qquad (21)$$

while its decay is ascribed to the back reaction:

$$AQDS + e_{aq}^- \rightarrow AQDS^- \qquad (22)$$

The fact that the spectrum of $AQDS^-$ is in emission is consistent with an electron transfer reaction involving the quinone triplet state in which the triplet spin polarization is carried over to the anion radical (TM CIDEP). Photoionization of $AQDS^-$ during the same laser pulse then generates e_{aq}^- with the same spin polarization as the anion radical precursor.

Attempts to verify the proposed mechanism with a two-laser, two-color experiment so far have not been successful [91]. It was confirmed that excitation (308 nm) of AQDS/MET solutions under the conditions described by Zubarev and Goez [90] yields the reported results. It was also found that excitation of the solutions with 355-nm laser light generates $AQDS^-$ which, as before, gives rise to an emissive FT-EPR spectrum. However, the change in wavelength resulted in the loss of the resonance from e_{aq}^-. This may be caused by the reduction in laser pulse width upon going from excimer (308 nm) to Nd-YAG (355 nm) laser which could preclude significant excitation of electron transfer product [Eq (21)]. Alternatively, it could signify that with 355-nm excitation the ionization threshold of AQDS$^-$ cannot be reached. In any case, 355-nm excitation produces spin-polarized AQDS$^-$. Therefore, it was anticipated that subsequent exposure to a pulse of 308-nm laser light would lead to formation of e_{aq}^- and that the FT-EPR signal would show a significant enhancement compared to that obtained with single-laser (308-nm) measurements. This turned out not to be the case, thus raising questions about the validity of the interpretation of the results obtained by Zubarev and Goez [90].

[It is noted that the two-laser experiment could have given interesting information on the effect of laser excitation on spin magnetization. According to Zubarev and Goez, the spin magnetization (M_z) created by TM CIDEP in the AQDS$^-$ radical is not affected by subsequent photoionization [Eq (21)] and is carried over to e_{aq}^-. The question is whether, with the pulse sequence 355-nm pulse–τ_1–$\pi/2$ microwave pulse–τ_2-308 nm pulse–FID acquisition, M_z produced in the electron transfer reaction [Eq (20)] and turned into the transverse plane by the microwave pulse is affected by subsequent excitation of AQDS$^-$.]

B. Excited-State Electron Transfer

1. Electron Transfer Mediated by $^3C_{60}$

C_{60} is a good photosensitizer because 1) its absorption spectrum extends from the UV through much of the visible, 2) *isc* to the relatively long-lived triplet state occurs with nearly 100% quantum yield, and 3) $^3C_{60}$ is an excellent electron acceptor [92–94]. For this reason, numerous spectroscopic studies have been concerned with the photophysics and photochemistry of C_{60} triplets.

$^3C_{60}$ has the unusual property that it gives rise to a narrow EPR signal [95–98]. This can be attributed to the fact that the dipole–dipole (zero field splitting, zfs) interaction between the unpaired electrons is averaged out by a dynamic Jahn–Teller effect [95,99–101]. Figure 8 illustrates the dramatic narrowing of the spectral width caused by this pseudorotation upon going from rigid matrices at 3 K and 123 K to a liquid solution at room temperature. The fact that $^3C_{60}$ gives a readily detectable EPR signal facilitates the study of excited-state electron transfer reactions involving the triplet [96,102,103]. Generally, TREPR measurements give information only through the spectra of double-radical redox products because of the fact that the signals from precursor triplets are too broad to be detected. (It is noted, however, that Yamauchi and co-workers recently reported the TREPR detection of porphyrin triplets in liquid solution [58,104].) In the case of $^3C_{60}$, electron transfer can be monitored via the triplet signal in cases where the reaction does not yield products that can be detected with EPR.

FT-EPR was first used to study the reductive quenching of $^3C_{60}$ by tritolylamine and hydroquinone [96]. This study and TREPR measurements by Gouds-

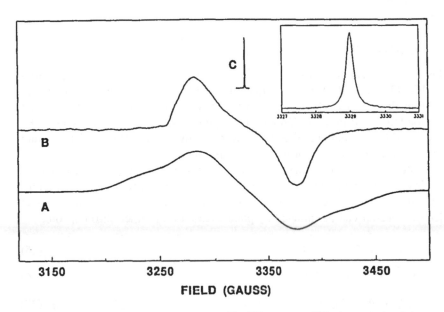

Figure 8 Rigid matrix continuous wave TREPR spectra of $^3C_{60}$ in methylcyclohexane at 3 K (A) and in toluene at 123 K (B), and FT-EPR spectrum of $^3C_{60}$ in toluene at room temperature (C). Inset shows the room temperature spectrum with an expanded-field scale. Delay between laser (532 nm) excitation and spectrum acquisition is about 1 μs.

mit and Paul [97] established that the triplets are "born" with 10–20% of the thermal equilibrium polarization. As a consequence, the $^3C_{60}$ resonance peak typically grows in over a period of 1 μs following excitation even though the triplets are generated within the time span of the laser pulse. That the rate of signal growth for this system reflects the T_1 of the triplet spin system was confirmed with the aid of inversion–recovery relaxation time measurements [100].

Oxidative quenching of $^3C_{60}$ has been investigated with TREPR as well [103,105]. However, since the triplet is a rather poor donor, quenching requires the presence of strong electron acceptors such as chloranil and tetracyanoethylene.

With transient optical absorption and FT-EPR measurements it has been shown that C_{60} can act as an efficient photocatalyst. In the presence of perylene (Pe) and chloranil (CA) in benzonitrile, excitation of C_{60} with visible light (532-nm) can trigger electron transfer from Pe to CA via two channels [105]:

$$C_{60} \xrightarrow{h\nu} {}^1C_{60}^* \xrightarrow{isc} {}^3C_{60}^* \xrightarrow{Pe} C_{60} + {}^3Pe^* \xrightarrow{CA} C_{60} + Pe^+ + CA^- \quad (23)$$

and

$$C_{60} \xrightarrow{h\nu} {}^1C_{60}^* \xrightarrow{isc} {}^3C_{60}^* \xrightarrow{Pc} C_{60}^- + Pc^- \xrightarrow{CA} C_{60} + Pc^+ + CA^- \qquad (24)$$

The quantum yield of ^{3}Pc formation [Eq. (23)] is 0.76, whereas that of C_{60}^- formation [Eq. (24)] is 0.26. The overall quantum yield of C_{60}-mediated electron transfer from Pc to CA is about 0.62. An interesting aspect of the reaction is that electron transfer from Pc to CA can be triggered by visible light whereas in the absence of C_{60} it would require UV or near-UV light. Furthermore, the quantum yield is increased because of the efficient production of $^3C_{60}$.

Studies have been made of the photophysics and photochemistry of porphyrin/fullerene systems. Flash photolysis measurements [106] on solutions of C_{60} or C_{70} and zinc tetraphenylporphyrin (ZnTPP) in a polar solvent show that electron transfer can be initiated by photoexcitation of either chromophore:

$$C_{60} \xrightarrow{h\nu_1} {}^1C_{60}^* \xrightarrow{isc} {}^3C_{60}^* \qquad (25)$$

$$P \xrightarrow{h\nu_2} {}^1P^* \xrightarrow{isc} {}^3P^* \qquad (26)$$

$$^3C_{60}^* + P \rightarrow C_{60}^- + P^+ \qquad (27)$$

$$C_{60} + {}^3P^* \rightarrow C_{60}^- + P^+ \qquad (28)$$

A cw TREPR study by Fujisawa et al. [107] demonstrated that photoexcitation of porphyrins in toluene in the presence of C_{60} also can lead to triplet energy transfer:

$$^3P^* + C_{60} \rightarrow P + {}^3C_{60}^*. \qquad (29)$$

The authors reported as well on electron and energy transfer from porphyrin triplets to C_{60} studied with TREPR [108].

The solvent effect on the relative importance of the two ^{3}P* quenching channels has been investigated with FT-EPR [109]. In the case of MgTPP/C_{60} and octaethylporphyrin (OEP)/C_{60} in toluene, triplet energy transfer [Eq. (29)] dominates. In Fig. 9a the time profile of the intensity of the $^3C_{60}$ resonance given by excitation of OEP (4×10^{-4} M) in a toluene solution containing C_{60} (2×10^{-4} M) is compared with that given by a solution containing C_{60} (2×10^{-4} M) only. That C_{60} triplets are produced by energy transfer is evident from the fact that the resonance peak is in emission at short delay times. It is known that the spin selectivity of the *isc* process in OEP generates triplets with emissive spin polarization [110]. This polarization is carried over in the triplet energy transfer process. The same effect is observed in the case of the MgTPP/C_{60}/toluene system. A quantitative analysis of the time profile shown in Fig. 9 shows that energy transfer occurs at a diffusion-controlled rate ($\sim 1.6 \times 10^{10}$ M^{-1} s^{-1}) and that the

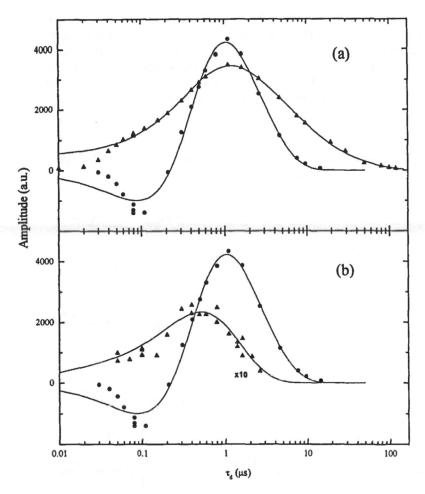

Figure 9 Time profiles of the intensity of the FT-EPR signal from $^3C_{60}$. (a) Solvent toluene; ▲ C_{60} alone, ● C_{60} with OEP. (b) ● C_{60} with OEP in toluene, ▲ C_{60} with OEP in benzonitrile. (Adapted with permission from Ref. 109.)

amount of $^3C_{60}$ formed is about two times larger than that given by excitation of a solution of C_{60} alone in toluene.

As shown in Fig. 9b with benzonitrile as solvent, formation of C_{60} triplets via the energy transfer route is negligible. Under the conditions of the experiments, a small fraction of the laser light is absorbed by C_{60} and this accounts for the weak $^3C_{60}$ signal observed. The dominant $^3P^*$ quenching channel (for both porphyrins) in this polar solvent is electron transfer [Eq. (28)]. This is reflected in the FT-EPR spectra given by a benzonitrile solution containing MgTPP and C_{60} displayed in Fig. 10 [109]. The spectra show resonance peaks due to the redox products $MgTPP^+$ and C_{60}^- in addition to the resonance from $^3C_{60}$. The pronounced solvent effect on the relative importance of triplet energy and electron transfer paths is attributed to the increase in driving force ($\Delta G°$) of the electron transfer reaction with increase in solvent dielectric constant.

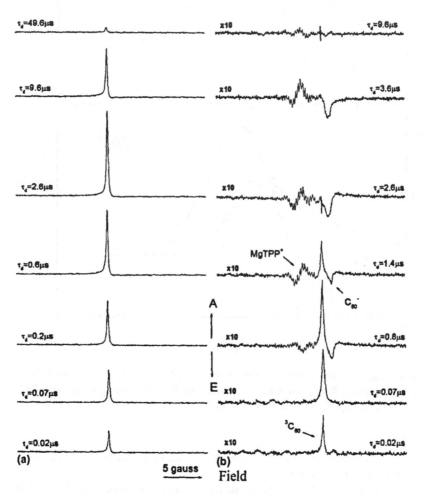

Figure 10 (a) Time evolution of the FT-EPR spectrum of $^3C_{60}$ generated by photoexcitation (532 nm) of a solution of C_{60} in benzonitrile. (b) Time evolution of the FT-EPR spectrum by paramagnetic species given by photoexcitation (532 nm) of a solution of C_{60} with MgTPP in benzonitrile. (Adapted with permission from Ref. 109.)

2. Porphyrin/Quinone Systems

Numerous cw TREPR and FT-EPR studies dealing with the oxidative quenching of porphyrin triplets by quinones have been published. Early work in this field has been reviewed previously [24]. Here the focus will be on some recent applications.

Yamauchi and co-workers applied electron spin-echo–detected FT-EPR in a study of photo-induced electron transfer from zinc-substituted myoglobin (ZnMb) to benzoquinone (BQ) [111]. Spectra obtained in this study show the five-line signal due to the BQ anion radical superimposed on the broad resonance due to the porphyrin cation radical. With FT-EPR measurements the latter signal is not detected because it is so broad that its contribution to the FID decays within the dead time of the spectrometer. With ESE-detected FT-EPR the dead time problem is removed. The time evolution of the FT-EPR spectrum from the anion radical, shown in Fig. 11, is characteristic for a radical formation mechanism in which both TM CIDEP and RPM CIDEP make strong contributions. As was found in earlier studies of Zn porphyrin/quinone systems [23,36,37], the BQ^- spectrum initially displays absorptive polarization due to TM CIDEP. Because the triplet T_1 is short compared to the reaction rate, transfer of triplet spin polarization plays an important role only during the first few hundred nanoseconds. At longer delay times (800 ns–5 μs), the *EA* polarization pattern generated by ST_0 RPM CIDEP makes the dominant signal contribution. The time evolution of the intensity of the hyperfine components in the BQ^- spectrum can be accounted for quantitatively by taking the following parameters into account: 1) initial triplet spin polarization, 2) triplet spin–lattice relaxation time, 3) rate of electron transfer, 4) spin sorting of the RPM, and 5) spin-lattice relaxation time of BQ^-. The data analysis [111] shows that the electron transfer rate constant in the ZnMb/BQ system ($\sim$6.6 $\times$ 10^8 M^{-1} s^{-1}) is an order of magnitude smaller than that of photo-induced electron transfer from zinc tetrakis(4-sulfonatophenyl)porphyrin (ZnTPPS) to BQ in aqueous solution. However, the electron transfer rates are high compared to those found for ZnMb modified by pentaammine–Ru bound to surface histidine [112]. It is proposed [111] that electron transfer occurs to BQ present at protein surface sites near the ZnTPP chromophore. From the fact that the T_1 of BQ^- in the ZnMb/BQ system is longer than that of BQ^- generated in the ZnTPPS/BQ system it is deduced that the anion radical must be loosely bound to the surface of the protein. According to the analysis, the triplet T_1 increases by nearly a factor of 3 upon going from ZnTPPS to ZnMb. The magnitude of this parameter is determined by the modulation of the electron spin–spin dipolar interaction by molecular rotation [35], and the increase is consistent with an increase in rotational correlation time on going from ZnTPPS to the porphyrin-containing protein.

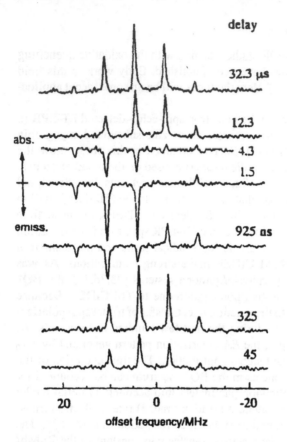

Figure 11 Time evolution of the FT-EPR spectrum of the benzoquinone anion radical formed by electron transfer quenching of zinc-substituted myoglobin. (Reprinted with permission from Ref. 111.)

The work described above illustrates the fact that TREPR measurements can be used to get an insight into molecular motion of paramagnetic species generated in photo-induced reactions. An FT-EPR study of photo-induced electron transfer from ZnTPPS to duroquinone (DQ) solubilized in cetyltrimethylammonium chloride (CTAC) micelles [113] also demonstrates that the technique can provide information that cannot be obtained with optical studies. The hyperfine components in the spectrum of the quinone anion radicals formed by photoexcitation of the ZnTPPS/DQ/CTAC system during the first few microseconds following the laser pulse exhibit the derivative-like lineshape (cf. Fig. 2) characteristic for the presence of a spin-correlated radical pair (SCRP). This is attributed to the fact that the electrostatic field created by the cationic head groups of the surfactant molecules inhibits the dissociation of the radical pairs formed by electron transfer from ZnTPPS to DQ [113].

The long-lived SCRP serves as an ideal model system for the testing of pulsed EPR methods that can be applied in studies of the (electronic) structure. Yamauchi et al. [114,115] used 2D pulsed EPR to determine that the average value of the exchange interaction between the unpaired electrons in the SCRP is about -0.3 MHz. This value is in close agreement with that obtained by simulating the (1D) FT-EPR spectrum of the SCRP. The research group also used the 2D FT-EPR method in a study of ketyl SCRPs generated by hydrogen-atom transfer from 2-propanol to acetone in the photoexcited triplet state [116].

As Fig. 12 shows, the relative intensity of contributions of FT-EPR signals from SCRPs and free DQ$^-$ depends on the power of the applied microwave pulse [48]. With a $\pi/2$ pulse the SCRP signal vanishes, so that only the contribution from the free anions is observed. When a $\pi/4$ pulse is applied, the SCRP signal is optimum and that of free DQ$^-$ is strongly attenuated. Weis et al. [49] used this power dependence to isolate the SCRP signal and investigate the dependence of radical pair formation and decay on DQ concentration. The rate of quenching of ^{3}ZnTPPS* by DQ, adjusted for the probability that a micelle contains both donor and acceptor, is about 1.0×10^6 s^{-1}. Taking into account that DQ is confined within the micelle and the microviscosity within the micelle, this value is close to what would be expected for a diffusion-controlled reaction. The rate of decay of the SCRPs shows a pronounced dependence on [DQ] that is attributed to the self-exchange reaction DQ$^-$ + DQ $\leftrightarrow$ DQ + DQ$^-$. Electron hopping from one quinone to another apparently is a mechanism that has a noticeable effect on the SCRP lifetime. In the case of ZnTPPS/BQ/CTAC, the FT-EPR spectrum at short delay time ($<$100 ns) is dominated by the resonance peaks from free BQ$^-$ [49]. Here the SCRP spectrum grows in gradually and the intensities of SCRP and free anion signals are of comparable magnitude. The difference in characteristics of the spectra obtained with DQ and BQ is attributed to the difference in partition over the lipophilic and aqueous phases. While DQ is mostly confined within the micelles, BQ is more equally distributed. Electron transfer to aqueous phase BQ accounts for the signal from free BQ$^-$, whereas the (slower) electron transfer to BQ captured within the micelles generates the SCRP spectrum. To be effective in reducing acceptors in both phases, ^{3}ZnTPPS* must be positioned at the micelle–water interface.

3. Quinone Photochemistry

Photoexcitation of quinones in the presence of amines can lead to excited-state electron transfer. Reductive quenching of quinone triplets can produce contact

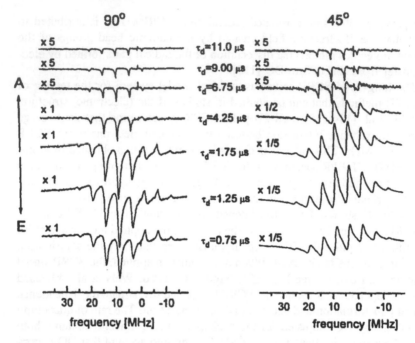

Figure 12 FT-EPR spectra from the duroquinone anion radical produced by photo-induced electron transfer from ZnTPPS in CTAC micellar solution: (left) $\pi/2$ microwave pulse, (right) $\pi/4$ microwave pulse. (Reprinted with permission from Ref. 49.)

and/or solvent-separated ion pairs or free ions depending on reaction conditions. In studies by Beckert et al. [117–120] it is shown that a careful analysis of the time evolution of the FT-EPR spectra makes it possible to identify signal contributions from loosely coupled ion pairs and monitor cage escape giving free radicals.

Photoexcitation (308-nm) of an alkaline (pH 11) aqueous solution of 1,5-(or 2,6-) anthraquinone disulfonate (AQDS, 10^{-3} M) and triethylamine (TEA, 3×10^{-2} M) produces the quinone anion radical and amine cation radical according to [118];

$$AQDS \xrightarrow{h\nu} {}^1AQDS* \xrightarrow{isc} {}^3AQDS* \xrightarrow{TEA} [AQDS^- \cdots TEA^+] \quad (30)$$

Figure 13a displays the FT-EPR spectrum of the 1,5-AQDS anion radical for a delay of 2 μs between laser excitation and microwave pulse. The spectrum is in emission because of TM CIDEP. Figure 13b gives the spectrum of the same

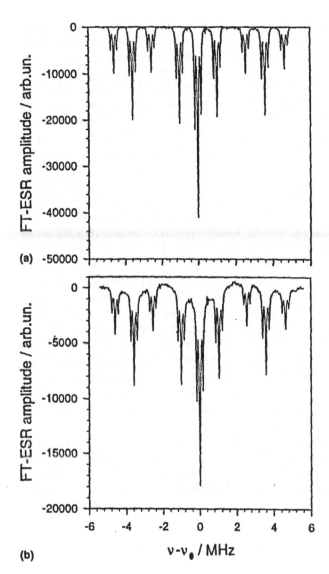

Figure 13   FT-EPR spectrum of the quinone anion radical generated by photoexcitation (308 nm) of an alkaline (pH 11) aqueous solution of 1,5-anthraquinone disulfonate (10^{-3} M) and triethylamine (3×10^{-2} M). (a) Delay between laser pulse and microwave pulse 2 μs; (b) delay between laser pulse and microwave pulse 32 ns. (Adapted with permission from Ref. 118.)

system, but now with a delay time setting of 32 ns. It is evident that in the latter spectrum each group of narrow lines sits on top of a broad resonance peak. With the aid of simulations, it was established [118] that the species responsible for the broad- and narrow-line spectra have the same g value and hyperfine splitting constants, but are distinguished by different linewidths (900 kHz vs. 84 kHz). The simulations also show that the rate of decay of the broad component matches a slow component of the rise time of the narrow-line spectrum. Furthermore, the broad-line spectrum of TEA$^+$, which falls outside the frequency range displayed in Fig. 13, decays with similar kinetics. From this it is concluded that the broad-line signals are due to the Coulomb-coupled, solvent-separated ion pair [AQDS$^-$ $\cdots$ TEA$^+$]. The lifetime of the ion pair shows an [OH$^-$] dependence with an increase in pH causing an increase in cage escape rate [118]. The lifetime of TEA$^-$ is pH-dependent as well because of the deprotonation reaction TEA$^+$ + OH$^-$ $\rightarrow$ Et$_2$N-C$\cdot$HCH$_3$. The buildup kinetics of this α-aminoalkyl radical can be derived from the time profile of its characteristic FT-EPR spectrum.

An interesting change was found in the time evolution of the FT-EPR spectrum of the anthraquinone anion radical (AQ$^-$) formed by photo-induced electron transfer from TEA in alcoholic solutions [121]. Here the emissively polarized AQ$^-$ spectrum develops over a period of about 100 ns, even though the time constant of reductive quenching of ^{3}AQ*, under the conditions of the measurements, is of the order of 1 ns. The difference with the results obtained with the AQDS/TEA system in aqueous solution is attributed [121] to formation of "EPR-silent," strongly coupled radical pairs in alcoholic solvents. The cage escape rate is a sensitive function of the alcohol used as solvent, which can be accounted for by considering the effect of dielectric shielding on the coulombic attraction between anion and cation radicals [121].

A FT-EPR study [122] of the reduction quenching of ^{3}AQ* by 4-methyl-2,6-di-*tert*-butylphenol shows the presence of two reaction channels, one producing AQ$^-$ and the other giving the neutral radical AQH$\cdot$. The time profiles of intensities of resonance peaks given by the radicals could be modeled in terms of Eqs. (31) and (32):

$$^3\text{AQ*} + \text{PhOH} \rightarrow \text{AQ}^- + \text{PhOH}^+ \tag{31}$$

$$^3\text{AQ*} + \text{PhOH} \rightarrow \text{AQH}\cdot + \text{PhO}\cdot \tag{32}$$

with an additional, slow, proton dissociation reaction establishing the equilibrium

$$\text{AQH}\cdot \leftrightarrow \text{AQ}^- + \text{H}^- \tag{33}$$

taking CIDEP and spin relaxation into account. The overall reaction scheme that formed the basis of computer analysis of the data is displayed in Fig. 14. Figure 15 shows experimental data for two different solvents with computer fits based on the model given in Fig. 14 [122]. Considering the complexity of this system

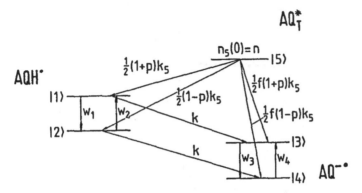

Figure 14 Kinetic scheme used for the analysis of the time evolution of the signals from AQH· and AQ⁻· formed by reductive quenching of ³AQ* by 4-methyl-2,6-di-*tert*-butylphenol. Triplet spin polarization is accounted for by the parameter p, the relative importance of the two radical formation paths by the parameter f, W_n denotes relaxation rate. (Reprinted with permission from Ref. 122.)

(the solution of the set of differential equations representing the model occupies a little over one page of Ref. 122), the agreement between experimental data and simulations is remarkable. Plüschau and Dinse [123] established the presence of the dynamic equilibrium [Eq. (33)] and determined the rate of AQH· ↔ AQ⁻ interconversion with 2D FT-EPR measurements.

Geimer et al. [120,124] investigated electron transfer from thymine, uracil, and methyluracil to AQDS triplets with FT-EPR. The spectra are highly resolved, which facilitates the identification of the doublet radicals formed, the study of structural changes accompanying changes in pH, and the extraction of kinetic parameters from time profiles of signal intensities. Specifically, in the case of the 2,6-AQDS/thymine system, the pH effect on the spectrum from the quinone anion radical could be interpreted in terms of an equilibrium between three protonation states [120]. Furthermore, the CIDEP pattern in the spectrum of the thymine-1-yl radical, which is formed by rapid deprotonation of the thymine cation radical, points to two distinct structures of the radical pair precursor of the free radicals [120].

C. Photochemistry of Transition Metal Complexes

Numerous studies dealing with the spin selectivity of photochemical reactions involving transition metal complexes have been published in recent years [27,125]. For the most part, the investigations have been based on data provided by optical spectroscopy and have focused on magnetic field effects on reaction kinetics. The number of applications of TREPR in this field of research remains

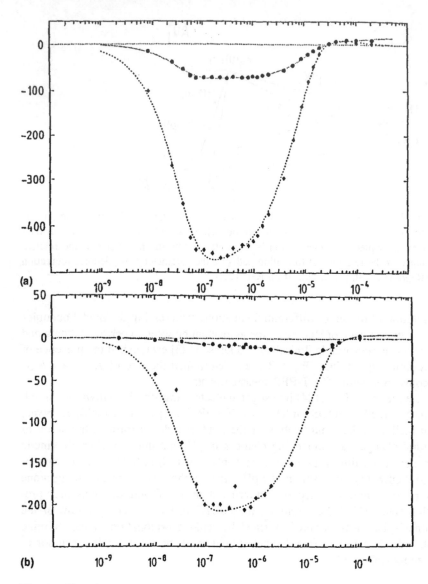

Figure 15 Time profiles of the intensities of resonance peaks from AQH˙ and AQ⁻˙ formed by reductive quenching of ³AQ* by 4-methyl-2,6-di-*tert*-butylphenol in (a) ethanol and (b) 2-propanol. The solid lines show the results of least-squares fits of the data to the model presented in Fig. 14. The horizontal (time) scale is in seconds. (Adapted with permission from Ref. 122.)

limited. The reason for this is that in most cases the detection of paramagnetic species is hampered by a poor signal-to-noise ratio because of short relaxation times. However, a number of TREPR studies have dealt with organic free radicals generated by photo-induced bond homolysis in organometallics. The results demonstrate that TREPR can be a valuable tool in studies of transition metal photochemistry.

The first example is the detection by cw TREPR of the methyl radical produced by pulsed laser excitation of methylaquocobaloxime [126]. The four-line spectrum of $CH_3^{\cdot}$ is in emission, which was attributed to TM CIDEP and was taken as evidence that bond fission proceeds through the triplet excited state of the complex. It should be noted, however, that the origin of the spin polarization also can be the spin-selective recombination reaction:

$$^3[LCo \cdots CH_3^{\cdot}] \xrightarrow{\quad risc \quad} {}^1[LCo \cdots CH_3^{\cdot}] \rightarrow LCo\text{-}CH_3 \qquad (34)$$

driven by spin-orbit coupling–induced reversed *isc* in a strongly coupled radical pair [60]. The finding that the free radical is "born" spin-polarized is of interest since it signifies that the spin-lattice relaxation time of paramagnetic precursor, possibly the triplet excited state of the Co complex, is long enough that bond cleavage can compete with establishment of thermal equilibrium of the spin system. That spin selectivity may play a role in the reactivity of this Co complex is of relevance to the question of whether magnetic fields can affect enzymatic reactions [127].

Yamauchi *et al.* [128,129] investigated the photodissociation of bis(*S*-benzyl-1,2-diphenyl-1,2-ethylenedithiolato)Me, where Me = Ni, Pd, or Pt with

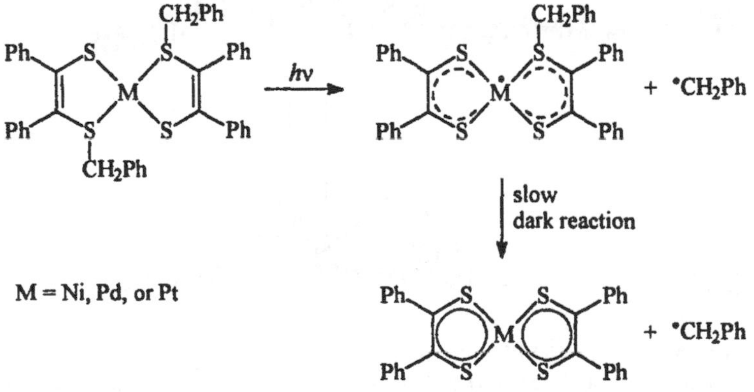

Figure 16 Reaction scheme proposed [129] for the photodissociation of bis(*S*-benzyl-1,2-diphenyl-1,2-ethylenedithiolato) M, where M = Ni, Pd, or Pt.

cw TREPR. It was previously postulated that bond cleavage producing the benzyl radicals and $Me(S_2C_2Ph_2)_2$ proceeds in a stepwise fashion according to the scheme given in Fig. 16.

TREPR spectra given by solutions of the dithiolato complexes following pulse laser excitation are shown in Fig. 17. In the case of the Ni and Pd complexes, the spectra display a complex pattern of narrow resonance peaks in the high-field region and a broad resonance at low field. The high-field resonances are due to the benzyl radical while the broad line is attributed to the metal com-

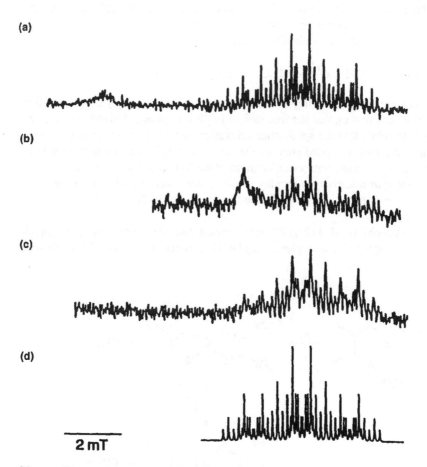

(a)

(b)

(c)

(d)

2 mT

Figure 17 Continuous wave TREPR spectra of the benzyl radical generated by photo-induced bond homolysis in bis(S-benzyl-1,2-diphenyl-1,2-ethylenedithiolato)Me with (a) Me = Ni, (b) Me = Pd, (c) Me = Pt, and (d) simulation of the benzyl radical EPR spectrum. (Reprinted with permission from Ref. 129.)

plex produced by removal of *one* benzyl group. With the Pt complex only the benzyl radical spectrum is detected: All spectra are in absorption but show some intensity asymmetry that could point to an *EA* CIDEP contribution. It is postulated that the dominant absorptive signal contribution is due to TM CIDEP. If correct, this means that the bond cleavage reaction involves the triplet excited state of the thiolato complexes. The intermediate paramagnetic complex is found to lose the remaining benzyl group in a dark reaction occurring on a time scale of 10 to 20 min.

Kleverlaan et al. [130–132] used FT-EPR to study the photo-induced metal–alkyl bond cleavage in a series of [Re(R)(CO)$_3$(α-diimine)] and [Ru(X)(R)(CO)$_2$(α-diimine)] complexes. Here R = alkyl (methyl, ethyl, isopropyl) or benzyl, X = I or SnPh$_3$, and the α-diimine ligand is 4,4'-dimethyl-2,2'-bipyridine (DMB) or *N,N*'-diisopropyl-1,4-diazabutadiene (iPr-DAB). The general structure of the compounds is given in Fig. 18. Detailed spectroscopic analyses (UV/Vis, IR, Raman, NMR, cw EPR) of these complexes [132] show that the quantum yield of metal–alkyl (benzyl) bond homolysis is a sensitive function of the identity of R and the coligands. The quantum yield can display a solvent dependence as well. In a few cases, the light-induced formation of free radicals could be demonstrated with conventional EPR measurements either by direct detection of the paramagnetic species or with the aid of the spin-trapping method. With FT-EPR the transient alkyl (benzyl) radicals formed upon pulsed laser excitation of the complexes could be detected. The following examples illustrate the insights provided by the FT-EPR data.

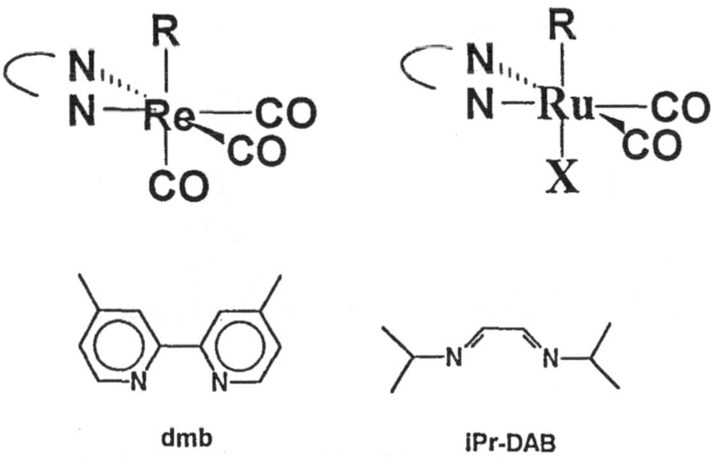

dmb   iPr-DAB

Figure 18 Structure of [Re(R)(CO)$_3$(α-diimine)] and [Ru(X)(R)(CO)$_2$(α-diimine)] complexes. R = alkyl (methyl, ethyl, isopropyl) or benzyl, X = I or SnPh$_3$.

Spectra from the methyl radical produced upon photoexcitation of solutions of [Re(CH$_3$)(CO)$_3$(DMB)] and [Re(CD$_3$)(CO)$_3$(DMB)] in toluene are given in Fig. 19 [131]. The spectrum of CH$_3$· displays overall net emissive polarization superimposed on an *EA* CIDEP pattern. The *EA* pattern points to a precursor triplet radical pair state generating RPM CIDEP and the net *E* contribution to TM CIDEP, suggesting the reaction sequence:

$$[Re(CH_3)(CO)_3(DMB)] \xrightarrow{h\nu} {}^1[Re(CH_3)(CO)_3(DMB)]^*$$

$$\xrightarrow{isc} {}^3[Re(CH_3)(CO)_3(DMB)]^*$$

$${}^3[Re(CH_3)(CO)_3(DMB)]^* \rightarrow {}^3\{[Re(CO)_3(DMB)]^\cdot \cdots CH^\cdot_3\}$$

$$\rightarrow [Re(CO)_3(DMB)]^\cdot + CH_3^\cdot \quad (35)$$

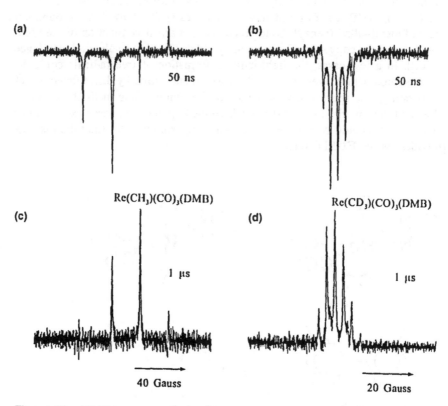

Figure 19   FT-EPR spectra from the methyl radical produced upon photoexcitation (355 nm) of solutions of [Re(CH$_3$)(CO)$_3$(DMB)] and [Re(CD$_3$)(CO)$_3$(DMB)] in toluene for delay times of 50 ns and 1 μs. (Reprinted with permission from Ref. 131.)

Figure 19 shows that deuteration almost completely removes the *EA* polarization. This is in agreement with its assignment to RPM CIDEP. The magnitude of the RPM CIDEP contribution is a function of the difference in resonance frequency (Δv) of the two radicals that make up the pair [Eqs (7) and (8)]. In the present case Δv apparently is determined virtually completely by the hfcc of the methyl protons. Deuteration reduces the splitting constant by about a factor of 6, which results in the pronounced reduction in *EA* signal contribution. The emissive spin polarization created by spin-selective *isc* is not affected by the deuteration.

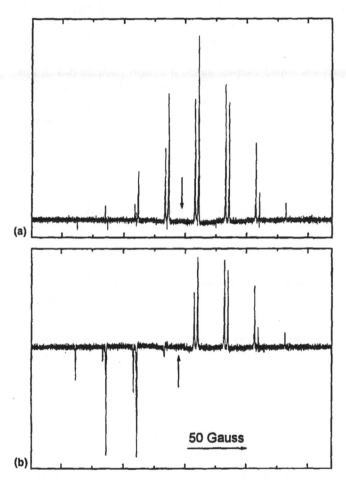

Figure 20   FT-EPR spectra from the isopropyl radical formed upon photoexcitation (355 nm) of $Ru(I)(iPr)(CO)_2(iPr\text{-}DAB)$ in toluene and 2-propanol. The arrows mark the center of the spectra. Delay 100 ns. (Reprinted with permission from Ref. 130.)

That TM polarization is observed is consistent with the proposal [132] that bond dissociation involves a $^3\sigma\pi^*$ state where σ represents the $\sigma(\text{Re-CH}_3)$ bonding orbital and π^* a diimine antibonding orbital. From the strong dependence of RPM CIDEP on the hyperfine interaction it can be deduced that the g value of the $[\text{Re(CO)}_3(\text{DMB})]^\bullet$ radical must be close to that of the methyl radical, so that the unpaired electron must reside in an orbital that has predominant diimine π^* character.

Spectra given by the isopropyl radical formed upon photoexcitation of $\text{Ru(I)(iPr)(CO)}_2(\text{iPr-DAB})$ in toluene and 2-propanol are presented in Fig. 20. Here as well an *EA* polarization contribution is observed. A striking feature is the strong solvent dependence of the spin polarization pattern. Its cause has not been determined, but several plausible explanations have been offered [130]. For instance, TM CIDEP can exhibit a solvent dependence because of a shift in relative energies of singlet and triplet excited states. It is also possible that in this case RPM CIDEP exhibits a pronounced solvent dependence because radical pair characteristics may depend critically on whether the coordination position vacated by benzyl is promptly occupied by a solvent molecule. Finally, the solvent

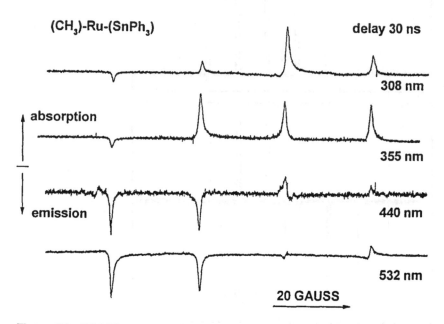

Figure 21 FT-EPR spectra of the methyl radical produced by photo-induced bond homolysis in the $[\text{Ru(SnPh}_3)(\text{CH}_3)(\text{CO})_2(\text{iPr-DAB})]$ complex in toluene. Excitation wavelengths ranging from 308 nm to 532 nm. Delay time 30 ns. (Reprinted with permission from Ref. 131.)

may determine whether or not the radical pair formed by bond homolysis is strongly or weakly coupled. In the former case, SOC-induced *risc* can give rise to net spin polarization.

A novel effect is found in the FT-EPR spectrum of the methyl radical produced by photoexcitation of [Ru(SnPh$_3$(CH$_3$)(CO)$_2$(iPr-DAB)]. As Figure 21 illustrates, in this case the CIDEP pattern observed is a sensitive function of the excitation wavelength (λ_{ex}). With visible light excitation (440- or 532-nm), the spectra show an *EA* pattern superimposed on a net emission signal. The pattern turns to *EA* plus net absorption for $\lambda_{ex} = 335$ or 308 nm. The effect is not found in the other complexes and may be related to the fact that the absorption spectrum of [Ru(SnPh$_3$)(CH$_3$)(CO)$_2$(iPr-DAB)] exhibits absorption bands centered at 525 and 350 nm assigned as two $\sigma \rightarrow \pi^*$ transitions which directly populate the dissociative $\sigma\pi^*$ state [132]. By contrast, in the other complexes the dissociative state is populated indirectly by excitation into the MLCT band. It suggests that bond homolysis in this case occurs in competition with relaxation to the lowest excited electronic state.

IV. CONCLUDING REMARKS

As anticipated in a review of applications of TREPR in studies of photochemical reactions published in 1993 [24], the introduction of commercial pulsed EPR spectrometers around that time has led to a rapid development of this field of research. As the present chapter shows, the number of research groups that use TREPR has grown and the range of applications has expanded. New mechanisms of spin polarization generation have been identified and this has been exploited in studies of reaction mechanisms.

The outcome of a photochemical reaction can depend to a significant extent on the spin-state dynamics of radical pair intermediates. Therefore, it is of particular interest that TREPR can provide direct information on the interaction between constituents of radical pairs as well as their formation and decay. With other spectroscopic techniques this information generally is not available. It is not surprising, therefore, that the study of strongly and weakly coupled radical pairs with TREPR is a particularly active area of research [28,30,48,49,118,121]. The interpretation of spectroscopic data is aided by theoretical analyses of the dependence of CIDEP on the interaction between radical pair constituents as modulated by relative diffusion [133,134].

The studies of photoionization reactions demonstrate that while it now is straightforward to monitor the time evolution of CIDEP patterns over the time regime extending from nanoseconds to tens of microseconds, the interpretation of these data in terms of reaction mechanisms can be a matter of dispute. Because of CIDEP signal enhancement, minor reaction paths can make dominant signal contributions. With the high laser intensities used (to optimize signal-to-noise) and relatively long laser pulses, a variety of two-photon processes can affect what is measured [74,84,90,110]. The improved instrumentation that has become available makes it possible to identify and study such processes.

It is of interest that processes such as photoionization [68–71] and photo-induced bond homolysis [131] can involve nonrelaxed excited states. In TREPR spectra of free-radical products this may be reflected in wavelength-dependent CIDEP effects. Therefore, the technique can serve as a probe of such processes.

Applications in the field of photochemistry of transition metal complexes suggest that TREPR studies can aid in the delineation of reaction mechanisms. However, the interpretation of CIDEP effects will be more difficult than in the case in organic photochemistry, among other things, because of the more important role played by spin orbit coupling and the greater variety in character of excited states involved.

As the discussion in Section II shows, spin polarization in products of photochemical reactions can stem from a variety of processes, some of which are not linked to the mechanism of radical formation. Even so, polarization patterns observed in TREPR spectra still are attributed almost exclusively to ST_0 RPM and TM CIDEP. Now that the evolution of spectra can be monitored with nanosecond time resolution, the identification of sources of spin polarization can be based on firm spectroscopic evidence. Moreover, it makes it possible to quantify the polarization generated in the reaction, which can provide information on the molecular motion of photoexcited triplet precursors and the spin-state mixing in radical pairs.

ACKNOWLEDGMENT

Financial support for this work was provided by the Division of Chemical Sciences, Office of Basic Energy Sciences of the U.S. Department of Energy (DE-FG02-84ER-13242).

REFERENCES

1. Budil, D. E.; Earle, K. A.; Lynch, W. B.; Freed, J. H. *Advanced EPR: Applications in Biology and Biochemistry*; Hoff, A. J., ed.: Elsevier: Amsterdam, 1989, pp. 307–340.
2. Lebedev, Y. S. *Modern Pulsed and Continuous-Wave Electron Spin Resonance*; Kevan, L.; Bowman, M. K., eds.; Wiley: New York, 1990, pp. 365–404.
3. Terazima, M.; Miura, Y.; Ohara, K.; Hirota, N. *Chem. Phys. Lett.* **1994**, *95*, 224.
4. Prisner, T. F.; Rohrer, M.; Möbius, K. *Appl. Magn. Reson.* **1994**, *7*, 167.
5. Gorcester, J.; Millhauser, G. L.; Freed, J. H. *Advanced EPR: Applications in biology and biochemistry*; Hoff, A. J., ed.; Elsevier: Amsterdam, 1989, pp. 177–242.
6. Bowman, M. K. *Modern Pulsed and Continuous-Wave Electron Spin Resonance*; Kevan, L.; Bowman, M. K., eds.; Wiley: New York, 1990, pp. 1–42.
7. Schweiger, A. *J. Chem. Soc. Farad. Trans.* **1995**, *91*, 177.

8. *Advanced EPR: Applications in biology and biochemistry*; Hoff A. J., ed.; Elsevier: Amsterdam, 1989.
9. *Modern Pulsed and Continuous-Wave Electron Spin Resonance*; Kevan, L.; Bowman, M. K., eds.; Wiley: New York, 1990.
10. Weil, J. A.; Bolton, J. R.; Wertz, J. E. *Electron Paramagnetic Resonance*; Wiley: New York, 1994.
11. Salikov, K. M.; Molin, Y. N.; Sagdeev, R. Z.; Buchachenko, A. L. *Spin Polarization and Magnetic Effects in Radical Reactions*; Molin, Y. N., ed.; Elsevier: Amsterdam, 1984.
12. Hore, P. J. *Advanced EPR: Applications in biology and biochemistry*, Hoff, A. J., ed.; Elsevier: Amsterdam, 1989; pp 405–440.
13. McLauchlan, K. A. *Modern Pulsed and Continuous-Wave Electron Spin Resonance*; Kevan, L.; Bowman, M. K., eds.; Wiley: New York, 1990; pp 285–364.
14. Hirota, N.; Yamauchi, S. *Dynamic Spin Chemistry*; Nagakura, S.; Hayashi, H.; Azumi, T., eds.; Wiley: New York, 1998; pp 187–248.
15. Paul, H. *Chem. Phys.* **1997**, *40*, 265; **1979**, *43*, 294.
16. Verma, N. C.; Fessenden, R. W. *J. Chem. Phys.* **1973**, *58*, 2501.
17. Kim, S. S.; Weissman, S. I. *J. Magn. Reson.* **1976**, *24*, 167.
18. Trifunac, A. D.; Thurnauer, M. C.; Norris, J. R. *Chem. Phys. Lett.* **1978**, *57*, 471.
19. Forbes, M. D. E.; Peterson, J.; Breivogel, C. S. *Rev. Sci. Instrum.* **1991**, *62*, 2662.
20. Koptyug, I.; Turro, N. J.; van Willigen, H.; McLauchlan, K. A. *J. Magn. Reson.* **1994**, *109*, 121.
21. Steren, C. A.; Levstein, P. R.; van Willigen, H.; Linschitz, H.; Biczók, L. *Chem. Phys. Lett.* **1993**, *204*, 23.
22. Massoth, R. J. Thesis, University of Kansas, 1987.
23. Prisner, T.; Dobbert, O.; Dinse, K. P.; van Willigen, H. *J. Am. Chem. Soc.* **1988**, *110*, 1622.
24. van Willigen, H.; Levstein, P. R.; Ebersole, M. H. *Chem. Rev.* **1993**, *93*, 173.
25. Trifunac, A. D.; Lawler, R. G.; Bartels, D. M.; Thurnauer, M. C. *Progr. React. Kinet.* **1986**, *14*, 43.
26. *Spin Polarization and Magnetic Effects in Radical Reactions*; Molin, Y. N., ed.; Elsevier: Amsterdam, 1984.
27. Steiner, U. E.; Ulrich, T. *Chem. Rev.* **1989**, *89*, 51.
28. *Dynamic Spin Chemistry*; Nagakura, S.; Hayashi, H.; Azumi, T., eds.; Wiley: New York, 1998.
29. McLauchlan, K. A.; Yeung, M.T. *Specialist Periodical Reports: Electron Spin Resonance*; Royal Society: London, **1994**; Vol. 14, pp. 32–62.
30. Clancy, C. M. R.; Tarasov, V. F.; Forbes, M. D. E. *Specialist Periodical Reports: Electron Spin Resonance*; Royal Society; London, **1998**, *16*, 50–78.
31. *The Photosynthetic Reaction Center*; Deisenhofer, J.; Norris, J. R., eds.; Academic Press: New York, 1993.
32. Levanon, H.; Möbius, K. *Annu. Rev. Biophys. Biomol. Struct.* **1997**, *26*, 495.
33. Wong, S. K.; Hutchinson, D. A.; Wan, J. K. S. *J. Chem. Phys.* **1973**, *58*, 985.
34. Atkins, P. W.; Evans, G. T. *Mol. Phys.* **1974**, *27*, 1633.
35. Carrington, A.; McLachlan, A. D. *Introduction to Magnetic Resonance*; Harper & Row: New York, 1967, p. 201.

36. Angerhofer, A.; Toporowicz, M.; Bowman, M. K.; Norris, J. R.; Levanon, H. *J. Phys. Chem.* **1988**, *92*, 7164.

37. van Willigen, H.; Vuolle, M.; Dinse, K. P. *J. Phys. Chem.* **1989**, *93*, 2441.

38. Levstein, P. R.; Ebersole, M. H.; van Willigen, H. *Proc. Indian Acad. Sci.-Chem. Sci.* **1992**, *104*, 681.

39. Kottis, P.; Lefebvre, R. *J. Chem. Phys.* **1963**, *39*, 393.

40. Ohara, K.; Hirota, N.; Martino, D. M.; van Willigen, H. *J. Phys. Chem. A* **1998**, *102*, 5433.

41. Kleverlaan, C. J.; Martino, D. M.; van Slageren, J.; van Willigen, H.; Stufkens, D. J.: Oskam, A. *Appl. Magn. Res.* **1998**, *15*, 203.

42. Kamata, Y.; Akiyama, K.; Tero-Kubota, S. *J. Phys. Chem. A* **1999**, *103*, 1714.

43. Levstein, P. R.; van Willigen, H. *J. Chem. Phys.*, **1991**, *95*, 900.

44. Kaptein, R.; Oosterhoff, L. J. *Chem. Phys. Lett.* **1969**, *4*, 214.

45. Adrian, F. *J. Chem. Phys.* **1971**, *54*, 3918.

46. Freed, J. H.; Pedersen, J. B. *Adv. Magn. Reson.* **1976**, *8*, 1.

47. Closs, G. L.; Forbes, M. D. E.; Norris, J. R. *J. Phys. Chem.* **1991**, *91*, 3592.

48. Hasharoni, K.; Levanon, H.; Tang, J.; Bowman, M. K.; Norris, J. R.; Gust, D.; Moore, T. A.; Moore, A. L. *J. Am. Chem. Soc.* **1990**, *112*, 6477.

49. Weis, V.; van Willigen, H. *J. Porphyrins & Phthalocyanines* **1998**, *2*, 353.

50. Thurnauer, M. C.; Norris, J. R. *Chem. Phys. Lett.* **1980**, *76*, 557.

51. Kroll, G.; Plüschau, M.; Dinse, K. P.; van Willigen, H. *J. Chem. Phys.* **1990**, *93*, 8709.

52. Imamura, T.; Onitsuka, O.; Obi, K. *J. Phys. Chem.* **1986**, *90*, 6741.

53. Blättler, C.; Jent, F.; Paul, H. *Chem. Phys. Lett.* **1990**, *166*, 375.

54. Kawai, A.; Okutsu, T.; Obi, K. *J. Phys. Chem.* **1991**, *95*, 9130.

55. Goudsmit, G.-H.; Paul, H.; Shushin, A. I. *J. Phys. Chem.* **1993**, *97*, 13243.

56. He, G.; Chen, C.; Yang, J.; Xu, G. *J. Phys. Chem. A* **1998**, *102*, 2865.

57. Goudsmit, G.-H.; Paul, H. *Chem. Phys. Lett.* **1993**, *208*, 73.

58. Fujisawa, J.; Ohba, Y.; Yamauchi, S. *J. Phys. Chem. A* **1997**, *101*, 434.

59. Kobori, Y.; Kawai, A.; Obi, K. *J. Phys. Chem.* **1994**, *98*, 6425.

60. Steiner, U.; Haas, W. *J. Phys. Chem.* **1991**, *95*, 1880.

61. Katsuki, A.; Akiyama, K.; Ikegami, Y.; Tero-Kubota, S. *J. Am. Chem. Soc.* **1994**, *116*, 12065.

62. Katsuki, A.; Akiyama, K.; Tero-Kubota, S. *Bull. Chem. Soc. Jpn.* **1995**, *68*, 3383.

63. Sasaki, S.; Katsuki, A.; Akiyama, K.; Tero-Kubota, S. *J. Am. Chem. Soc.* **1997**, *119*, 1323.

64. Sasaki, S.; Kobori, Y.; Akiyama, K.; Tero-Kubota, S. *J. Phys. Chem. A* **1998**, *102*, 8078.

65. Grossweiner, L. I. *Curr. Top. Radiat. Res. Q.* **1976**, *11*, 141.

66. Feitelson, J.; Hayon, E.; Treinin, A. *J. Am. Chem. Soc.* **1973**, *95*, 1025.

67. Bent, D. V.; Hayon, E. *J. Am. Chem. Soc.*, **1975**, *97*, 2599.

68. Grabner, G.; Köhler, G.; Zechner, J.: Getoff, N. *Photochem. Photobiol.* **1977**, *26*, 449.

69. Grabner, G.; Köhler, G.; Zechner, J.; Getoff, N. *J. Phys. Chem.* **1980**, *84*, 3000.

70. Zechner, J.; Köhler, G.; Grabner, G.; Getoff, N. *Can. J. Chem.* **1980**, *58*, 2006.

71. Grabner, G.; Köhler, G.; Marconi, G.; Monti, S.; Venuti, E. *J. Phys. Chem.* **1990**, *94*, 3609.
72. Jeevarajan, A. S.; Fessenden, R. W. *J. Phys. Chem.* **1992**, *96*, 1520.
73. Bussandri, A.; van Willigen, H.; Nakagawa, K. *Appl. Magn. Reson.* **1999**, *17*, 577.
74. Clancy, C. M. R.; Forbes, M. D. E. *Photochem. Photobiol.* **1999**, *69*, 16.
75. Avdievich, N. I.; Jeevarajan, A. S.; Forbes, M. D. E. *J. Phys. Chem.* **1996**, *100*, 5334.
76. Murai, H.; Kuwata, K. *Chem. Phys. Lett.* **1989**, *164*, 567.
77. Shkrob, I. A.; Trifunac, A. D. *Chem. Phys.* **1996**, *202*, 117.
78. Ishiwata, N.; Murai, H.; Kuwata, K. *J. Phys. Chem.* **1993**, *97*, 7129.
79. Säuberlich, J.; Beckert, D. *J. Phys. Chem.* **1995**, *99*, 12520.
80. Säuberlich, J.; Brede, O.; Beckert, D. *J. Phys. Chem.* **1996**, *100*, 18101.
81. Dixon, W. T.; Murphy, D. *J. Chem. Soc. Perkin Trans.2* **1976**, 1823.
82. Alkaitis, S. A.; Beck, G.; Grätzel, M. *J. Am. Chem. Soc.* **1975**, *97*, 5723.
83. Nakagawa, K.; Katsuki, A.; Tero-Kubota, S. *J. Am. Chem. Soc.* **1996**, *118*, 5778.
84. Turro, N. J.; Khudyakov, I. V.; van Willigen, H. *J. Am. Chem. Soc.* **1995**, *117*, 12273.
85. Gauduel, Y.; Berrod, S.; Migus, A.; Yamada, N.; Antonetti, A. *Biochemistry* **1988**, *27*, 2509.
86. Smith, A. G.; McGimpsey, W. G. *J. Phys. Chem.* **1994**, *98*, 2923.
87. van Willigen, H.; Levstein, P. R.; Martino, D. M.; Ouardaoui, A.; Tassa, C. *Appl. Magn. Reson.* **1997**, *12*, 395.
88. Lech, J. *Thesis*, University of Massachusetts, Boston, 1999.
89. Rozenshtein, V.; Zilber, G.; Rabinovitz, M.; Levanon, H. *J. Am. Chem. Soc.* **1993**, *115*, 5193.
90. Zubarev, V.; Goez, M. *Angew. Chem.* **1997**, *109*, 2779.
91. Bussandri, A.; van Willigen, H. Unpublished results.
92. Terazima, M.; Hirota, N.; Shinohara, H.; Saito, Y. *J. Phys. Chem.* **1991**, *95*, 9080.
93. Arborgast, J. W.; Foote, C. S.; Kao, M. *J. Am. Chem. Soc.* **1992**, *114*, 2277.
94. Biczók, L.; Linschitz, H.; Walter, R. I. *Chem. Phys. Letters* **1992**, *195*, 339.
95. Closs, G. L.; Gautam, P.; Zhang, D.; Krusic, P. J.; Hill, S. A.; Wasserman, E. *J. Phys. Chem* **1992**, *96*, 5228.
96. Steren, C. A.; Levstein, P. R.; van Willigen, H.; Linschitz, H.; Biczók, L. *Chem. Phys. Lett.* **1993**, *204*, 23.
97. Goudsmit, G.-H.; Paul, H. *Chem. Phys. Lett.* **1993**, *208*, 73.
98. Zhang, D.; Norris, J. R.; Krusic, P. J.; Wasserman, E.; Chen, C.-C.; Lieber, C. M. *J. Phys. Chem.* **1993**, *97*, 5886.
99. Regev, A.; Gamliel, D.; Meiklyar, V.; Michaeli, S.; Levanon, H. *J. Phys. Chem.* **1993**, *97*, 3671.
100. Steren, C. A.; van Willigen, H.; Dinse, K.-P. *J. Phys. Chem.* **1994**, *98*, 7464.
101. Bennati, M.; Grupp, A.; Mehring, M. *J. Chem. Phys.* **1995**, *185*, 221.
102. Bennati, M.; Grupp, A.; Bäuerle, P.; Mehring, M. *Chem. Phys.* **1994**, *102*, 9457.
103. Michaeli, S.; Meiklyar, V.; Schulz, M.; Möbius, K.; Levanon, H. *J. Phys. Chem.* **1994**, *98*, 7444.
104. Fujisawa, J.; Ohba, Y.; Yamauchi, S. *J. Am. Chem. Soc.* **1997**, *119*, 8736.
105. Steren, C. A.; van Willigen, H.; Biczók, L.; Gupta, N.; Linschitz, H. *J. Phys. Chem.* **1996**, *100*, 8920.

106. Nojiri, T.; Watanabe, A.; Ito, O. *J. Phys. Chem. A* **1998**, *102*, 5215.
107. Fujisawa, J.; Yasunori, O.; Yamauchi, S. *Chem. Phys. Lett.* **1998**, *282*, 181.
108. Fujisawa, J.; Yasunori, O.; Yamauchi, S. *Chem. Phys. Lett.* **1998**, *294*, 248.
109. Martino, D. M.; van Willigen, H. *Proceedings of the Symposium on Recent Advances in the Chemistry and Physics of Fullerenes and Related Materials*, Kadish, K. M.; Ruoff, R. S., Eds.; Electrochemical Society: Pennington, 1998, pp. 338–352.
110. Levanon, H.; Regev, A.; Galili, T.; Hugerat, M.; Chang, C. K.; Fajer, J. *J. Phys. Chem.* **1993**, *97*, 13198.
111. Satoh, R.; Ohba, Y.; Yamauchi, S.; Iwaizumi, M.; Kimura, C.; Tsukahara, K. *J. Chem. Soc. Faraday Trans.* **1997**, *93*, 537.
112. Winkler, J. R.; Gray, H. B. *Chem. Rev.* **1992**, *92*, 369.
113. Levstein, P. R.; van Willigen, H. *Chem. Phys. Lett.* **1991**, *187*, 415.
114. Hanaishi, R.; Yamamoto, K.; Ohba, Y.; Yamauchi, S.; Iwaizumi, M. *Appl. Magn. Reson.* **1996**, *10*, 55.
115. Hanaishi, R.; Ohba, Y.; Yamauchi, S.; Iwaizumi, M. *Bull. Chem. Soc. Jpn.* **1996**, *69*, 1533.
116. Hanaishi, R.; Ohba, Y.; Akiyama, K.; Yamauchi, S. *J. Chem. Phys.* **1995**, *103*, 4819.
117. Beckert, D.; Fessenden, R. W. *J. Phys. Chem.* **1996**, *100*, 1622.
118. Säuberlich, J.; Brede, O.; Beckert, D. *J. Phys. Chem. A* **1997**, *101*, 5659.
119. Säuberlich, J.; Brede, O.; Beckert, D. *Acta Chim. Scand.* **1997**, *51*, 602.
120. Geimer, J.; Beckert, D. *Chem. Phys. Lett.* **1998**, *288*, 449.
121. Beckert, D.; Plüschau, M.; Dinse, K. P. *J. Phys. Chem.* **1992**, *96*, 3193.
122. Plüschau, M.; Kroll, G.; Dinse, K. P.; Beckert, D. *J. Phys. Chem.* **1992**, *96*, 8820.
123. Plüschau, M. *Thesis*, University of Dortmund, **1992**.
124. Geimer, J.; Brede, O.; Beckert, D. *Chem. Phys. Lett.* **1997**, *280*, 353.
125. Khudyakov, I. V.; Serebrennikov, Y. A.; Turro, N. J. *Chem. Rev.* **1993**, *93*, 537.
126. Sakaguchi, Y.; Hayashi, H.; I'Haya, Y. *J. Phys. Chem.* **1990**, *94*, 291.
127. Grissom, C. B. *Chem. Rev.* **1995**, *95*, 3.
128. Ohtani, M.; Ohkoshi, S.; Kajitani, M.; Akiyama, T.; Sugimori, A.; Yamauchi, S.; Ohba, Y.; Iwaizumi, M. *Inorg. Chem.* **1992**, *31*, 3873.
129. Ohkoshi, S.; Ohba, Y.; Iwaizumi, M.; Yamauchi, S.; Ohkoshi-Ohtani, M.; Tokuhisa, K.; Kajitani, M.; Akiyama, T.; Sugimori, A. *Inorg. Chem.* **1996**, *35*, 4569.
130. Kleverlaan, C. J.; Martino, D. M.; van Willigen, H.; Stufkens, D. J.; Oskam, A. *J. Phys. Chem.* **1996**, *100*, 18607.
131. Kleverlaan, C. J.; Martino, D. M.; van Slageren, J.; van Willigen, H.; Stufkens, D. J.; Oskam, A. *Appl. Magn. Reson.* **1998**, *15*, 203.
132. Kleverlaan, C. J.; Stufkens, D. J.; Clark, I. P.; George, M. W.; Turner, J. J.; Martino, D. M.; van Willigen, H.; Vlcek Jr., A. *J. Am. Chem. Soc.* **1998**, *120*, 10871 and references cited therein.
133. Shushin, A. I. *Chem. Phys. Lett.* **1990**, *170*, 78.
134. Norris, J. R.; Morris, A. L.; Thurnauer, M. C.; Tang, J. *J. Chem. Phys.* **1990**, *92*, 4239.

6

Photochemical Generation and Studies of Nitrenium Ions

Daniel E. Falvey
University of Maryland, College Park, Maryland

I. INTRODUCTION

This chapter describes recent contributions of photochemical methods to the study of nitrenium ions. This family of reactive intermediates includes any species with a positively charged, divalent nitrogen atom (RRN^+). Because the nitrogen atom has only six valence electrons, nitrenium ions are typically very strong electrophiles and tend to be very short-lived at room temperature in solution. Arylnitrenium ions (i.e., nitrenium ions which have a benzene or other aromatic ring directly attached to the nitrenium ion center) have received the most attention due to their postulated role in DNA damage mechanisms and cancer formation. This aspect of nitrenium ion chemistry has been reviewed elsewhere. Most of the biologically relevant reactions are thermal rather than photochemical. Consequently, this chapter will not focus on this aspect of nitrenium ion chemistry other than to note the contributions of photochemistry to these studies.

Photochemistry has long been recognized as an indispensible tool in the study of reactive intermediates. Indeed, significant progress in the study of carbenes, diradicals, and nitrenes that has been made in the past two decades can be traced to the application of laser flash photolysis and low-temperature matrix photochemistry to their generation and detection. Of course, these methods can be applied to a given reactive intermediate only when there are established and well-understood photochemical procedures for their generation. For example, it has been firmly established that carbenes are formed cleanly in the photolysis of diazo compounds or, alternatively, from diazirines [1–4]. Likewise diradicals are accessible through a variety of photochemical routes including the photochemical extrusion of CO from cyclic ketones [4]. Nitrenes are accessible through the photolysis of azides [5–7].

The study of nitrenium ions has long lagged behind the progress made on these neutral intermediates largely because the photochemical routes to the nitrenium ions had until recently been in the earliest stages of development. Photochemical generation of nitrenium ions presents several interesting challenges that were not faced in the generation of the neutral intermediates. One obstacle has been the challenge of designing a photochemical process that creates appropriately charged fragments. For the case of nitrenium ions this requires leaving a positive charge on a fairly electronegative nitrogen atom.

It will become apparent in the discussion below that many of the technical obstacles have been surmounted. There are now several well-characterized photochemical procedures for generating nitrenium ions. These methods have been used to unravel many of the mysteries surrounding these intermediates. Photochemical methods have been used to differentiate chemistry of the singlet and triplet states of nitrenium ions. Photochemical generation has also made it possible to apply laser flash photolysis (LFP) to the detection and kinetic characterization of nitrenium ions. Information from these experiments has provided the first insights into the lifetimes and reactivities of these intermediates.

II. ELECTRONIC STRUCTURES OF NITRENIUM IONS

A complete survey of the rapidly expanding theoretical and computational literature on nitrenium ions and related species is beyond the scope of this chapter. However, theory has been influential in the interpretation of nitrenium ion behavior. Therefore, it is appropriate to consider several key examples in detail.

The simplest nitrenium ion is the parent system, NH_2^+ (**1**). As a general rule, this ion and all higher nitrenium ions have a bent rather than a linear structure. Thus NH_2^+ has two nonbonding orbitals, designated here σ and p (Fig. 1) as well as two nonbonding electrons. There are four possible electronic configurations. In the singlet states the two electrons have antiparallel spins, allowing for the configurations $(\sigma)^2$, $(p)^2$, and (σp). Of these the $(\sigma)^2$ is considered to be the lowest in energy. The triplet state can only have the configuration σp. High-level, correlated calculations on the NH_2^+ system place the lowest singlet state $(\sigma)^2$ 30 kcal/mol higher in energy than the triplet [8–11]. This quantitative prediction was later verified by gas phase electron photodetachment spectroscopy experiments [12].

To date, **1** is the only nitrenium ion for which there is an experimentally verified singlet-triplet energy gap (ΔE_{st}). For the higher homologs, the ΔE_{st}'s are available only through various theoretical calculations. Fortunately, a number of

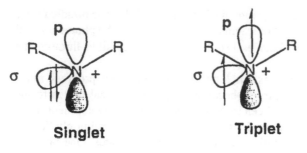

Singlet   **Triplet**

Figure 1 Electronic configurations of nitrenium ions.

trends in ΔE_{st} are seen consistently with every theoretical procedure that has been applied to nitrenium ions. Table 1 shows ΔE_{st} values for some key examples. Alkyl substitution, exemplified by methylnitrenium ion (CH_3NH^-, **2**) and the dimethylnitrenium ion ((CH_3)$_2N^+$, **3**), stabilizes the singlet state relative to the triplet and thus reduces ΔE_{st} [13,14]. Alkyl groups donate electron density into the nitrenium p orbital through hyperconjugation. This tends to increase the energy splitting between the two orbitals, favoring the closed-shell singlet state.

Aromatic substitution shows an even more pronounced stabilization of the singlet state [15–18]. For example, substitution of a benzene ring at the nitrenium center (e.g., **4**) [19,20] reverses the state ordering, making the singlet state lower in energy by -18.8 kcal/mol. As with the alkyls, donation from the filled π orbitals of the aromatic ring raise the energy of the out-of-plane p orbital relative to the in-plane σ, favoring electron pairing. The calculated geometry for singlet $PhNH^-$ indicates considerable alternation in the bond lengths of the benzene ring

Table 1 Theoretical Predictions of Nitrenium Ion Singlet–Triplet Energy Gaps (ΔE_{st} in kcal/mol) and Central Bond Angles (<R-N-R' in Degrees)

Nitrenium ion	ΔE_{st}	Singlet	Triplet	Method	Ref.
NH_2^+ (**1**)	+30.0	106.1	154.5	CASSCF	9
CH_3NH^- (**2**)	+7.8	112.1	150.4	BLYP/cc-pVDZ	13
(CH_3)$_2N^-$ (**3**)	+2.3	119.7	144.3	BLYP/cc-pVDZ	13
$PhNH^+$ (**4**)	−18.8	111.2	131.8	BPW91/cc-pVDZ	20
$PhNCH_3^-$ (**5**)	−14.1	123.7	150.7	BPW91/cc-pVDZ	24
Ph_2N^- (**6**)	−11.6	125.0	147.9	BPW91/cc-pVDZ	25
Ar_2N^+ (**7**)	+7.3	141.0	180.0	AM1	25
(Ar = 2,6-di-*tert*-butylphenyl)					

[15,21,22]. In other words, this species has a geometry that could be well described as 4-imine-2,5-cyclohexadienyl cation. This has been confirmed by some recent time-resolved infrared (IR) measurements [23].

It is interesting that substitution of an additional aromatic ring or alkyl group actually destabilizes the singlet relative to the triplet. This is seen with N-methyl-N-phenylnitrenium ion 5 [24] and diphenylnitrenium ion 6 [25], which have $\Delta E_{st} = -14.1$ kcal/mol and -11.6 kcal/mol, respectively. This can be attributed to steric interactions between the substituents which require a wider central bond angle. Increasing the bond angle imparts more p character to the σ nonbonding orbital and thus raises its energy, bringing it closer to the p nonbonding orbital. Reducing the energy difference between the two nitrenium nonbonding orbitals tends to favor the triplet state. This is further illustrated by bis(2,6-di-*tert*-butylphenyl)nitrenium ion 7 where the steric bulk of the ortho *tert*-butyl groups forces the central bond angle to be linear (at least in the triplet state) [25]. This creates a nitrenium ion with a triplet ground state.

In summary, the unsubstituted nitrenium ion NH_2^+ (1) is a ground-state triplet. The singlet–triplet energy gap can be controlled by the nature of the substituents. Aryl substitution and, to a lesser extent, alkyl substitution stabilize the singlet state through electron donation to the nonbonding p orbital on the nitrogen. In fact, to date all calculations indicate that typical arylnitrenium ions have singlet ground states. Bulky substituents tend to favor the triplet by forcing a more obtuse bond angle.

III. CHEMICAL BEHAVIOR OF NITRENIUM IONS: GENERAL TRENDS

There are few comprehensive reviews of nitrenium ion chemistry. Gassman's treatment is several decades old [26] but should be consulted by anyone interested in the history of the topic. The most recent reviews are by Abramovitch [27] and Simonova [28]. Other reviews have focused on more specific aspects of nitrenium ion chemistry. For example, Kadlubar [29] discusses the role of arylnitrenium ion intermediates in cellular DNA damage and carcinogenesis. McClelland [30–32] has written several accounts that are focused specifically on laser flash photolysis and kinetic studies of nitrenium ions. An account of singlet and triplet state nitrenium ion behavior has also appeared [33].

Most early investigations of nitrenium ions employed thermal generation of these species from neutral precursors. Under such conditions, the nitrenium ions are consumed much more rapidly than they are formed, and only steady-state methods such as decomposition kinetics and chemical trapping can be used to probe their properties. Extensive studies by the late Paul Gassman [34–39] showed that N-chloroamines and N-hydroxylamine esters generate nitrenium ions by polar solvents. One way in which this was confirmed was by substituent effects on the rates of solvolysis. Rate constants for the solvolysis various substituted *N*-chloroaniline derivatives (**8**, Scheme 1) were measured and analyzed using the Hammett relationship [37]. This revealed a strong dependence of these rates on the electronegativity of the substituents ($\rho = -6.35$) indicating that a positively charged intermediate (i.e., nitrenium ion **9**) is formed on solvolysis. Subsequent investigations in the laboratories of Boche [40–42] and Novak [43–47] have used similar experimental strategies to study the reactions of arylnitrenium ions that are believed to be involved in carcinogenic DNA damage.

Arylnitrenium ions have been the focus of most experimental studies because of the interest in DNA damage by activated nitrogenous mutagens. Therefore, much is known about the chemical behavior of this particular subset of nitrenium ions. In most cases, simple nucleophiles (i.e., water, alcohols, N_3^-, and halide ions) are added to the aromatic ring, giving products such as **10**, rather than being added to nitrogen [35,44,48,49]. This can be readily understood on the basis of the ab initio calculations, described in the previous section, which predict considerable charge delocalization from the nitrogen to the aromatic ring [20,21]. These nucleophiles are simply attracted to the sites of the highest positive charge density which are on the ortho and para ring carbon atoms.

While most nucleophiles add to the ring carbons, there are several cases where addition to the nitrogen atom is observed. Electron rich π systems, such as *N,N*-dimethylaniline [50,51], the DNA base guanosine [18,49,52–56], and glutathione have been observed to add to the nitrogen atom of various arylnitrenium ions. The reason why these nucleophiles add to the nitrenium center rather than the more charged sites is not clear at this time.

In the case of guanosine (**12**, Scheme 2), it has been postulated that the addition first involves complexation of the N-7 nitrogen lone pair with the empty orbital on the nitrenium center, producing an N-7–amino purinium ion intermediate, **13** [55]. This intermediate is thought to rearrange and deprotonate to give the observed product **15**. However, more recent experiments by Novak and McClelland have questioned this pathway. Novak showed that an C-8 adduct **14**

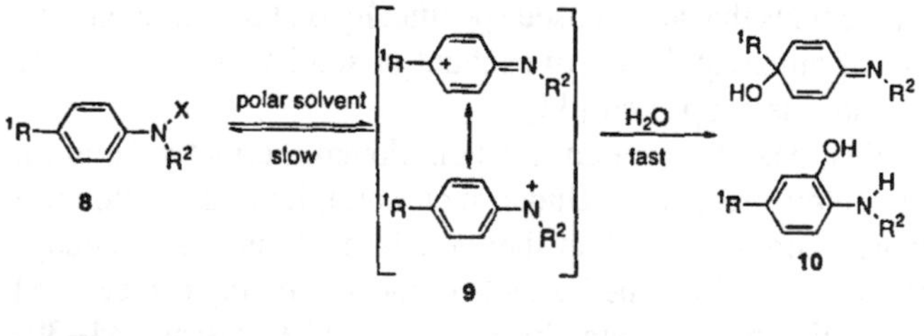

Scheme 1

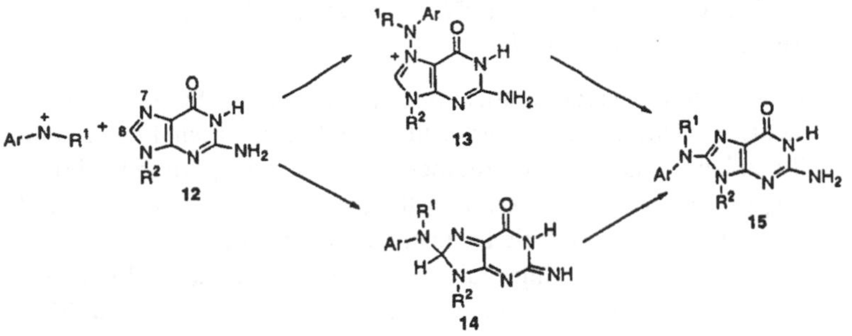

Scheme 2

could be detected when 8-methylguanosine was used as the trap [45]. McClelland's [31,57] laser flash photolysis study also provides evidence for direct formation of initial C-8 adduct **14**. It was also shown that the rate of initial addition does not correlate with the basicity of N-7 nitrogen, as would be expected if bond formation at N-7 were rate determining.

There is little that is known with certainty about the behavior of the alkylnitrenium ions. Their singlet states apparently undergo rapid 1,2 shift reactions of β-alkyl groups or hydrides resulting in an closed shell and presumably more stable iminium ion (e.g., **17** in Scheme 3). A problem that vexed both Gassman [26,34,36] and subsequent investigators [58] was distinguishing iminium ions that formed concertedly from those that originated from bona fide free nitrenium ions. In fact, even alkylaryl nitrenium ions **16** undergo this rearrangement [59,60]. However, the aromatic ring spreads the positive charge more diffusely throughout the ion, and as a result the rearrangement is slower and the intermediate nitrenium ions can be trapped through competing addition reactions.

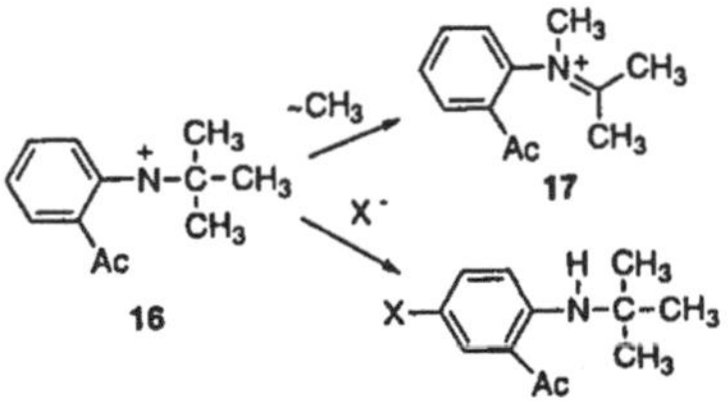

Scheme 3

IV. PHOTOCHEMICAL GENERATION OF NITRENIUM IONS:
GENERAL CONSIDERATIONS

There are five general strategies for photochemically generating charged intermediates (X⁻, Scheme 4). The first is to start with a neutral compound (X-Y) and
photochemically induce a heterolytic fragmentation of the appropriate bond.
Within this strategy, there are three possible pathways to consider: a) the desired
heterolytic fragmentation, b) heterolytic fragmentation in the opposite sense, and
c) homolytic fragmentation generating radicals, rather than the desired ion. It is
possible to cause the desired fragmentation to occur if the two fragments differ
sufficiently in electronegativity and the medium is sufficiently polar. Of course,
these are precisely the sorts of conditions required for thermal heterolysis. The
experimentalist is then confronted with the challenge of finding a system where

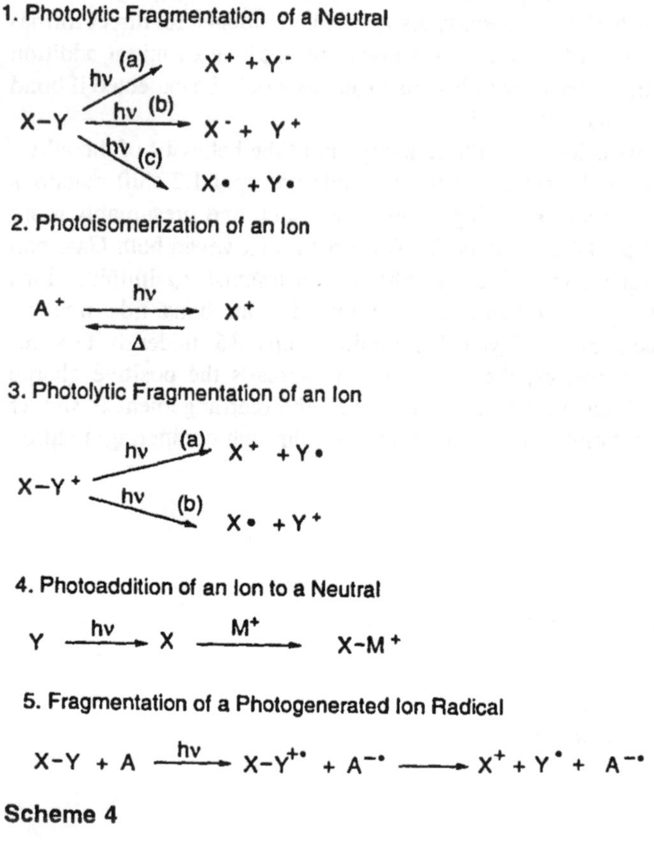

1. Photolytic Fragmentation of a Neutral

$$X-Y \xrightarrow{\substack{h\nu \ (a) \\ h\nu \ (b) \\ h\nu \ (c)}} \begin{array}{l} X^+ + Y^- \\ X^- + Y^+ \\ X\bullet + Y\bullet \end{array}$$

2. Photoisomerization of an Ion

$$A^+ \underset{\Delta}{\overset{h\nu}{\rightleftharpoons}} X^+$$

3. Photolytic Fragmentation of an Ion

$$X-Y^+ \xrightarrow{\substack{h\nu \ (a) \\ h\nu \ (b)}} \begin{array}{l} X^+ + Y\bullet \\ X\bullet + Y^+ \end{array}$$

4. Photoaddition of an Ion to a Neutral

$$Y \xrightarrow{h\nu} X \xrightarrow{M^+} X-M^+$$

5. Fragmentation of a Photogenerated Ion Radical

$$X-Y + A \xrightarrow{h\nu} X-Y^{+\bullet} + A^{-\bullet} \longrightarrow X^+ + Y^\bullet + A^{-\bullet}$$

Scheme 4

the prospective fragments are sufficiently polarized to allow for clean fragmentation but are not so polarized that they react prior to photolysis.

A second approach is to start with a photoreagent that is already appropriately charged (A^+) and create the desired intermediate (X^+) through photoisomerization. Since the overall charge is maintained, this circumvents the problem of charge separation inherent in the photoheterolysis strategy. However, it is almost inevitable that an intermediate generated through photoisomerization of a stable reactant will ultimately decay by reisomerizing back to ground-state reactant. The key issue becomes whether the barrier to reversion will be sufficiently high to allow for observation and trapping of the intermediate.

A third approach is to create the desired intermediate through a bond fragmentation of an appropriately charged starting material (X-Y^+). Such a reagent would be designed to expel a neutral leaving group upon photolysis giving the desired charged intermediate. As with the isomerization strategy, beginning with a charged starting material obviates the need to separate charges and such a route can be adapted to media of low polarity. Since the intermediate is created though a dissociation reaction, recombination to the starting material possesses a significant entropic barrier. Thus rapid reversal to the starting material should be less troublesome than in the isomerization approach. There does, however, remain the problem of electron apportionment in the initial dissociation. The charge may end up on the wrong fragment following photolysis.

Photoaddition reactions represent a fourth general strategy for creating a charged intermediate. In this case, photolysis of a neutral photoreagent (Y) creates either an excited state or a neutral intermediate (X) that in turn combines with a charged trap (M^+), creating the desired charged intermediate (X-M^+). The most straightforward implementation of this approach would be to protonate the species formed upon excitation ($M^+ = H^+$). However, in principle this could be adapted to any other charged trap.

Ion radical fragmentation represents a fifth potentially general route to ionic intermediates. Such fragmentations generally create a neutral radical and an ionic fragment. Methods for generating ion radicals via photo-induced electron transfer are well known. A number of fragmentation pathways for these high-energy ion radicals have been documented. Most of the recent work has emphasized the creation of interesting free-radical species through such processes [61]. The exploitation of such reactions for the creation of interesting ionic species has received little attention.

V. PHOTOCHEMICAL ROUTES TO NITRENIUM IONS

A. Photoisomerization of Anthranilium Ions

The first reported photochemical method for generating arylnitrenium ions was through photoisomerization of anthranilium ions **20** (also called 2,1-benzoisox-

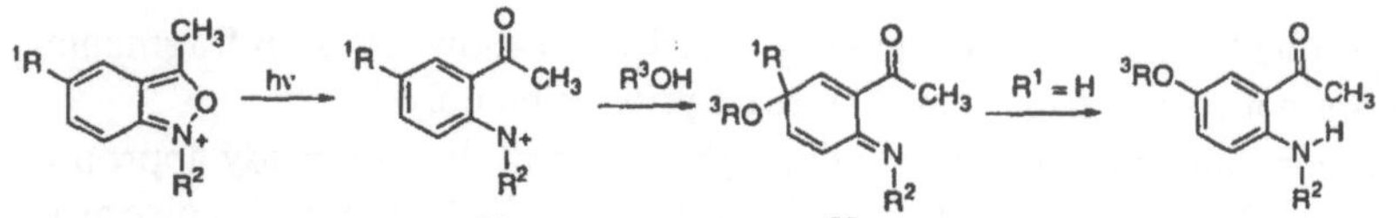

Scheme 5

azolium ions) [59]. As illustrated in Scheme 5, irradiation of these ions results in scission of the N–O bond and formation of an arylnitrenium ion **21**. This reaction was first reported by Haley who examined the N-alkylanthranilium ion derivatives. This study revealed a number of interesting decay pathways for aryl-nitrenium ions that were not apparent from the solvolytic studies. For example, it was found that water and methanol attack the benzene ring of the arylnitrenium ion giving substituted 2-acylaniline derivatives **22**.

Hansen et al. [62–64] as well and Giovannini and de Sousa [65] carried out extensive studies of anthranil **23** photochemistry in strongly acidic media (e.g., 98% H_2SO_4) (Scheme 6). Under these conditions, the anthranil is proton-ated, giving **24**. Consequently, photoisomerization leads to formation of a pri-mary nitrenium ion **25**. In accordance with Haley's observations, products **27** derived from addition of water or other nucleophiles to the ring were observed [64,65]. Attempts to sterically inhibit trapping at the positions ortho or para to the nitrenium center lead to products **29** resulting from addition to a 1,4-imine quinonemethide intermediate **28**. When photolysis was carried out in the presence of HBr, the corresponding reduction products **32** (the parent amine of the ni-trenium ion) were detected [66]. The latter were attributed to reactions of the triplet state nitrenium ion formed from heavy-atom-induced relaxation of the sin-glet. However, this mechanism assumes that the triplet state is the ground state of the nitrenium ion. This probably should be reevaluated in light of more recent

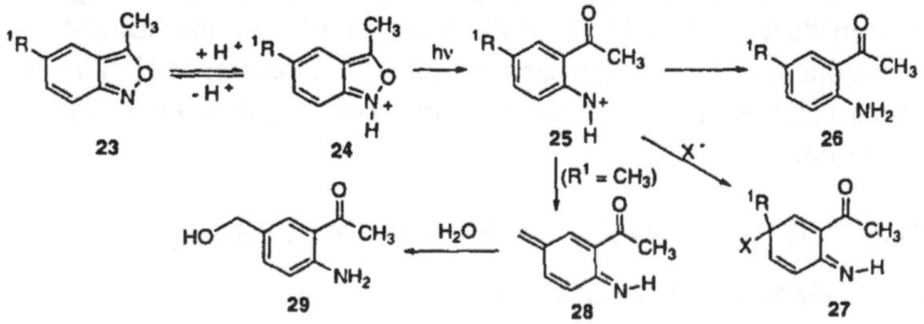

Scheme 6

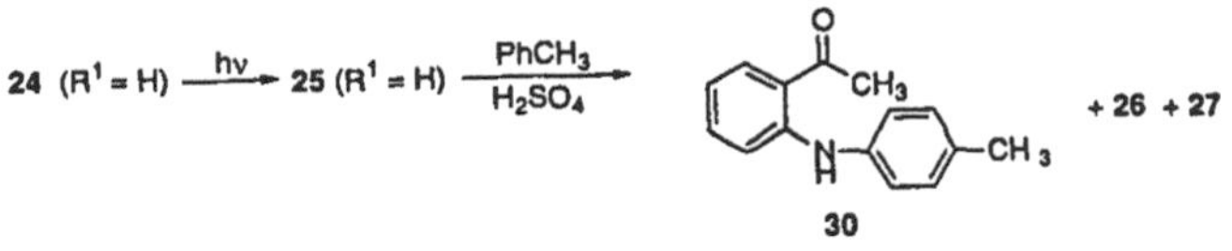

Scheme 7

ab initio calculations (Table 1), which suggest that arylnitrenium ions are more typically ground-state singlets. In fact, by analogy to the mechanism in Scheme 9 and the discussion below, the possibility that 32 forms from intersystem crossing in the excited state of the precursor ought to be considered. Photolysis of 2-indazoles gave an analogous N–N bond scission, in this case providing a 2-iminoarylnitrenium ion [62]. It was also shown in this series of studies that the singlet nitrenium ions could be trapped in modest yields with electron-rich aromatics. For example, photolysis of 24 in the presence of toluene and sulfuric acid gives the aromatic substitution product N-(2-acetylphenyl)-N-(4-tolyl)amine 30 as well as products arising from attack of water on the aromatic ring 27 (Scheme 7) [64].

Anderson et al. [60] also examined the photochemistry of N-tert-butyl-3-methylanthranilium ion 20 in CH$_3$CN (Scheme 8). Photolysis of this ion leads to three types of products: 1) the products of nucleophilic addition to the aryl ring on the nitrenium ion, 31; 2) the parent amine that results from a net two-electron reduction of the nitrenium ion, 32; and 3) a rearrangement product 17 resulting from a 1,2 shift of a methyl group from the tert-butyl substituent to the nitrenium nitrogen.

This work and subsequent investigations of various derivatives lead to the general mechanism depicted in Scheme 9. The key prediction of this mechanism is that there are two reactive states of the nitrenium ion: a singlet state ^{1}16 as well as a triplet state ^{3}16. These nitrenium ion spin states are distinct not only in their electronic structure but also in their characteristic products. The triplet reacts through sequential hydrogen atom abstractions to give the reduction prod-

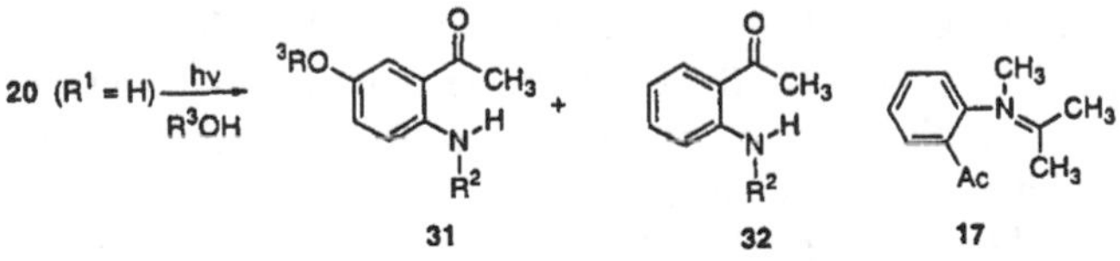

Scheme 8

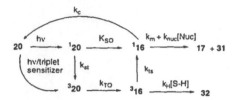

Scheme 9

uct (**32**). The singlet is responsible for the addition and rearrangement products **17** and **31**.

This spin-selective reactivity mechanism is supported by trapping studies using both triplet sensitization, as well as direct irradiations. These experiments were designed following the reasoning that direct irradiation would populate the singlet excited state of the precursor anthranilium ion ^{1}**20** and the latter would suffer N–O bond scission giving the nitrenium ion also in its singlet state ^{1}**16**. Likewise, triplet sensitization was expected to lead to the triplet excited state of the anthranilium ion ^{3}**20**, which in turn would open to give the same nitrenium ion in its triplet state ^{3}**16**.

The yields of both addition and rearrangement products **17** and **31** are highest under direct irradiation, whereas the yield of the reduction product **32** is highest under direct irradiation. The triplet state of the nitrenium ion decays by two pathways, H-atom abstraction from the solvent (k_H) to give (following a second H-atom abstraction) the parent amine **32**, and relaxation to the lower energy singlet state (k_{TS}). Thus, triplet-sensitized irradiation in a solvent such as CH_2Cl_2 (a relatively poor H-atom donor) gives significant yields of the singlet products **17** and **31**. However, triplet-sensitized irradiation in THF (a relatively strong H-atom donor) gives nearly quantitative yields of the parent amine **32**. In the latter case, the triplet is kinetically trapped prior to intersystem crossing. In the former case, H-atom transfer and intersystem crossing occur at comparable rates. The fact that the two modes of irradiation lead to different product distributions supports the assertion that there are at least two separate states responsible for the products.

Substitution of the anthranilium ion in the position para to the incipient nitrenium center leads to similar chemistry. The *para*-halogen, phenyl, and cyano substituents all lead to formation of ortho substitution products as well as the corresponding iminium ion derived from the 1,2-methyl shift [67–70]. In contrast, the methyl [68] and methoxy derivatives [70] are never observed to undergo this reaction. Apparently, in the latter cases the 1,2 shift takes place far more slowly than reaction with adventitious water.

Significant yields of the triplet product **32** are observed upon direct irradiation. This is attributed to a partitioning event that occurs in the excited state of the precursor anthranilium ion. Intersystem crossing from the singlet to the triplet competes (k_{ST}) with ring opening (k_{SO}) to give the singlet nitrenium ion. The ratio of these two rate constants, k_{SO}/k_{ST}, depends on the substituents present in the starting material. For the 5-Br [67] and 5-NO$_2$ [71]-derivatives this is relatively low, and significant amounts of the reduction product are observed. For the 5-CH$_3$ derivative the ratio is relatively high and virtually no reduction product is observed upon direct irradiation [68].

A 4-nitro-substituted arylnitrenium ion **33** was shown by compelling, albeit indirect, evidence to be a ground-state triplet [71]. This nitrenium ion shows qualitatively different behavior from that of other members of this series. For one thing, all attempts to trap this species with hydroxylic nucleophiles (CH$_3$OH, H$_2$O) were unsuccessful. The only two products detected were the reduction product **35** and the iminium ion **34** that results from the migration of a methyl group.

Direct photolysis of the anthranilium ion precursor to this nitrenium ion (**36**) gives relatively high yields of the triplet product **35** suggesting that the k_{isc}/k_{so} ratio is relatively high for this photoreagent. In fact, LFP experiments show a strong signal for the excited triplet state of the anthranilium ion immediately after photolysis. This signal decays in 300 ns, and, in the absence of trap, no other signals are detected.

When triphenylmethane (an H-atom donor) is added to the LFP reaction mixtures, a transient absorption signal corresponding the triphenylmethyl radical is observed to grow in after the decay of the excited triplet state of the anthranilium ion. Both qualitative inspection of the kinetic trace (Fig. 2) and quantitative kinetic analysis show that the growth of the trap radical does not coincide with the decay of the excited triplet state. Rather, there is a delay between the decay of the triplet and the onset of the growth of the trap radical, corresponding to the lifetime of the intermediate triplet nitrenium ion. A conservative estimate places the lifetime of the triplet nitrenium ion at about 2 μs in the absence of trap.

There are two possibilities that are consistent with these observations. 1) The triplet is the ground state. In this case, the 2-μs lifetime corresponds to decay of the triplet through ring closure and H-atom abstraction from the solvent

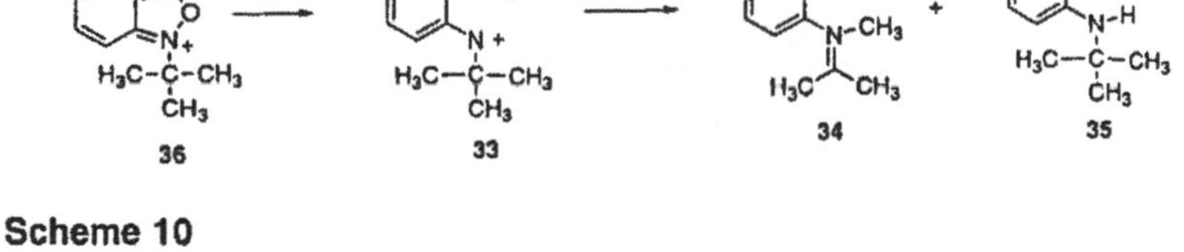

Scheme 10

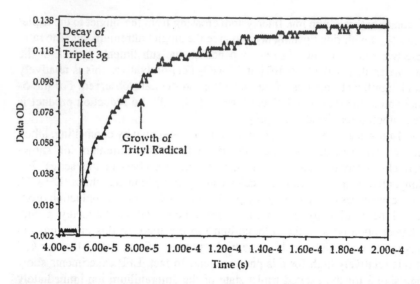

Figure 2 Transient absorption waveform from laser flash photolysis of **36** in the presence of 0.1 M Ph₃CH in CH₃CN. The wavelength is 380 nm where both the excited triplet state of **36** and the trityl radical absorb.

(CH₃CN). 2) The singlet is the ground state and the 2-μs lifetime corresponds to the sum of rate constants for singlet-to-triplet intersystem crossing as well as the aforementioned other decay pathways for the triplet. In this case, the rate constant for intersystem crossing from triplet to singlet nitrenium ion would be less than $5 \times 10^5 \text{ s}^{-1}$. This value is several orders of magnitude lower than that observed for the corresponding carbenes and therefore seems unreasonable. For this reason the first explanation is favored (Scheme 11).

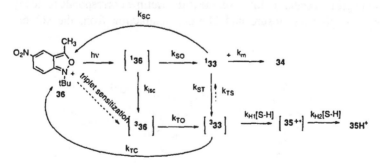

Scheme 11

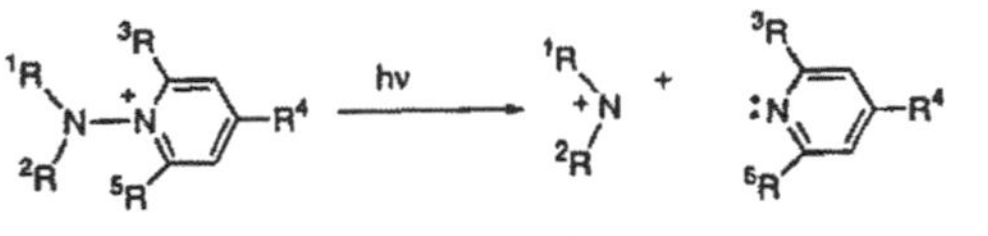

Scheme 12

B. Photofragmentation of *N*-Aminopyridinium Ions

While much was learned from the anthranilium ions, these substrates are obviously limited. Only arylnitrenium ions with a keto or formyl group in the position ortho to the nitrenium center can be generated with these substrates. Thus, we sought a photoreagent that would have broader applicability. Fortunately, Takeuchi [72–76] and Abramovitch [77,78] had shown that photolysis of *N*-aminopyridinium salts **37** can be photolyzed to give products from reactions of the corresponding nitrenium ion (Scheme 12). The nitrenium arises from N–N bond heterolysis giving the neutral pyridine **38** as a byproduct. The nitrenium ions thus generated are trapped by many of the same reactions outlined above. For example, the Takeuchi photolyzed several *N*-aminopyridinium salts and was able to trap the parent nitrenium ion **1** with various aromatic compounds (e.g., anisole, toluene) [74–76]. The yields of the trapped products **39** are modest–good and there is little selectivity between ortho, para, and meta adducts. These products were attributed to reactions of the singlet state nitrenium ion with the aromatic ring. Dimers of the aromatic trap (**40–42**) were attributed to triplet state H-atom abstraction processes. It was further proposed that the singlet state of the nitrenium ion could be stabilized through some unspecified interaction with the counterion or the nitrogen lone pair on the pyridine leaving group (Scheme 13).

Additional studies [79] of 1-amino-2,4,6-triphenylpyridinium ion (**37**, R^1, R^2 = H) photochemistry in the presence of toluene were largely in accord with the earlier study by Takeuchi. However, several interesting new findings emerged. First, it was found that methanol and water trapped the singlet nitrenium ion forming, respectively, methoxyamine and hydroxylamine (**46**). Second, a

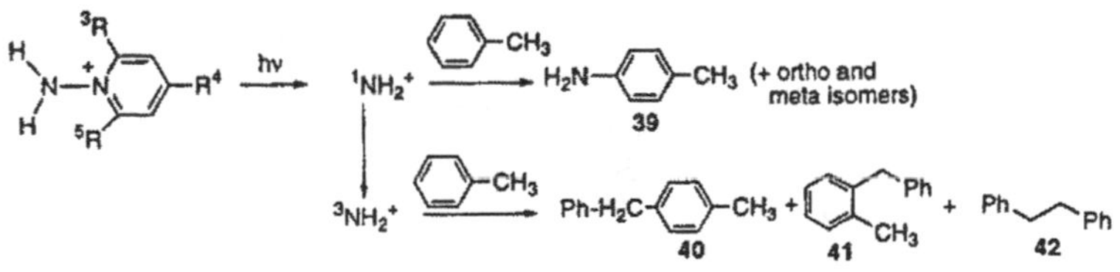

Scheme 13

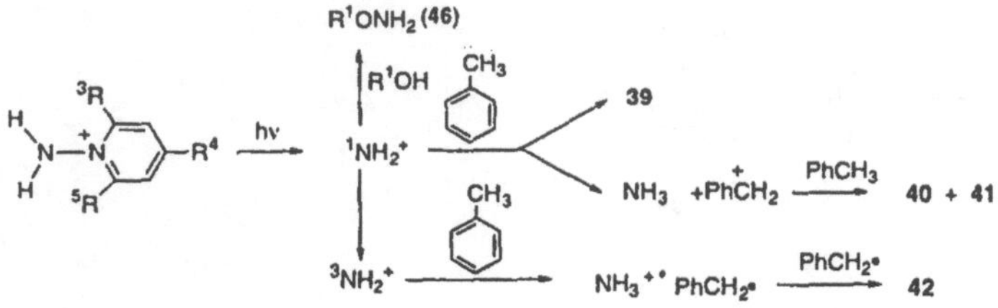

Scheme 14

closer investigation of the toluene dimers showed that only 1,2-diphenylethane **42**, the product of head-to-head dimerization of the benzyl radicals formed via the H-atom transfer pathway in Scheme 14. The benzyltoluene derivatives (**40** and **41**) actually originate from the *singlet* nitrenium ion, which abstracts a hydride from toluene. The resulting benzyl cation adds, in a Friedel–Crafts fashion, to an unreacted toluene.

The above mechanism is supported by two observations. 1) Triplet sensitized irradiation reduces the yield of benzyltoluenes (**40** and **41**) while it increases the yield of 1,2-diphenylethane (**42**). If both types of products had formed from a common intermediate, the ratios would be independent of the mode of generation. 2) Adding nonreactive diluents to the toluene also increases the yield of the triplet product (**42**). This is expected because the triplet state is the ground state and addition of the diluent allows the singlet state of **1** more time to relax to the lower energy triplet rather than abstract a hydride ion. This study also showed that the aromatic substitution products (**39**) were formed in accord with Takeuchi's results.

Abramovitch et al. have long had an interest in the synthetic utility of nitrenium ion chemistry and in a number of papers these workers have examined the trapping of nitrenium ions formed thermolytically [78,80–83]. Photolysis of 1-(*N*-phthalimido)-2,4,6-triphenylpyridinium tetrafluoroborate (**47**) in the presence of mesitylene (**49**) gave an adduct (**50**) of the corresponding nitrenium ion (**48**) with mesitylene in good yields (Scheme 15) [78]. Additional minor products

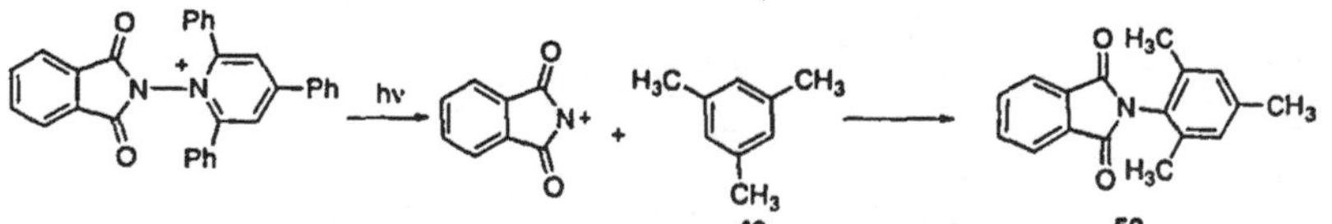

Scheme 15

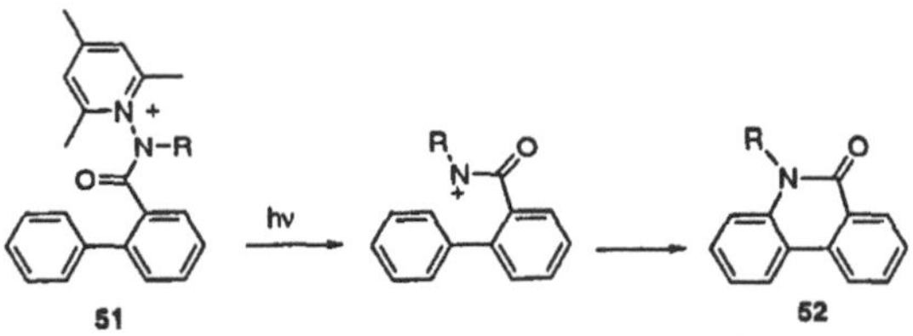

Scheme 16

were derived from a similar attack of the nitrenium ion on the 2,4,6-triphenylpyri-
dine leaving group.

Photolysis of an *N*-aroylaminopyridinium tetraflouroborate **51** gave an
aryl–acyl nitrenium ion having an ortho-phenyl substituent. The latter efficiently
traps the nitrenium ion at nitrogen giving a cyclic product **52** (Scheme 16) at
good yield [77]. Although the spin dependence of trapping was not investigated
in this case, the aromatic substitution process is likely to occur via a singlet
pathway by analogy to the mechanism in Scheme 13.

The aminopyridinium route was extended to the study of diarylnitrenium
ions. Moran and Falvey [84,85] investigated the photolysis of 1-(*N,N*-diphen-
ylamino)-2,4,6-trimethylpyridinium tetrafluoroborate (**55**) under various conditions (Scheme 17). In the absence of trapping reagents, a complex mixture of
poorly characterized products is observed, including the parent amine **56** and
carbazole **57**. Carbazole **57** is thought to arise from an electrocyclization of diphe-
nyl nitrenium ion **6**, similar to the Nazarov cyclization of carbocations. Formation
of **57** also requires a deprotonation and proton shift (Scheme 18).

Alcohols, water, and halide ions all trap **6** by the expected addition to the
ring carbons giving adducts such as **58** and **59**. The more reactive (vide infra)
chloride ion gives both para (**58**) as well as ortho adducts (**59**), whereas the less
reactive methanol gives para adducts with no detectable (<5%) products from
ortho addition. Addition of either nucleophile suppresses the yield of carbazole,
indicating that **57** as well as **58** and **59** arise from the same intermediate: the
singlet state of Ph_2N^+ [84].

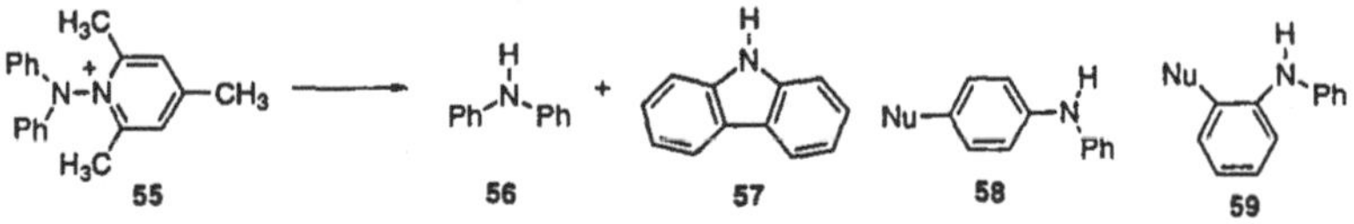

Scheme 17

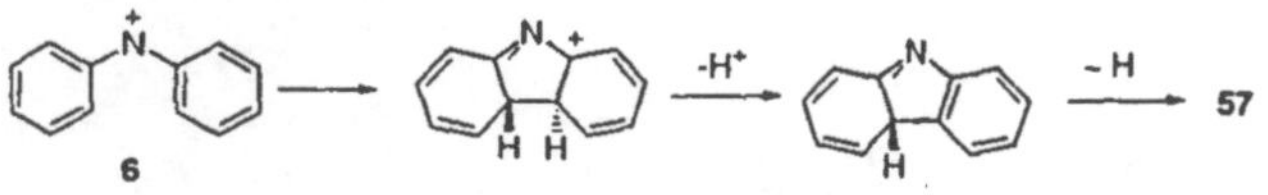

Scheme 18

Trapping of Ph_2N^- with electron-rich alkenes and arenes (i.e., π-nucleophiles) was also explored [85]. Simple alkenes trapped this nitrenium ion inefficiently, if at all. For example, a complex mixture, similar to that seen in the absence of traps, was found when Ph_2N^+ was photogenerated in the presence of 2,3-dihydropyran or β-methylstyrene. However, when similar experiments were carried out with various trimethylsiloxy-substituted alkenes (**61**, Scheme 19), the nitrenium ion **6** was successfully intercepted giving the set of products shown in Scheme 19. The major product, in all cases, was that from addition of the nucleophile to the ring position para to the nitrenium center (**62**). Among the two minor products are an *N*-phenylindole derivative (**63**) and a tertiary amine (**64**). Adduct **63** is apparently derived from an initial ortho attack of the alkene on the nitrenium ion's phenyl ring followed by attack of the nitrogen on the incipient α-siloxycarbocation concerted with deprotonation (Scheme 20). Loss of the elements of $(CH_3)_3SiOR$ gives indolones (**63**) or indoles (**65**) from the silyl ketene acetals and silylenol ethers, respectively.

The factors responsible for these differences in regiochemistry are not clear at this time. However, several can be ruled out. First, frontier molecular orbital coefficients do not seem to provide any guidance for predicting the relative rates or regiochemistry of addition. DFT calculations indicate that LUMO coefficients are similar at the nitrogen as well as the ortho and para ring positions [85]. Second, steric effects alone cannot account for the differences. In fact, 1-trimethylsilyloxy-1-methoxy-2-methylpropene gives the highest proportion of N-addition products, yet it is arguably the most sterically encumbered of the alkenes exam-

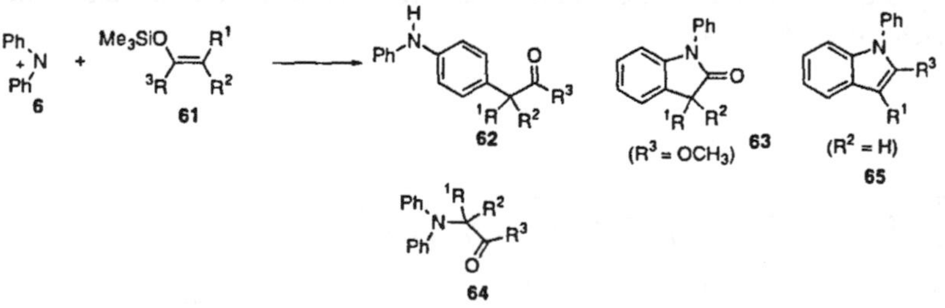

Scheme 19

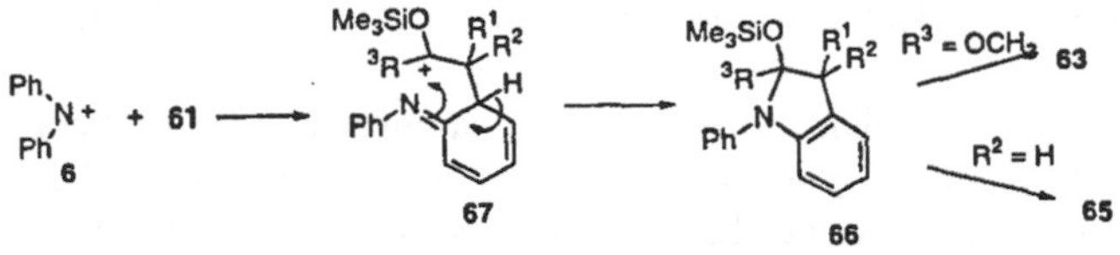

Scheme 20

ined. In contrast, 2-trimethylsilyloxypropene gives no detectable N adducts, despite its lack of steric encumbrance. Further studies of the regiochemistry of addition to arylnitrenium ions, especially for the π nucleophiles, will be necessary to obtain a clear predictive model for these reactions.

C. Photoheterolysis of Hydroxylamine Esters and N-Chloroamines

One of the more straightforward routes to arylnitrenium ions involves heterolysis of the same precursors that have been successfully used in the thermal (i.e., ground-state) generation of these species. Davidse et al. [86] examined the photochemistry of the N-chloro, N-sulfanoyloxy, and N-pivaloyloxy acetanilide derivatives (68) shown in Scheme 21. These compounds undergo heterolysis slowly under thermal conditions and more rapidly under photolysis. It was shown that similar, but not identical, product distributions were obtained from photochemistry and through thermolysis. The major difference between the two routes is that photolysis produces significant amounts of the parent amine (69) along with products derived from nucleophilic attack on the aromatic ring (70). Parent amine 69 was not detected in the thermal experiments.

The origin of the parent amine 69 in these experiments was not investigated in detail. However, two plausible mechanisms have been put forward. Photohomolysis might occur in competition with photoheterolysis. This would be consistent with the observation that nitrenium ion generation (as judged by its appear-

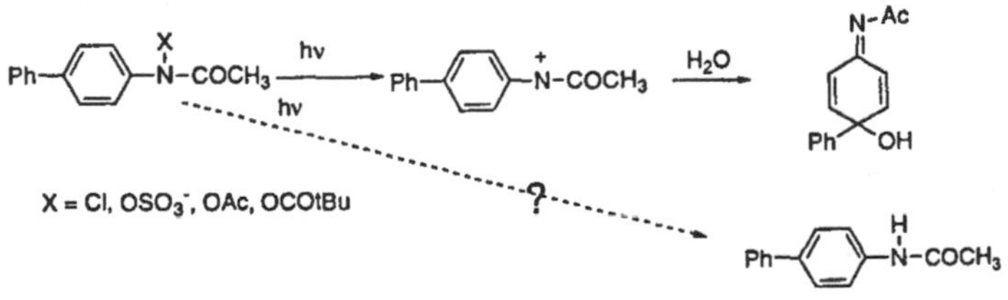

Scheme 21

ance in laser flash photolysis experiments) occurs in water, but not in the less polar solvent, acetonitrile.

An alternative possibility is that formation of the parent amine may reflect excited-state intersystem crossing. By analogy with the anthranilium ion mechanism outlined in Scheme 9, it could be that the *N*-chloro derivative suffers an excited-state intersystem crossing event competitively with direct heterolysis to form the singlet nitrenium ion. The triplet state photoprecursor would eliminate chloride to generate the triplet nitrenium ion.

Perhaps the instability of these precursors and the delicate balance of conditions required for photogeneration has discouraged further investigations into these photolysis mechanisms. Picosecond laser studies in particular would be very valuable in sorting out the details of the primary photoprocesses.

D. Nitrenium Ions Through Addition Reactions: Protonation of Nitrenes Generated from Azide Photolysis

The acid-catalyzed thermal decomposition of aryl azides has been known for some time [27,87]. Product trapping data indicate that the initial decomposition leads to an arylnitrenium ion intermediate though a net replacement of N_2 with H^+ (Scheme 22). Doppler et al. [63] compared the photolysis of 2-azidoacetphenone (75) in sulfuric acid with added toluene with to that of 3-methylanthranil under the same conditions. The same products, specifically the those resulting from attack of the nitrenium ion arene (76) as well as those from addition of water to the nitrenium ion (30), were derived from both experiments. This lead them to conclude that both processes proceed through a common intermediate, namely, the primary nitrenium ion (25) (Scheme 23).

McClelland et al. [31,32,57,88–90] report comprehensive LFP and product studies of arylnitrenium ions formed from photolysis of azides in protic media. Scheme 24 summarizes their general conclusions. It was found that the singlet nitrenes ^{1}79 were rapidly protonated both by acids as well as by water itself. The protonation reaction leading to 82 occurs competitively with ring expansion of singlet nitrene and with intersystem crossing to give the triplet state nitrene ^{3}79. Carrying out the photolysis in acetonitrile (containing trace amounts of water) gives products characteristics of the neutral phenylnitrene, namely, azepinone 80

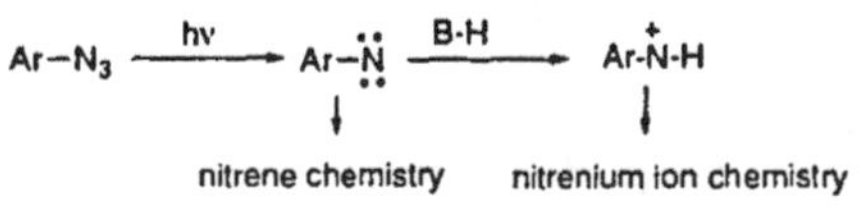

$$Ar-N_3 \xrightarrow{h\nu} Ar-\ddot{N} \xrightarrow{B-H} Ar-\overset{+}{N}-H$$

nitrene chemistry   nitrenium ion chemistry

Scheme 22

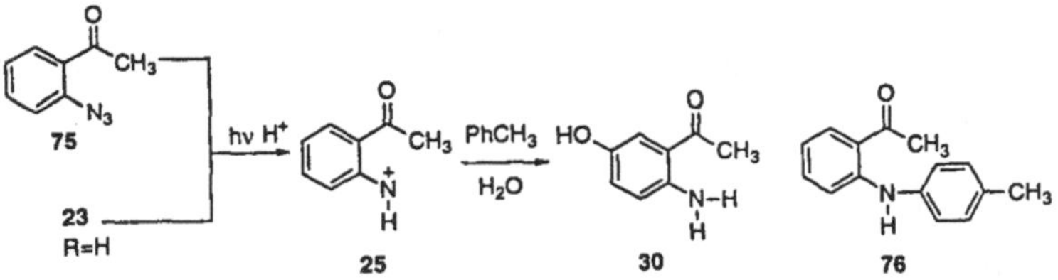

Scheme 23

and aniline **81**. Addition of increasing amounts of acid acts to suppress the yields of the nitrene products (such as **80** and **81**) and increase the yield of products derived from the singlet nitrenium ion, such as **83**.

Similar results were obtained with 4-biphenylyl azide (**78**, R = Ph). Under aprotic conditions products characteristic of 4-biphenylynitrene were detected. Under protic conditions nitrenium ion derived products (4-hydroxy-4-phenyl-2,5-cyclohexadienone from hydrolysis of the corresponding quinol imine, **77**, Scheme 24) [32,88,89] were detected. It is interesting to note that protonation of this nitrene does not require a strongly acidic medium. The nitrenium ion formed in neutral aqueous solution and even in the presence of hydroxide. For example, addition of 0.5 M NaOH diminishes the formation of nitrenium products by only about 75% relative to pH 7 solutions.

The apparent strong basicity of the arylnitrenes can be traced to an unusual electronic spin barrier to proton transfer, illustrated in Fig. 3. Both the nitrene and nitrenium ion possess two low-energy spin states: a singlet and a triplet. As described above, phenylnitrenium ion is a ground state singlet with the triplet state lying approximately 18 kcal/mol higher in energy [19,20]. In the case of

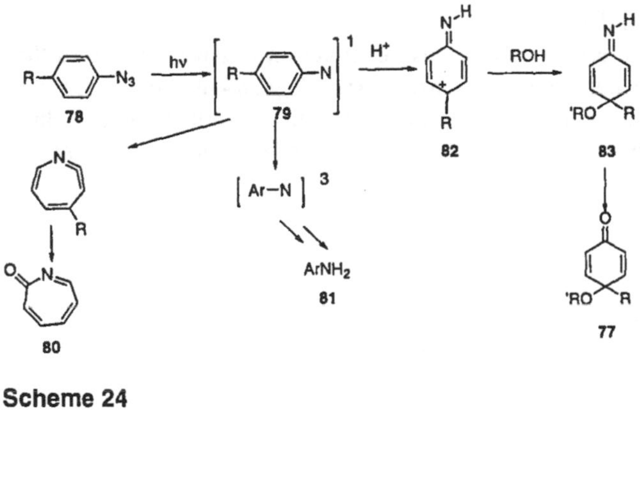

Scheme 24

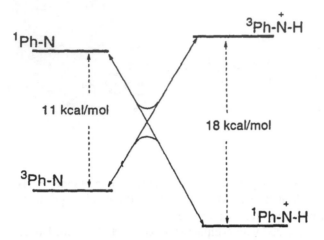

Figure 3 State energy diagram for the protonation of singlet and triplet phenylnitrene to give singlet and triplet phenylnitrenium ion.

the phenylnitrene, the state ordering is reversed with the singlet state now 11 kcal/mol above the triplet [7,91,92]. Rapid protonation of singlet phenylnitrene (the excited state) by water occurs adiabatically to give the singlet (ground-state) arylnitrenium ion.

It is interesting to consider how the reverse reaction would proceed. Would it proceed adiabatically to the singlet state of the nitrenium ion, or would there be a curve crossing to the triplet surface, giving the lower energy triplet nitrene? Likewise, is it possible to protonate the triplet nitrene through a nonadiabatic pathway? Clearly, the rapid decomposition pathways for all four states would make this a very difficult experiment with simple arylnitrenium ions. However, it may become possible to investigate such spin-controlled acid–base chemistry using highly stabilized derivatives.

It was found that arylnitrenium ions generated in this way can be further protonated to form the nitrenium dications, RNH_2^{2+} (e.g., **84** in Scheme 25) [89]. This finding was based on the observation that the rate constants for nucleophilic attack on the nitrenium ion increase with added acid and show saturation behavior

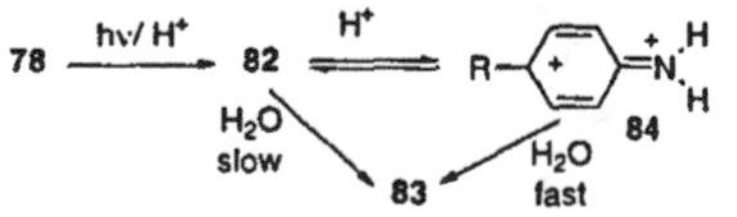

Scheme 25

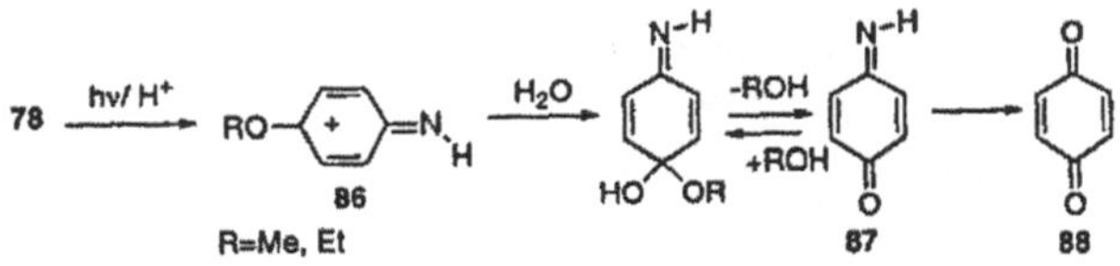

Scheme 26

at very high acid concentrations where most of the nitrenium ions would be protonated. The pK$_a$ values for the protonated biphenylyl nitrenium ion and protonated 2-fluorenylnitrenium were determined to fall in the range 0–1. This explains why it has not been possible to detect simple arylnitrenium ions under superacid conditions: the acidity required to stabilize them to nucleophilic attack is sufficient to protonate them.

The nitrene protonation studies have lately been expanded to include the several derivatives including 4-methoxy-and 4-ethoxyphenylnitrenium ions (**86**). The fate of these in aqueous solution was shown to be hydrolysis, first to benzoquinoneimine (**87**) and then to 1,4-benzoquinone itself (**88**, Scheme 26) [90,93].

Concurrently with McClelland's work, Platz et al. [94] discovered that photolysis of polyfluorinated phenyl azides in acidic media also produced the corresponding nitrenium ions. In this case strong acids, such as H_2SO_4, were required to protonate the corresponding nitrenes. In this report the nitrenium ions were characterized primarily by their kinetic and spectroscopic behavior. Likewise in the course of an investigation of (2,4,6-tribromophenyl)nitrene, Wirz et al. [95] detected a species formed in protic media which was attributed to the (2,4,6-tribromophenyl)nitrenium ion. Under aqueous conditions, this species eventually reacts to produce 2,6-dibromobenzoquinone. This presumably occurs through a mechanism analogous to that of the 4-alkoxyphenylnitrenium ions.

VI. SPECTROSCOPIC AND KINETIC PROPERTIES OF NITRENIUM IONS

Nitrenium ions are typically reactive and short-lived. For this reason, their direct detection has generally required flash photolysis (LFP) methods. However several highly stabilized nitrenium ions have been characterized spectroscopically without the use of fast kinetic techniques. Early electrochemical studies by Svanholm and Parker [96] as well as Serve [97,98] showed that various diarylnitrenium ions could be generated through anodic oxidation (CH_3CN, Pt electrode) of a diarylamine in the presence of a moderately strong base such as lutidine (Scheme 27). This causes a formal loss of two electrons and one proton from the amine, giving persistent diarylnitrenium ions. Some of these were sufficiently stable to

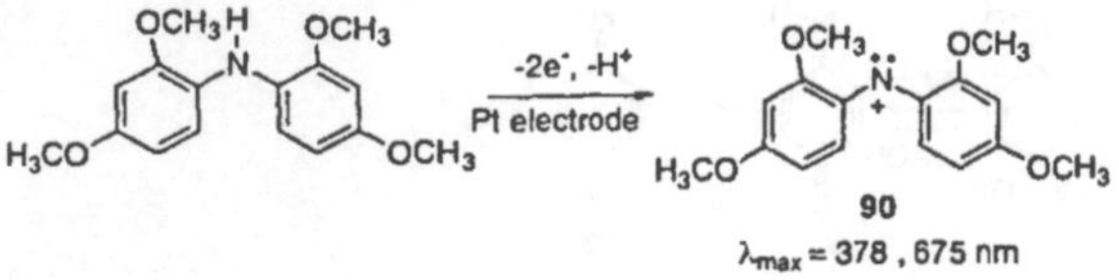

90
$\lambda_{max} = 378, 675$ nm

Scheme 27

be characterized by steady-state UV-Vis absorption spectroscopy. Bis(2,4-di-methoxyphenyl)nitrenium ion (**90**), for example, shows two strong absorption bands, one at 378 nm (log $\varepsilon = 4.27$) and the other at 675 nm (log $\varepsilon = 4.58$) [97].

Following similar procedures Rieker and Speiser [99,100] generated the 4'-dimethylamino-4-(2,6-di-*tert*-butyl)biphenylylnitrenium ion (**91**) through anodic oxidation of its parent amine. This species is sufficiently persistent to permit characterization by UV-Vis spectroscopy ($\lambda_{max} = 502$ nm) as well as ^{1}H NMR. Addition of HClO$_4$ protonates the nitrenium ion, giving the dication (**92**) (Scheme 28).

Recently, Boche et al. [101] argued that 1,2,3-triazolium salts (e.g., **93**), stable compounds which have been known for decades [102], can be considered to be nitrenium ions on the basis of resonance hybrid **93b**, which assigns a positive charge to the central nitrogen. The spectrum of the dimethylbenzotriazolium ion (**93**) in acetonitrile is typical for heterocyclic molecules; this compound shows a transition in the UV-B region of the spectrum, having $\lambda_{max} = 280$ nm [103] (Scheme 29).

This raises an interesting question about the nature of nitrenium ions: when is the positive charge so delocalized from a divalent nitrogen atom that the species in question cannot justifiably be termed a nitrenium ion? A rigorous definition of nitrenium ion would probably only apply to NH$_2^+$ itself. Most of the theoretical and experimental evidence suggests that arylnitrenium ions are closer to 4-imino-

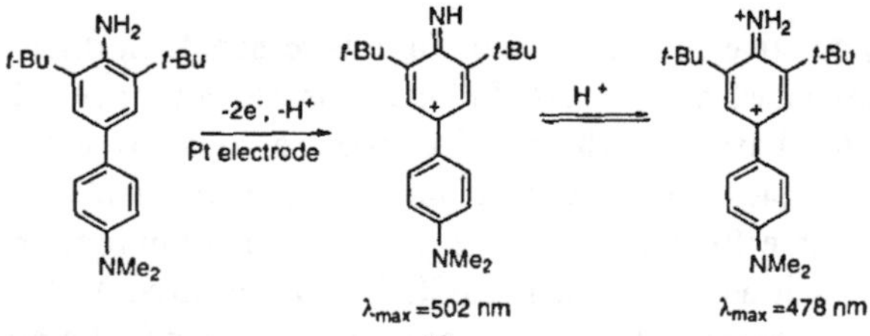

91
$\lambda_{max} = 502$ nm

92
$\lambda_{max} = 478$ nm

Scheme 28

cyclohexadienyl cations in terms of their structure and charge distributions [16,17,20,21,25]. The significant stabilization of the singlet state by simple alkyl groups (Table 1) shows that even these ligands act to delocalize charge through hyperconjugation [13]. Thus, one can only view this definition as a matter of degree of nitrenium ion character. In any case, it is clear that the triazolium species are so stable and charge-delocalized that studies of these species will not provide any meaningful information regarding the behavior of the more reactive species such as PhNH$^+$.

Aside from the aforementioned examples, spectroscopic and kinetic characterization of nitrenium ions requires LFP or other fast kinetic methods. In the LFP experiment, a stable photochemical precursor is excited with light from a pulsed laser. The laser pulse initiates a photochemical process that creates the short-lived intermediate (e.g., the nitrenium ion). The latter is then detected via its UV-Vis absorption spectrum using a probe beam from a continuous lamp and fast detector.

In principle, the LFP technique can be applied to the detection of any sort of reactive intermediate. There are, however, several limitations to the sort of species that can be examined in this way. First, the intermediate of interest must be accessible through a photochemical process which generates it from a stable precursor, rapidly and with a high quantum efficiency. Second, the intermediate must live longer than the time resolution of the instrument (generally this is determined by the time duration of the laser pulse). Third, the intermediate must have a strong absorption, distinct from that of the photochemical precursor.

Compiled in Table 2 are the absorption maxima (λ_{max}), solution lifetimes (τ), and representative rate constants for the reactions of the nitrenium ion with trapping agents (k_{trap}). Also indicated is the method used for photogeneration: azide photolysis under protic conditions (A), photolysis of N-aminopyridinium ion derivatives (P), and photochemical rearrangement of anthranilium ions (R).

When considering the LFP data, the reader should understand the limitations and potential distortions inherent in these sorts of measurements. The LFP experiment provides difference spectra where the positive signal due to the intermediate is combined with a negative signal due to depletion of the precursor. The change in optical density (ΔOD) observed in the LFP experiment is determined by the concentration change induced by the laser pulse (Δc), the molar absorptivity

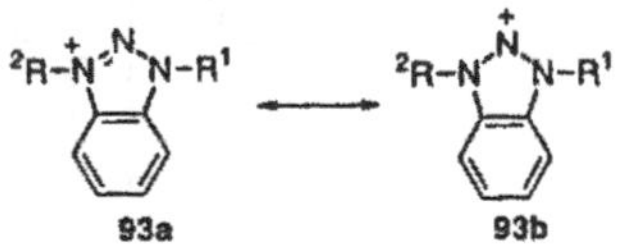

Scheme 29

Table 2 Lifetimes (τ) UV-Vis Absorption Maxima (λ_{max}) and Rate Constants for Trapping Reactions (k_{trap}) for Various Nitrenium Ions Determined by LFP

Ion		Route[a]	τ(μs)/conditions	λ_{max} (nm)	k_{trap} (M^{-1} s^{-1})/trap	Ref.
101	(X = Cl)	R	0.14/CH$_3$CN	385	3.2 × 10^6/H$_2$O	67
102	(X = Br)	R	0.13/CH$_3$CN	395	1.3 × 10^6/H$_2$O	67
103	(X = CH$_3$)	R	0.40/CH$_3$CN	460	3.1 × 10^7/H$_2$O	69
104	1f (X = Ph)	R	30/CH$_3$CN	470	8.5 × 10^5/Br$^-$ 6.2 × 10^7/H$_2$O	69
105	1e (X = OCH$_3$)	R	600/CH$_3$CN	500	2.1 × 10^7/H$_2$O	69
106	1g (X = NO$_2$)	R	2/CH$_3$CN	n.d.		71
107	1h (X = CN)	R	<0.1/CH$_3$CN	n.d.		69
108	1a (X = H)	R	<0.1/CH$_3$CN	n.d.		69
109		S	0.16/H$_2$O	450	5.1 × 10^9/N$_3^-$	86
110		S	13/H$_2$O	450	4.2 × 10^9/N$_3^-$	86
111		A	435,000/H$_2$O	430		107
112		A	0.2/H$_2$O/MeCN	420,600		95

[a] Indicates photochemical route for generation of the nitrenium ion: R = photoisomerization of anthranilium ion (Scheme 5); S = photosolvolysis (Scheme 21); A = protonation of nitrene generated from azide photolysis (Scheme 22); P = photofragmentation of pyridinium ion (Scheme 12).

Table 2 Continued

Ion		Route[a]	$\tau(\mu s)$/conditions	λ_{max} (nm)	k_{trap} (M^{-1} s^{-1})/trap	Ref.
113	R = OCH$_3$	A	0.53/H$_2$O	300	2.9 × 10^9/N$_3$$^-$	93
114	R = OEt	A	0.8/H$_2$O	300	3.0 × 10^9/N$_3$$^-$	93
115	R = OPh	A	0.05/H$_2$O	295, 375, 425		93
116	R = F	A	>1000/CH$_3$CN	375		94
117	R = CO$_2$CH$_3$	A	> 1000/CH$_3$CN	375		94
118	R = CH$_2$Br	A	> 1000/CH$_3$CN	380		94
119	R = CN	A	> 1000/CH$_3$CN	365		94
120	R = CH$_2$OH	A	> 1000/CH$_3$CN	380		94
121	R = CH$_2$N(CH$_3$)$_2$	A	> 1000/CH$_3$CN	370		94
122	Y = H	A	0.35/H$_2$O	465	9.6 × 10^9/N$_3$$^-$	106
123	Y = N(Ac)CH$_3$	A	26/H$_2$O	380, 560		106
124	Y = 4'-MeO	A	633/H$_2$O	500		106
125	Y = 4'-CH$_3$	A	3.7/H$_2$O	490		106
126	Y = 4'-F	A	0.76/H$_2$O	470		106
127	Y = 3'-CH$_3$	A	0.67/H$_2$O	470		106
128	Y = 3'-CH$_3$O	A	0.41/H$_2$O	475		106

No.	Structure	Route	Value 1	Value 2	Value 3	Ref.
129	Y = 4'-Cl	A	0.40/H_2O	480		106
130	Y = 3'-Cl	A	0.08/H_2O	475		106
131	Y = 4'-CF_3	A	0.03/H_2O	—		106
132	(fluorenyl-N^H)	A	75/H_2O	460	3×10^4/H_2O	89
133	(imidazolium, $N-CH_3$)	A	100,000/H_2O	235		109
134	(naphthyl $\overset{+}{N}:CH_3$)	P	0.21/CH_3CN	475	1.2×10^9/MeOH	110
135	(H_3CO–$\overset{+}{N}:CH_3$)	P	8.7/CH_3CN	330		110
136	(Cl–$\overset{+}{N}:CH_3$)	P	0.90/CH_3CN	350	1.9×10^7/H_2O	110
137	(biphenylyl $\overset{+}{N}:CH_3$)	P	0.59/H_2O (95%) 24/CH_3CN	460	2.4×10^{10}/Cl^- 9.8×10^4/H_2O	104
138	(X–$\overset{+}{N}:$–X)	P	1.2/CH_3CN	425, 680	6.1×10^5/H_2O	84
139	X = H / X = CH_3	P	280/CH_3CN	425, 690	1.9×10^7/H_2O	108
140	X = Br	P	10/CH_3CN	450, 690		108

[a] Indicates photochemical route for generation of the nitrenium ion: R = photoisomerization of anthranilium ion (Scheme 5); S = photosolvolysis (Scheme 21); A = protonation of nitrene generated from azide photolysis (Scheme 22); P = photofragmentation of pyridinium ion (Scheme 12).

of the reactive intermediate (ε_{int}) and the molar absorptivity of the photoprecursor (ε_{pre}) [Eq (1)]. This can distort and in some cases null out the spectrum of the intermediate in situations where it and the photoprecursor have overlapping absorption bands. Thus the observed λ_{max} might differ from the true λ_{max} of the isolated species.

$$\Delta OD(\lambda) = \Delta c[\varepsilon_{int}(\lambda) - \varepsilon_{pre}(\lambda)] \tag{1}$$

While the τ values provide a rough indication of the stability of the nitrenium ion in question, these can be highly sensitive to changes in the medium and even trace impurities in the solvent. For example, the lifetime of N-methyl-N-(4-biphenylyl)nitrenium ion (135, in Table 2) is 24 μs in CH_3CN, but only 590 ns in H_2O [104]. Thus, a comparison between τ values is meaningful only when they are measured under identical conditions.

The third parameter, k_{trap}, is the bimolecular rate constant measured for the reactions of the nitrenium ions with various nucleophiles. This is derived from the dependence of the nitrenium ion pseudo-first-order decay rate constant (k_{obs}) at various concentrations of nucleophile [trap (Eq. (2)].

$$k_{obs} = \frac{1}{\tau} + k_{trap}[trap] \tag{2}$$

As a general rule, all but the most stable nitrenium ions react with anionic nucleophiles at the diffusion limited rate. In aqueous solution this rate constant is $k_{diff}(H_2O) = 5 \times 10^9 \text{ M}^{-1} \text{ s}^{-1}$. Acetonitrile is less viscous and the diffusion-limited rate constant $k_{diff}(CH_3 CN)$ is 1.9×10^{10}.

The simplest monoarylnitrenium ion, $PhNH^+$, has not yet been characterized by direct detection. Fishbein and McClelland [105] have estimated its lifetime at 0.1 ns in aqueous solution. This was done by a competition experiment where azide ion (N_3^-) was assumed to react at the diffusion limit. This low lifetime, coupled with an absorption band predicted to be at low wavelengths (cf. 113), is what has probably precluded its LFP characterization.

The only monoarylnitrenium ions that have been detected possess substituents in the 4-position of the phenyl ring. These include 101–105, which were generated by photolysis of anthranilium ions (Scheme 5); 112–115, which were generated through azide photolysis in protic media (Scheme 22); and 135–136, which were generated via photolysis of N-aminopyridinium derivatives (Scheme 12).

First we consider the simplest of the monoaryls, 113, 114, 135, and 136. These each show a single absorption maximum in the UV region. 4-Methoxynitrenium ion (113), for example, absorbs at 300 nm [90,93]. The effect of N-methylation (135) is to shift this absorption band to the red by about 30 nm [104]. As might be expected, changing the 4-methoxy group to a 4-ethoxy group (114) has no discernible effect on absorption [90]. The effect of 4-chlorination (136) is to shift the absorption band to yet higher wavelengths (350 nm) relative to the 4-methoxy group [104]. Resonance electron donation from the alkoxy groups stabilizes the nitrenium ion relative to the 4-chloro derivative. For example, the lifetime of 135 is 8.7 μs compared with 900 ns for its 4-chloro analog (136).

The additional conjugation of 4-phenoxy group (115) creates several additional bands. In addition to the UV peak at 295 nm, there are peaks at 375 nm and 425 nm [93]. The additional phenyl ring attenuates electron donation from the oxygen nonbonding pair. This diminishes the stability of the nitrenium ion by a factor of approximately 10 relative to the alkoxy species 113 and 114.

Multiple halogenation of the phenyl ring shows some fascinating effects on the lifetime and absorption spectrum. Consider first the series of 2,3,5,6-tetra-fluorophenylnitrenium ions studies by the Platz group (116–121) [94]. These species have absorption maxima between 360 and 380 nm, significantly to the red of their 4-alkoxy nonfluorinated analogs 113 and 114. More interestingly, these species are extremely stable in the CH_3CN/H_2SO_4 media where they are generated. In fact, their lifetimes are over 1 ms—too long to be accurately measured by LFP. Furthermore, the initial study found them to be unreactive to a number of simple nucleophiles. This high stability is attributed to the resonance electron donation by the nonbonding electron pairs on the fluorine atoms. This is expected to become especially important in cases where the phenyl ring is substituted with a strong-resonance electron-withdrawing group, such as the nitrenium center. Further examination of this interesting effect is clearly warranted.

The 2,4,6-tribromophenylnitrenium ion 112 also shows some unusual properties [95]. Unlike the other simple monoaryl nitrenium ions, 112 shows two visible absorption bands (420 and 600 nm). As with the fluorinated series, one might attribute this to the π donation of the halogen nonbonding pairs. However, this effect does not stabilize the tribromo nearly as much as tetrafluoro substitution. Nitrenium ion 112 lives for only about 200 ns in CH_3CN/H_2O.

The anthranilium ion-derived arylnitrenium ions (101–105) all have an acetyl group in the 2-position of the phenyl ring. This appears to cause a slight red shift in the absorption maxima of the nitrenium ions and in some cases gives rise to an additional, weaker band at higher wavelengths. The transient absorption spectrum of N-tert-butyl-N-(2-acetyl-4-bromo)nitrenium ion (102) is shown in Fig. 4 [67]. This particular species shows a strong band with an apparent absorption maximum at 380 nm. In this experiment, the anthranilium ion precursor has an absorption that extends to about 370 nm. Undoubtedly, the depletion of the starting material diminishes any nitrenium ion absorption below this wavelength. Therefore, it is possible that the true absorption maximum for this band actually occurs at somewhat lower wavelengths. In addition to the intense absorption in the UV, there is also a weak, diffuse band that extends to the visible region of spectrum. In the case of 103 and 105, depletion of the starting material apparently

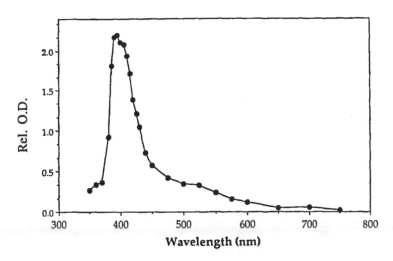

Figure 4 The transient absorption spectrum at *N-tert*-butyl-*N*-(2-acetyl-4-bromo-phenyl)nitrenium ion in CH$_3$CN.

obscures the intense UV band and the observed maximum is thought to corre-spond to the weak, broad visible absorption [69].

The derivatives of *N*-(4-arylphenyl)nitrenium ion (**104, 109–111,** and **122–132**) have been studied extensively by LFP [106]. This is no doubt due to their relatively long lifetimes in aqueous solution and because of their relevance in carcinogenesis. The 4-arylphenyl species are characterized by strong absorption bands in the region 400–500 nm. The parent of this series (**122**) has an absorption maximum at 465 nm. Addition of an alkyl group (Fig. 5, **137**) or a acyl group (**109**) to the remaining nitrogen position has a minimal effect on absorption. Electron-donating substituents in the 4′-position of the distal phenyl ring tend to shift this band to higher wavelengths (**111, 123–126**) and, as expected, substan-tially increase the lifetime of the nitrenium ion. Inductive electron-withdrawing substituents in the 3′-position (**128–131**) diminish the lifetime but have little effect on the position of the absorption band, other than perhaps to shift it to slightly higher wavelengths.

An interesting exception to this trend is the *N*-acetyl-*N*-(4′-amino-4-biphe-nyl)nitrenium ion shown as entry **111** [107]. This has an apparent absorption maximum at 430 nm, approximately 20 nm lower than the simple acetylated system in entry **109** [106]. There are two possible reasons for this. First, the observed band might not represent the lowest energy transition. We note that a similar 4′-aminobiphenylynitrenium ion, **125**, shows two absorption maxima, at 380 and 560 nm. It is possible that in **111** both of the corresponding transitions are shifted to higher wavelength and that only the lower one (430 nm) appears

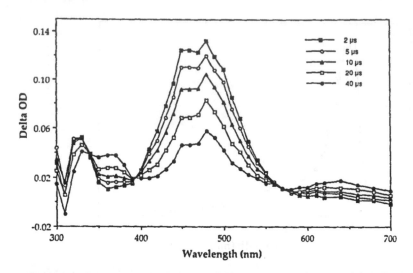

Figure 5 The transient absorption spectrum of *N*-methyl-*N*-(4-biphenylyl)nitrenium ion obtained by LFP of the corresponding 2,4,6-trimethylpyridine adduct in CH₃CN.

in the accessible region of the spectrum. Alternatively, the *N*-acetyl group and the 4′-amino group may interact in an unexpected way, shifting the observed transition to lower wavelengths.

The 2-fluorenylnitrenium ions, **110** [86] and **132** [89], have much longer lifetimes in aqueous solution than the corresponding biphenylyl systems (cf. **122** and **109**). The bridging methylene group apparently improves conjugation between the two phenyl rings, and this in turn improves the charge delocalization. Interestingly, this improved conjugation does not significantly change the position of the visible absorption band.

The *N*-methyl-*N*-(1-naphthyl)nitrenium ion **134** is interesting in that it is the only monoarylnitrenium ion to be detected that does not have some sort of substituent in the position para to the nitrenium ion center [104]. The additional conjugation of the naphthyl group extends the absorption maximum out to 475 nm. However, it is still much less stable than the 4-biphenylylnitrenium ion. Its lifetime in acetonitrile is only about 200 ns.

Three *N*,*N*-diarylnitrenium ions have been detected by LFP (**138–140**) [84,108]. These all show two visible absorptions, one in the 400- to 470-nm region and one in the 640- to 720-nm region. Figure 6 shows the spectrum for diphenylnitrenium ion **6** [103]. The observed absorption bands correspond well to the absorptions of the highly stabilized diarylnitrenium ions detected in the electrochemical experiments (see **90** in Scheme 27, above). As expected the additional *N*-aryl substituent substantially increases the lifetime and stability toward

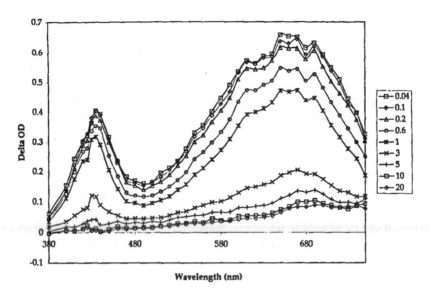

Figure 6 Transient absorption spectrum of diphenylnitrenium ion **6** obtained by LFP of **55**.

hydration. For example, the *N,N*-bis-(4-methyl)phenylnitrenium ion **139** reacts with water almost three orders of magnitude more slowly then its monoaryl analog **102**.

The only non–phenylnitrenium ion that has been studied using flash photolysis is the 1-methyl-(2-imidazolyl)nitrenium ion **133** [109]. This species is especially stable, undoubtedly due to the large degree of charge delocalization throughout the ring. Its lifetime in aqueous solution is about 0.1 s. However, this delocalization does not lead to a high-wavelength absorption band. This nitrenium ion absorbs only in the far-UV at 235 nm.

In summary, LFP experiments have provided the first direct information concerning nitrenium ion UV-Vis spectra, lifetimes, and reaction rates constants. Most studies to date have focused on arylnitrenium ions and, more specifically, those arylnitrenium ions that possess either an additional phenyl ring or an alkoxy group in the position para to the nitrenium ion center because these groups provide sufficient stability to allow its detection in solution.

VII. CONCLUSIONS

Recent photochemical studies have provided a wealth of information about nitrenium ion stabilities and lifetimes, as well as spin-state dependent chemical behavior. Several photochemical routes to nitrenium ions have been demonstrated. These include the photoisomerization of anthranilium ions (Scheme 5), photofragmentation of *N*-aminopyridinium ions (Scheme 12), and protonation of photogenerated singlet nitrenes (Scheme 22). The anthranilium ion route has the advantage of high quantum yields. However, it is limited to a rather narrow range of nitrenium ion structures, namely, those nitrenium ions that possess a 2-acylphenyl substituent. The nitrene protonation route is attractive because the precursors (alkyl or aryl azides) are readily accessible through very straightforward synthetic procedures. This route is much more flexible than the anthranilium route in terms of the types of nitrenium ions that can be created. However, it is confined to primary nitrenium ions generated in acidic (or at least protic) media. The route which has the highest potential for generality is the pyridinium fragmentation. In principle, this can accommodate any primary or secondary nitrenium ion; moreover, it can be applied to nonprotic, nonnucleophilic media where some of the less stable nitrenium ions ought to be longer lived. However, synthesis of the precursors is more time consuming, and in many cases the photoproduct mixtures can be complex due to the potential involvement of the pyridine leaving group in the decay reactions of the nitrenium ion.

REFERENCES

1. Kirmse, W. *Carbene Chemistry*; Academic: New York, 1971.
2. *Carbenes*; Moss, R. A.; Jones, M., Ed.; Wiley: New York, 1983.
3. Schuster, G. B. *Adv. Phys. Org. Chem.* **1986**, *22*, 311–361.
4. *Kinetics and Spectroscopy of Carbenes and Biradicals*; Platz, M. S., Ed.; Plenum: New York, 1990.
5. Leyva, E.; Platz, M. S.; Persy, G.; Wirz, J. *J. Am. Chem. Soc.* **1986**, *108*, 3783–3790.
6. Platz, M. S.; Leyva, E.; Haider, K. In *Organic Photochemistry*; A. Padwa, Ed.; Marcel Dekker: New York, 1991; Vol. 11; pp 367–429.
7. Platz, M. S. *Acc. Chem. Res.* **1995**, *28*, 487–492.
8. Peyerimhoff, S. D.; Buenker, R. J. *Chem. Phys.* **1979**, *42*, 167–176.
9. van Huis, T. J.; Leininger, M. L.; Sherrill, C. D.; Schaefer, H. F., III *J. Comput. Chem.* **1998**, *63*, 1107–1142.
10. Jensen, P.; Bunker, P. R.; McLean, A. D. *Chem. Phys. Lett.* **1987**, *141*, 53–57.
11. Cramer, C. J.; Dulles, F. J.; Storer, J. W.; Worthington, S. E. *Chem. Phys. Lett.* **1994**, *218*, 387–394.
12. Gibson, S. T.; Greene, J. P.; Berkowitz, J. *J. Chem. Phys.* **1985**, *83*, 4319–4328.
13. Lim, M. H.; Worthington, S. E.; Dulles, F. E.; Cramer, C. J. In *Chemical Applications of Density Functional Theory. ACS Symposium Series 629*; B. B. Laird, R. B. Ross and T. Ziegler, Ed.; American Chemical Society: Washington, DC, 1996.
14. Ford, G. P.; Herman, P. S. *J. Am. Chem. Soc.* **1989**, *111*, 3987–3996.
15. Ford, G. P.; Scribner, J. D. *J. Am. Chem. Soc.* **1981**, *103*, 4281–4291.

16. Ford, G. P.; Herman, P. S.; Thompson, J. W. *J. Comput. Chem.* **1999**, *20*, 231–243.
17. Ford, G. P.; Herman, P. S. *J. Mol. Struct. (THEOCHEM)* **1991**, *236*, 269–282.
18. Ford, G. P.; Thompson, J. W. *Chem. Res. Toxicol.* **1999**, *12*, 53–59.
19. Cramer, C. J.; Dulles, F. J.; Falvey, D. E. *J. Am. Chem. Soc.* **1994**, *116*, 9787–9788.
20. Sullivan, M. B.; Brown, K.; Cramer, C. J.; Truhlar, D. G. *J. Am. Chem. Soc.* **1998**, *120*, 11778–11783.
21. Cramer, C. J.; Dulles, F. J.; Falvey, D. E. *J. Am. Chem. Soc.* **1994**, *116*, 9787–9788.
22. Falvey, D. E.; Cramer, C. J. *Tetrahedron Lett.* **1992**, *33*, 1705–1708.
23. Srivastava, S.; Toscano, J. P.; Moran, R. J.; Falvey, D. E. *J. Am. Chem. Soc.* **1997**, *119*, 11552–11553.
24. Cramer, C. J.; Truhlar, D. J.; Falvey, D. E. *J. Am. Chem. Soc.* **1997**, *119*, 12338–12339.
25. Cramer, C. J.; Falvey, D. E. *Tetrahedron Lett.* **1997**, *38*, 1515–1518.
26. Gassman, P. G. *Acc. Chem. Res.* **1970**, *3*, 26–33.
27. Abramovitch, R. A.; Jeyaraman, R. In *Azides and Nitrenes: Reactivity and Utility*; E. F. V. Scriven, Ed.; Academic: Orlando, FL, 1984; pp 297–357.
28. Simonova, T. P.; Nefedov, V. D.; Toropova, M. A.; Kirillov, N. F. *Russ. Chem. Rev* **1992**, *61*, 584–599.
29. Kadlubar, F. F. In *DNA Adducts Identification and Significance*; K. Hemminki, A. Dipple, D. E. G. Shuker, F. F. Kadlubar, D. Segerbäck and H. Bartsch, Ed.; University Press: Oxford, UK, 1994; pp 199–216.
30. McClelland, R. A. *Tetrahedron* **1996**, *52*, 6823–6858.
31. McClelland, R. A.; Gadosy, T. A.; Ren, D. *Can. J. Chem.* **1998**, *76*, 1327–1337.
32. McClelland, R. A.; Kahley, M. J.; Davidse, P. A. *J. Phys. Org. Chem.* **1996**, *9*, 355–360.
33. Falvey, D. E. *J. Phys. Org. Chem.* **1999**, *12*, 589–596.
34. Gassman, P. G.; Cryberg, R. L. *J. Am. Chem. Soc.* **1969**, *91*, 2047–2052.
35. Gassman, P. G.; Campbell, G. A. *J. Chem. Soc. Chem. Commun.* **1970**, *1970*, 427.
36. Gassman, P. G.; Hartman, G. D. *J. Am. Chem. Soc.* **1973**, *95*, 449–454.
37. Gassman, P. G.; Campbell, G. A. *J. Am. Chem. Soc.* **1972**, *94*, 3891–3896.
38. Gassman, P. G.; Granrud, J. E. *J. Am. Chem. Soc.* **1984**, *106*, 2448–2449.
39. Gassman, P. G.; Granrud, J. E. *J. Am. Chem. Soc.* **1984**, *106*, 1498–1499.
40. Meier, C.; Boche, G. *Tetrahedron Lett.* **1990**, *31*, 1685–1688.
41. Ulbrich, R.; Famulok, M.; Bosold, F.; Boche, G. *Tetrahedron Lett.* **1990**, *31*, 1689–1692.
42. Bosold, F.; Boche, G. *Angew. Chem. Int. Ed. Engl.* **1990**, *29*, 63–64.
43. Novak, M.; Martin, K. A.; Heinrich, J. L. *J. Org. Chem.* **1989**, *54*, 5430–5431.
44. Novak, M.; Kahley, M. J.; Lin, J.; Kennedy, S. A.; James, T. G. *J. Org. Chem.* **1995**, *60*, 8294–8304.
45. Kennedy, S. A.; Novak, M.; Kolb, B. A. *J. Am. Chem. Soc.* **1997**, *119*, 7654–7664.
46. Panda, M.; Novak, M.; Magonski, J. *J. Am. Chem. Soc.* **1989**, *111*, 4524–4525.
47. Novak, M.; Kahley, M. J.; Lin, J.; Kennedy, S. A.; Swanegan, L. A. *J. Am. Chem. Soc.* **1994**, *116*, 11626–11627.

48. Gassman, P. G.; Campbell, G. A.; Frederick, R. C. *J. Am. Chem. Soc.* **1972**, *94*, 3884–3891.
49. Novak, M.; Kennedy, S. A. *J. Am. Chem. Soc.* **1995**, *117*, 574–575.
50. Helmick, J. S.; Martin, M. A.; Heinrich, J. L.; Novak, M. *J. Am. Chem. Soc.* **1991**, *113*, 3459–3466.
51. Helmick, J. S.; Novak, M. *J. Org. Chem.* **1991**, *56*, 2925–2927.
52. Famulok, M.; Bosold, F.; Boche, G. *Angew. Chem. Int. Ed. Engl.* **1989**, *28*, 337–338.
53. Famulok, M.; Boche, G. *Angew. Chem. Int. Ed. Engl.* **1989**, *28*, 468–469.
54. Turesky, R.; Markovic, J. *Chem. Res. Toxicol.* **1994**, *7*, 752–761.
55. Humphreys, W. G.; Kadlubar, F. F.; Guengerich, F. P. *Proc. Natl. Acad. Sci. U.S.A.* **1992**, *89*, 8278–8282.
56. Scribner, J. D.; Naimy, N. K. *Cancer Res.* **1975**, *35*, 1416–1421.
57. McClelland, R. A.; Ahmad, A.; Dicks, A. P.; Licence, V. *J. Am. Chem. Soc.* **1999**, *121*, 3303–3310.
58. Hoffman, R. V.; Kumar, A.; Buntain, G. A. *J. Am. Chem. Soc.* **1985**, *107*, 4731–4736.
59. Haley, N. F. *J. Org. Chem.* **1977**, *42*, 3929–3933.
60. Anderson, G. B.; Yang, L. L.-N.; Falvey, D. E. *J. Am. Chem. Soc.* **1993**, *115*, 7254–7262.
61. Su, Z.; Mariano, P. S.; Falvey, D. E.; Yoon, U. C.; Oh, S. W. *J. Am. Chem. Soc.* **1998**, *120*, 10676–10686.
62. Georgarakis, E.; Schmid, H.; Hansen, H.-J. *Helv. Chim. Acta* **1979**, *62*, 234–269.
63. Doppler, T.; Schmid, H.; Hansen, H.-J. *Helv. Chim. Acta* **1979**, *62*, 271–303.
64. Doppler, T.; Schmid, H.; Hansen, H.-J. *Helv. Chim. Acta* **1979**, *62*, 304–313.
65. Giovaninni, E.; Sousa, B. F. S. E. d. *Helv. Chim. Acta* **1979**, *62*, 185–197.
66. Giovaninni, E.; Sousa, B. F. S. E. d. *Helv. Chim. Acta* **1979**, *62*, 198–204.
67. Anderson, G. B.; Falvey, D. E. *J. Am. Chem. Soc.* **1993**, *115*, 9870–9871.
68. Robbins, R. J.; Falvey, D. E. *Tetrahedron Lett.* **1994**, *35*, 4943–4946.
69. Robbins, R. J.; Laman, D. M.; Falvey, D. E. *J. Am. Chem. Soc.* **1996**, *118*, 8127–8135.
70. Robbins, R. J.; Yang, L. L.-N.; Anderson, G. B.; Falvey, D. E. *J. Am. Chem. Soc.* **1995**, *117*, 6544–6552.
71. Srivastava, S.; Falvey, D. E. *J. Am. Chem. Soc.* **1995**, *117*, 10186–10193.
72. Takeuchi, H.; Watanabe, K. *J. Phys. Org. Chem.* **1998**, *11*, 478–484.
73. Takeuchi, H.; Koyama, K. *J. Chem. Soc. Perkin Trans I* **1988**, 2277–2281.
74. Takeuchi, H.; Hayakawa, S.; Tanahashi, T.; Kobayashi, A.; Adachi, T.; Higuchi, D. *J. Chem. Soc. Perkin Trans. 2* **1991**, 847–855.
75. Takeuchi, H.; Higuchi, D.; Adachi, T. *J. Chem. Soc. Perkin Trans. 2* **1991**, 1525–1529.
76. Takeuchi, H.; Hayakawa, S.; Murai, H. *J. Chem. Soc. Chem. Commun.* **1988**, 1287–1289.
77. Abramovitch, R. A.; Shi, Q. *Heterocycles* **1994**, *37*, 1463–1466.
78. Abramovitch, R. A.; Beckert, J. M.; Chinnasamy, P.; Xiaohua, H.; Pennington, W.; Sanjivamurthy, A. R. V. *Heterocycles* **1989**, *28*, 623–628.

79. Srivastava, S.; Kercher, M.; Falvey, D. E. *J. Org. Chem.* **1999**, *64*, 5853–5857.
80. Abramovitch, R. A.; Evertz, K.; Huttner, G.; Gibson, H. H.; Weems, H. G. *J. Chem. Soc. Chem. Commun.* **1988**, 325–327.
81. Li, Y.; Abramovitch, R. A.; Houk, K. N. *J. Org. Chem.* **1989**, *54*, 2911–2914.
82. Abramovitch, R. A.; Gibson, H. H.; Nguyen, T.; Olivella, S.; Sole', A. *Tetrahedron Lett.* **1994**, *35*, 2321–2324.
83. Abramovitch, R. A.; Beckert, J. M.; Pennington, W. T. *J. Chem. Soc. Perkin Trans I* **1991**, 1761–1762.
84. Moran, R. J.; Falvey, D. E. *J. Am. Chem. Soc.* **1996**, *118*, 8965–8966.
85. Moran, R. J.; Cramer, C. J.; Falvey, D. E. *J. Org. Chem.* **1996**, *61*, 3195–3199.
86. Davidse, P. A.; Kahley, M. J.; McClelland, R. A.; Novak, M. *J. Am. Chem. Soc.* **1994**, *116*, 4513–4514.
87. Abramovitch, R. A.; Kyba, E. P. *The Chemistry of Azide Group*; Interscience: New York, 1971, p. 234.
88. McClelland, R. A.; Davidse, P. A.; Hadzialic, G. *J. Am. Chem. Soc.* **1995**, *117*, 4173–4174.
89. McClelland, R. A.; Kahley, M. J.; Davidse, P. A.; Hadzialic, G. *J. Am. Chem. Soc.* **1996**, *118*, 4794–4803.
90. Sukhai, P.; McClelland, R. A. *J. Chem. Soc. Perkin Trans. 2* **1996**, 1529–1530.
91. Hrovat, D. A.; Waali, E. E.; Borden, W. T. *J. Am. Chem. Soc.* **1992**, *114*, 8698–8699.
92. Travers, M. J.; Cowles, D. C.; Clifford, E. P.; Ellison, G. B. *J. Am. Chem. Soc.* **1992**, *114*, 8699–8701.
93. Ramlall, P.; McClelland, R. A. *J. Chem. Soc. Perkin Trans. 2* **1999**, 225–232.
94. Michalak, J.; Zhai, H. B.; Platz, M. S. *J. Phys. Chem.* **1996**, *100*, 14028–14036.
95. Born, R.; Burda, C.; Senn, P.; Wirz, J. *J. Am. Chem. Soc.* **1997**, *119*, 5061–5062.
96. Svanholm, U.; Parker, V. D. *J. Am. Chem. Soc.* **1974**, *96*, 1234–1236.
97. Serve, D. *J. Am. Chem. Soc.* **1975**, *97*, 432–434.
98. Caquis, G.; Delhomme, H.; Serve, D. *Electrochim. Acta* **1976**, *21*, 557–565.
99. Rieker, A.; Speiser, B. *Tetrahedron Lett.* **1990**, *35*, 5013–5014.
100. Rieker, A.; Speiser, B. *J. Org. Chem.* **1991**, *1991*, 4664–4671.
101. Boche, G.; Andrews, P.; Harms, K.; Marsch, M.; Rangappa, K. S.; Schimeczek, M.; Willecke, C. *J. Am. Chem. Soc.* **1996**, *118*, 4925–4930.
102. Finley, K. T. In *Heterocyclic Compounds* Wiley: New York, 1981; Vol. 39; pp 292–311.
103. McIlroy, S.; Falvey, D. E. **1999**, unpublished results.
104. Chiapperino, D.; Falvey, D. **1999** (manuscript in preparation).
105. Fishbein, J. C.; McClelland, R. A. *J. Am. Chem. Soc.* **1987**, *109*, 2824–2825.
106. Ren, D.; McClelland, R. A. *Can. J. Chem.* **1998**, *76*, 78–84.
107. Dicks, A. P.; Ahmad, A.; D'Sa, R.; McClelland, R. A. *J. Chem. Soc. Perkin Trans. 2* **1999**, 1–3.
108. Moran, R. J. Ph. D. Thesis, University of Maryland, College Park, 1997.
109. Gadosy, T. A.; McClelland, R. A. *J. Am. Chem. Soc.* **1999**, *121*, 1459–1465.
110. Chiapperino, D.; Falvey, D. **1999**, unpublished results.

7

Photophysical Properties of Inorganic Nanoparticle–DNA Assemblies

Catherine J. Murphy
University of South Carolina, Columbia, South Carolina

I. INTRODUCTION

Some of the most interesting and important advances in science occur at the boundaries between subdisciplines; the area of supramolecular photochemistry itself is an example. Technological advances and scientific expertise in different areas—synthetic chemistry, molecular modeling, and photophysics in the case of supramolecular photochemistry—are a prerequisite for good experimental design and rational interpretation of results.

The 1980s saw great advances in two areas that seem far apart but in fact have come together in recent years: 1) DNA synthesis and 2) the synthesis and materials science of nanometer-size inorganic particles. Commercial DNA synthesizers became available in that decade and for the first time the chemist at the bench was no longer obliged to synthesize oligonucleotides by hand or isolate oligonucleotides by cutting genomic DNA with restriction enzymes. Nowadays many companies sell oligonucleotides made to order that can be delivered in several days. The general scale of the synthesis is 0.2 μmol to 10 μmol of single strands, up to about 50 bases long. Procedures describing the chemistry involved and the purification of the resulting oligonucleotides have been worked out in detail [1,2].

Also in the 1980s, chemists and materials scientists began to make significant advances in the synthesis and characterization of inorganic particles with
dimensions on the nanometer scale. The work in general has focused on metal
nanoparticles and semiconductor nanoparticles (although other materials such as
silica, an insulator, and organic polymer beads are also being examined). One
of the driving forces behind these investigations has been the predicted unusual
electronic properties of metals and semiconductors in the size range between
molecular and bulk (see below). Another driving force has been the use of heavy-
metal colloids for the visualization of biological macromolecular structure in
electron and light microscopy [3,4].

The dimensions of inorganic nanoparticles and biological molecules overlap to some extent (Fig. 1). Molecular dimensions are typically on the order of
10 Å (1 nm), while semiconductor nanoparticles generally have diameters of 2–
10 nm, and metal nanoparticles have diameters of 5–100 nm. Protein sizes are
in the 5- to 20-nm range, and DNA can be considered as a rod of 2 nm diameter
and length up to about 1 m, depending on the number of base pairs it contains;
there is 0.34 nm between adjacent base pairs (bp), so a 30-bp oligonucleotide
would have a length of 10 nm.

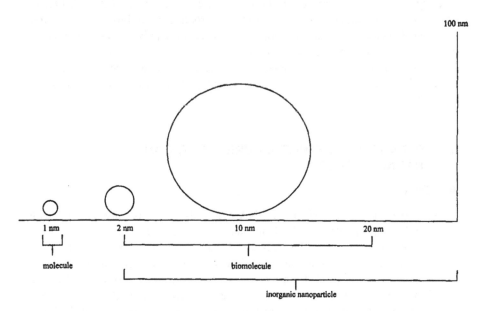

Figure 1 Comparison of molecular and biomolecular dimensions. Beyond approximately 100 nm, inorganic particles, in general, behave as bulk materials.

With better synthetic methods and better means to purify and characterize oligonucleotides and inorganic nanoparticles, research groups are now combining the accumulated expertise in oligonucleotide and materials synthesis to:

1. Image DNA transcription [5]
2. Visualize DNA hybridization in a color-coded spot test [6,7]
3. Model protein–DNA interactions, with the inorganic nanoparticle functioning as an "inorganic protein" [8–11]
4. Introduce DNA into cells via adsorption to inorganic nanoparticles (which for this application can be up to 1 μm in diameter) for gene therapy [12–16]
5. Learn about the in vivo behavior of biologically generated inorganic particles [17]
6. Develop new diagnostic techniques in DNA quantitation and drug–DNA interactions [18–20]
7. Use DNA as a template to organize nanoparticles for ordered nanoscale materials [21–24]

In all of these methodologies, the signal that is monitored arises from the nanoparticle; selectivity comes from the DNA, which is either adsorbed noncovalently or covalently attached to the nanoparticle. The focus of this chapter is on spectroscopic signals, such as visible absorbance, photoluminescence, and surface-enhanced Raman spectroscopy, which are intimately related to the nanometer scale of particle size. Related work, not covered in this chapter, includes DNA adsorption to clay and organic nanoparticles for environmental tracer purposes [25–28] and DNA bound to magnetic nanoparticles for magnetic resonance imaging [29].

II. PREPARATIVE METHODS, PURIFICATION AND CHARACTERIZATION

A. DNA

As mentioned in the Introduction, DNA synthesis per se is ably handled with commercial instrumentation. The synthesis is done on a solid support using phosphoramidite chemistry with the protected 3' base attached covalently to the support. Subsequent deprotection and coupling steps allow the growth of the DNA single strand from 3' to 5'. Here we remind the reader of the primary structure of DNA (Fig. 2) and the Watson–Crick base pairing (Fig. 3) that provides the selectivity in hybridization. It is worth noting that many derivatives of the bases are commercially available as well. In addition, the backbone can be modified

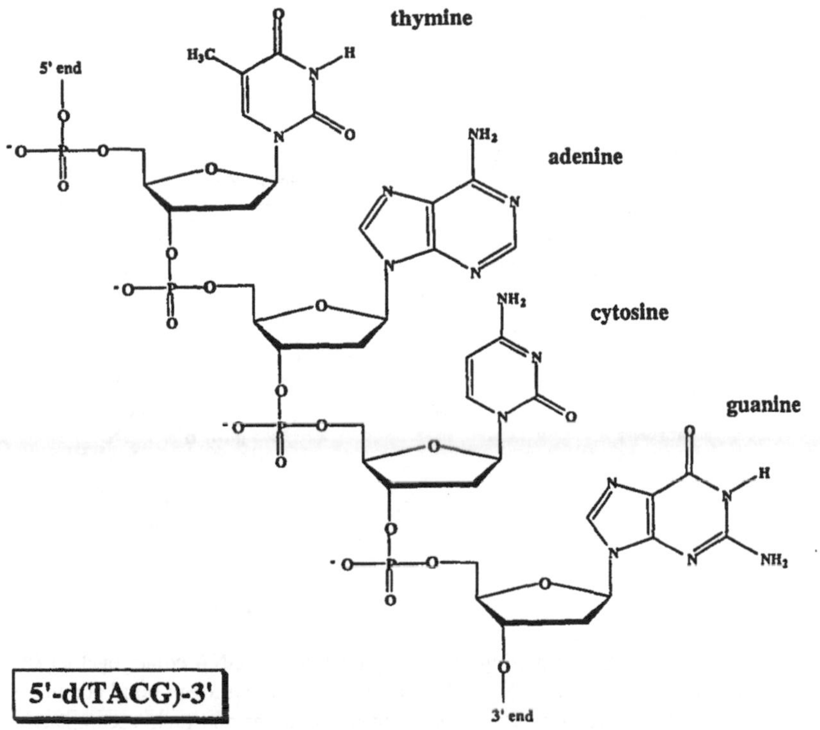

Figure 2 The primary structure of DNA, deoxyribonucleic acid. The sugar–phosphate backbone is negatively charged at physiological pH. The abbreviation for the structure is given at bottom.

to produce sulfur-containing or neutral phosphate analogs; fluorescent probes and conjugatable groups can be linked to the 5′ or 3′ end of oligonucleotides.

Oligonucleotides are, in general, only soluble in solutions that are aqueous or, at most, 30% organic/70% aqueous. Purification is generally achieved by high-pressure liquid chromatography [2]. The double helix can be melted back to the single strands under high temperature (90°C at the most) or low-salt conditions; the melting temperature depends on DNA concentration and ionic strength. Double-helix formation is favored at high DNA concentration (millimolar in bases), long DNA length, high ionic strength, and low temperature. Also, the presence of mismatched bases (e.g., A–C, G–T) will in general destabilize the double helix. Many hybridization schemes depend on this lowering of the melting temperature by imperfect duplexes compared to perfect duplexes of the same length under the same conditions.

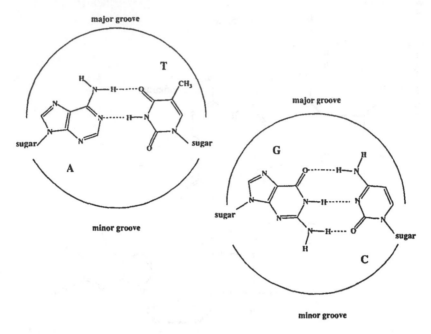

Figure 3 Watson–Crick base pairing. Two single strands of DNA come together (hybridize) in an antiparallel fashion ($5' \rightarrow 3'$ with $3' \rightarrow 5'$) to hydrogen-bond adenine to thymine and guanine to cytosine. The orientation of the bases and their accompanying backbones, in three dimensions, gives the familiar double helix of DNA, which is defined by a large major groove and a somewhat smaller minor groove (for standard "B-form" DNA).

B. Metal Nanoparticles

The most common metal nanoparticles used in DNA-based applications are silver and gold, although copper has also been examined. The syntheses generally involve reduction of a metal salt to the element under appropriate solution conditions to produce elemental nanoparticles. Gold nanoparticles of various sizes (5–20 nm or larger) and surface groups (e.g., with proteins) are also available commercially (e.g., Sigma).

A specific example of a silver nanoparticle preparation [30] is the addition of one equivalent of a 1.0 mM aqueous silver nitrate solution to six equivalents of a sodium borohydride solution, cold, with vigorous stirring. The resulting solution is pale yellow.

A specific example of a preparation that produces 13-nm-diameter gold nanoparticles is given in [7]. A 1 mM aqueous solution of $HAuCl_4$ is brought to reflux with stirring, and 3.5 equivalents of trisodium citrate is added quickly. The solution color changes from pale yellow to deep red; the solution is refluxed for an additional 15 min and filtered through a 0.45-μm filter to remove large particles. The 13-nm nanoparticles bear a slight net negative charge from coordinated citrate.

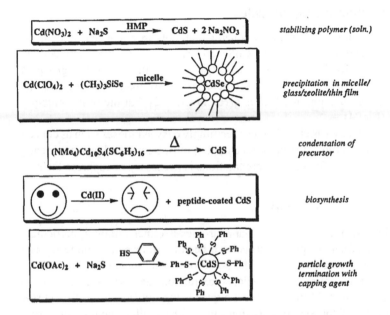

$Cd(NO_3)_2$ + Na_2S $\xrightarrow{\text{HMP}}$ CdS + 2 Na_2NO_3	*stabilizing polymer (soln.)*
$Cd(ClO_4)_2$ + $(CH_3)_3SiSe$ $\xrightarrow{\text{micelle}}$ CdSe	*precipitation in micelle/ glass/zeolite/thin film*
$(NMe_4)Cd_{10}S_4(SC_6H_5)_{16}$ $\xrightarrow{\Delta}$ CdS	*condensation of precursor*
Cd(II) + peptide-coated CdS	*biosynthesis*
$Cd(OAc)_2$ + Na_2S $\xrightarrow{\text{HS-}}$ Ph-S-CdS-S-Ph	*particle growth termination with capping agent*

Figure 4 Synthetic methods to make CdS and CdSe nanoparticles. In the first (top), a soluble cadmium salt is added to a polar solvent in the presence of a stabilizing polymer such as a polyphosphate, polyamine, or polyalcohol (HMP is hexametaphosphate in the example). Cd(II) coordinates to the polymer and is thus somewhat protected from solvent. The subsequent addition of sulfide produces CdS nanoparticles whose growth is arrested due to steric and electrostatic hindrance (depending on pH) of the polymer. The second method is similar to the first, except that the stabilizing agent has a predetermined cavity size that the nanoparticle can fill; in the particular example shown here, the inverse micelle has a pool of water inside that decomposes the selenium reagent. The third method, not as elegant as the first two, thermolyzes atomically well-defined clusters to make larger but less well-defined nanoparticles. The fourth method is a detoxification response on the part of a eukaryotic organism to the presence of Cd(II); CdS particle size is controlled by a peptide rich in cysteine residues. (In vitro experiments can use the same peptide as the stabilizing polymer, identical in principle to the first method.) The final method is to compete for sulfide with a capping group such as a thiol; the ratio of capping agent to sulfide determines the size and also produces tailored surfaces.

As is clear from these examples, purification is minimal in these syntheses. Characterization of nanoparticle size distribution is best done by transmission electron microscopy (TEM) and can be confirmed spectrophotometrically; the color of the gold nanoparticles in particular has been well correlated with average size over certain ranges. In the 13-nm gold nanoparticle case described above, a characteristic absorbance peak at 518–520 nm is observed [7].

C. Semiconductor Nanoparticles

The most commonly studied semiconductor materials in nanoparticle synthesis are the II–VI materials, such as CdS, CdSe, and ZnS [31–42]. Other semiconductors, such as the III–V materials (GaAs, GaP, GaN, InP, etc.) and group IV materials (Si, Ge), will not be discussed here. There are many procedures in the literature, only a selection of which is given in [31–42]; but the basic principles are similar (Fig. 4).

There are many variables in these seemingly easy syntheses such as concentration, pH, temperature, stirring time, etc., that must be worked out empirically. In particular, high pH for aqueous preparations is necessary to suppress the reaction:

$$S^{2-} + H_2O \rightarrow HS^- + OH^- \tag{1}$$

Purification of semiconductor nanoparticles is either minimal, particularly if stabilizing polymers or porous hosts have been used, or can be more rigorous if capping agents have been used. The solubility of the capping agent dictates the solubility of the entire nanoparticle, so extractions and washings can be employed to good effect. More recent techniques using hot injection of precursors and "focusing" of the size distribution allow for excellent control of particle size [40,43,44]. Characterization of particle size is best done by TEM, although spectrophotometric methods are also used (see below).

III. PHOTOPHYSICAL PROPERTIES OF INORGANIC NANOPARTICLES

A. Metals

1. Visible Absorbance

It has been known for centuries that colloidal solutions of metals (in particular, gold) are visibly colored [45]. The color of these solutions arises from resonant excitation of the collective normal modes of the conduction electrons in the metal particles (a "plasmon") [46]. Scattering of light also contributes to the color, and Mie theory describes the dependence of absorption and scattering effects on the particle size and dielectric function of the metal [46]. These effects are optimal in the visible region of the spectrum for silver and gold due to their dielectric functions [46], and nanoparticle solutions of these metals have been the best studied. One of the most interesting features of the visible absorbance is that it strongly depends on the aggregation state of the nanoparticles (which in turn is affected by salt concentration, temperature, nature of any surface groups, etc.). For example, gold nanoparticles whose dimensions are substantially smaller than the wavelengths of visible light can appear red to red–purple when well separated but upon aggregation turn blue to blue–gray. These color changes are readily monitored with a conventional spectrophotometer; distinct peaks are easily observed. Extensive aggregation, of course, affords precipitation of the entire ensemble.

2. Surface-Enhanced Raman Spectroscopy

Surface-enhanced Raman spectroscopy (SERS) is a technique in which the Raman scattering of molecular vibrations is enhanced up to a million fold by adsorption of the molecule to metal surfaces that have roughness on the 10- to 100-nm scale [47,48]. The three most common surfaces used are roughened electrodes, metal nanoparticles, or metal island films [47,48]. There are two general mechanisms that contribute to the enhancement: electromagnetic and chemical [47,48]. The intensity of Raman scattering is proportional to the square of the electric field–induced dipole moment P, where $P = \alpha E$; α is the molecular polarizability (a tensor) and E is the electric field incident on the molecule [47]. Obviously, increases in either α or E will result in enhanced Raman scattering. In the electromagnetic mechanism, the electric field incident on the molecule depends on the laser and on the dielectric function of the metal [48]:

$$E_{induced} = \{[\varepsilon_1(\omega) - \varepsilon_2]/[\varepsilon_1(\omega) + 2\varepsilon_2]\}E_{laser} \qquad (2)$$

where $\varepsilon_1(\omega)$ is the complex, frequency-dependent dielectric function of the metal and ε_2 is the relative permittivity of the ambient phase. This function is resonant at the frequency for which $\varepsilon_1 = -2\varepsilon_2$; it turns out that for the visible laser light used in most SERS experiments, the noble metals satisfy this condition. Thus, the metal nanoparticle surface upon laser excitation acts as an antenna for the Raman scattering of the adsorbed molecule.

In the chemical mechanism, the polarizability of the molecule is altered upon adsorption to the metal nanoscale surface. This can arise either from shifting and broadening of the molecular electronic states by interaction with the surface, or by the production of new resonant electronic states (akin to a charge transfer transition) that arise from the adsorbed molecule on the metal surface [48].

A major advantage of SERS is that the incident laser light need not resonantly excite the adsorbed molecule itself; thus, in principle any molecule that is adsorbed to the appropriate surface can be examined [49].

B. Semiconductors

1. Visible Absorbance

Electronically, a semiconductor is characterized by a filled valence band, consisting of very closely spaced occupied orbitals, separated by the bandgap energy E_g from the (nearly) empty conduction band, which consists of very closely spaced unoccupied orbitals. The bandgap E_g is then analogous to the familiar HOMO–LUMO gap of chemistry. Semiconductors have bandgaps of about 0.5– 3.5 eV (1 eV = 1.602 × 10^{-19} J), which corresponds to wavelengths of light from the near-infrared through the visible, just into the ultraviolet (UV). Thus, semiconductors absorb with a characteristic "edge" in the UV-Vis region in the electromagnetic spectrum. When a semiconductor absorbs light, an electron is promoted from the valence band to the conduction band, leaving a positively charged "hole" behind. This hole is the absence of an electron but can be treated as a particle with its own mass and positive charge. The spatial separation of the photogenerated electron–hole pair (an "exciton") can be estimated using an adaptation of the Bohr model:

$$r = \varepsilon h^2 / \pi m_r e^2 \tag{3}$$

where r is the radius of the sphere defined by the three-dimensional separation of the electron–hole pair; ε is the dielectric constant of the semiconductor; and m_r is the reduced mass of the electron–hole pair. For CdS, the Bohr excitonic radius is about 30 Å. If the physical size of the semiconductor approaches that of the excitonic Bohr radius, unusual quantum confinement effects are observed akin to the familiar particle-in-a-box notion of physical chemistry; these effects are manifested as quantization of the energy levels in the semiconductor and as an increase in the bandgap of the semiconductor particle [50–60]. Alternatively, quantum confinement can be viewed as the electronic intermediate between bulk solid (bands) and molecule (bonds) (Fig. 5). Semiconductor nanoparticles in this size regime are therefore also known as "quantum dots."

Experimentally, electronic absorption spectra of semiconductor nanoparticle solutions are good indicators of average particle size. As the particle decreases in size, the band edge blue-shifts and the spectrum contains more structure. Many excellent reviews are available that describe these effects and others in more detail [35–37,39,50–60]. One early and popular calculation is Brus' effective mass model for the quantum confinement effect [61]:

$$E_g(\text{nanoparticle}) = E_g(\text{bulk}) + (h^2 / 8R^2)(1/m_e^* + 1/m_h^*) \tag{4}$$
$$- 1.8e^2 / 4\pi\varepsilon_0\varepsilon R$$

(neglecting spatial correlation effects), where E_g is the bandgap energy of nanoparticle or bulk solid, R is the nanoparticle radius, m_e and m_h are the effective

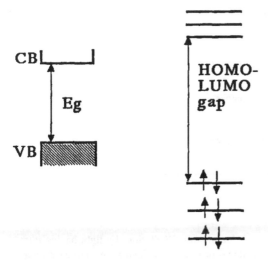

Figure 5 A comparison of bulk semiconductor electronic structure (left) to molecular electronic structure (right). VB is valence band, CB is conduction band, Eg is the energy gap between valence band edge and conduction band edge, HOMO–LUMO gap is the energy gap between the highest occupied molecular orbital (HOMO) and the lowest unoccupied molecular orbital (LUMO). The semiconductor nanoparticle electronic structure is size-dependent and is between the band picture and the bond picture.

masses of the electron and hole in the solid, respectively, and ε is the dielectric constant of the solid. The middle term on the right-hand side of the equation is a particle-in-a-box-like term for the exciton, while the third term on the right-hand side of the equation represents the electron–hole pair coulombic attraction mediated by the solid. Implicit in this equation is that the nanoclusters are spherical and that both the effective masses of charge carriers and the dielectric constant of the solid are constant as a function of size.

2. Photoluminescence

Upon excitation, a semiconductor can emit light that is analogous to the fluorescence/phosphorescence observed in molecules (Fig. 6). The electron–hole pair recombines to emit photons of bandgap (or subbandgap, if appropriate impurities are added to produce energy levels within the bandgap) energy. This photoluminescence is far more sensitive to the presence of adsorbates than the absorption spectrum. Molecules adsorbed to the semiconductor surface may, at one extreme, be oxidized by the photogenerated hole or reduced by the photogenerated electron, quenching the luminescence [53,62]. Adsorbates may also form adducts with the surface that influence the rate of electron-hole pair recombina-

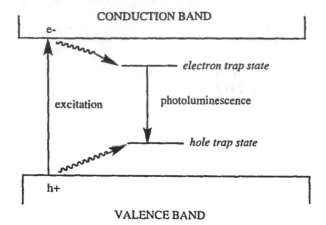

CONDUCTION BAND

Figure 6 Generation of light from a semiconductor. Excitation (by light having ultra-bandgap energy, or by application of a voltage) promotes an electron (e−) to the conduction band, leaving a hole (h+) in the valence band. The presence of adsorbates, or dangling bonds at the surface, can produce trap states within the bandgap that capture either charge carrier (squiggly arrows). One frontier in semiconductor nanoparticle synthesis is to produce materials that are well passivated and do not have trap states. Electron–hole pair recombination from these trap states (or from the band edges themselves, should the trap states be avoided) can result in photoluminescence visible to the naked eye at room temperature. For typical solution syntheses of CdS nanoparticles, the emitted light can be blue, green, yellow, orange, or red depending on synthesis conditions, particle size, and degree of surface passivation [8,9,31,42].

tion or alter the near-surface electronic structure of the semiconductor [63–65]. If the surface is well passivated, the photoluminescence wavelength maximum tracks the absorbance edge with particle size: smaller nanoparticles have blue-shifted emission spectra compared to larger ones. Careful treatment of the nanoparticle surface can produce high emission quantum yields [31,39,40].

IV. NONCOVALENT DNA–INORGANIC NANOPARTICLE ASSEMBLIES

A. Metals

Silver nanoparticles have been deposited on DNA by reducing DNA-associated Ag(I) counterions [24]. The purpose of this study was to make an extremely small silver wire between two electrodes, and DNA was used as a template [24]. The authors concluded that DNA could be useful in the making of electronic circuits on the nanometer scale [24].

Silver nanoparticles have been used as SERS substrates to examine DNA adsorption [66–73]. These studies have for the most part focused on detection of DNA and its constituent bases in solution; color changes due to DNA-induced aggregation of nanoparticles have also been reported [66]. Some of the studies are contradictory, which is most likely due to different nanoparticle preparations. It is worth reminding the reader that only those chemical species that are very close to the surface will give a signal in SERS.

Silver colloids prepared by sodium borohydride reduction of silver nitrate that are 42 nm on average have an absorption band at 380 nm, which shifts to longer wavelengths upon addition of free guanine at acidic pH, according to Sequaris et al. [66]. Micromolar detection of alkylated guanine derivatives by SERS was found by this group [66]. Sequaris et al. found that calf thymus DNA (about 10,000 bp long; commercially available) adsorbs to roughened silver electrodes differently compared to methylated calf thymus DNA as judged by SERS (0.15 M NaCl, 2 mM Tris, pH 8); methylation increases the SERS signals assigned to adenine at 735 and 1330 cm^{-1} and to 7-methylguanine at 656, 700, and 1360 cm^{-1}. Conversely, methylation decreases the signals assigned to phosphate at 236 cm^{-1} and cytosine at 1200 cm^{-1} and 1300 cm^{-1} [66]. The implication is that methylation induces a conformational change, such as unwinding, that makes adenine and methylguanine more accessible to the surface.

A similar study was performed by Kneipp and Flemming [67] with silver colloids prepared via reduction of silver nitrate by sodium citrate. These nanoparticles were polydisperse, with diameters measured by TEM ranging from 30 to 150 nm. Calf thymus DNA solutions were prepared that were 1 M NaCl ("native") or boiled in water for 20 min, followed by rapid cooling and mixing with a salt stock solution to achieve 1 M NaCl ("denatured"). One worrisome feature of these solutions is the lack of buffer and pH control; DNA will degrade in pure water that is slightly acidic. In any case, Kneipp and Flemming found that denatured DNA gave very similar SERS peaks to those found by Sequaris et al. [66], while native DNA was distinctly different and in particular lacked the dominant 735 cm^{-1} band that is assigned as a ring-breathing mode of adenine; their conclusion was that native DNA did not unwind/denature on their surface to give this 735 cm^{-1} signal.

Nabiev et al. examined calf thymus DNA and a plasmid DNA in both its supercoiled and linear forms with two different silver colloidal nanoparticles [68]. All of the silver nanoparticles were prepared by citrate reduction of silver nitrate; "nonactivated" nanoparticles were left as is, and "activated" nanoparticles were obtained by further treatment with NaClO$_4$, which presumably aggregated the nanoparticles to some extent. No details about the DNA solution buffers and pH were provided. Nabiev et al. found that the activated nanoparticles selectively enhanced the 735 cm^{-1} adenine ring-breathing vibration upon DNA denaturation, suggesting that these nanomaterials may selectively detect local regions of unwinding that contain adenines. The plasmid data were preliminary, but they indicated that supercoiled and relaxed circular DNAs did give different SERS signatures [68].

Other workers have noted that the time it takes to adsorb nucleic acids to nanoparticles in solution may be quite long (days), which may reflect the relative rigidity of the macromolecules [69]. Gold and copper nanoparticles are beginning to be studied more thoroughly as SERS substrates for the DNA bases [72]. One frontier that is being vigorously explored is the use of near-infrared SERS for single-molecule detection of the DNA bases, which may be useful for rapid sequencing analysis [73].

Other SERS work in the literature has focused on drug–DNA interactions [18–20,74]. In general, the drug is first adsorbed to the inorganic nanoparticle and its SERS spectrum recorded. The drugs are much smaller molecules than the DNA and frequently have well-defined vibrational spectra. The addition of DNA reduces the intensity of certain drug vibrations, implying that these portions of the drug are interacting with the DNA rather than the surface. Visualization of how the DNA "swallows" the surface-bound drug can be garnered from the intensity and positions of DNA-specific bands. Comparisons can also be made to preincubated drug–DNA complexes.

B. Semiconductors

The signal that is followed in DNA–semiconductor nanoparticle interactions has been, by and large, photoluminescence from the nanoparticle [8–11,75–79], although microscopic methods for characterization of DNA-assembled nanoparticles have also been used [23,79]. The Coffer group has focused on using DNA as a polyphosphate to stabilize CdS nanoparticle formation [23,75–79]; our group has focused on using CdS nanoparticles as photoluminescent "proteins" to learn about DNA sequence–directed structure [8–11]. Both groups rely on the physisorption of DNA to the nanoparticles; thus, the net charge on the nanoparticles and environmental conditions such as salt and temperature influence the binding of DNA to the inorganic surface. The DNA–nanoparticle assemblies, moreover, are not well defined on the nanometer scale; rather, they can be pictured as having "spaghetti-and-meatball" morphology.

Coffer's group has found that the base composition of the DNA stabilizer influences the light output of the resulting DNA–nanoparticle assembly [75,78]. Viscosity measurements of bulk solutions containing the assemblies support the notion that calf thymus DNA is "balled up" upon interaction with the CdS nanoparticles [77]. Photoluminescence titrations of preformed CdS nanoparticles with different DNAs [*E. coli* double-helical DNA and single-stranded poly (A)] revealed that the observed quenching of the emission is influenced by the nucleic acid, and the quenching follows a Perrin "sphere-of-action" model; in this model, the DNAs must be within a certain critical distance of the nanoparticles to cause photoluminescence quenching [76]. This behavior is contrasted with the more common Stern–Volmer quenching behavior, in which emitter and quencher collide at a diffusion-limited rate [76]. Measurements of photoluminescence quenching alone, however, cannot give detailed information about the mechanism of DNA adsorption to the CdS surface; hydrogen bonding has been suggested as one means of coordinating DNA to the heterogeneous nanoparticle surface [76].

   Our own group has examined both long calf thymus DNA and short, well-defined oligonucleotides as adsorbates onto CdS nanoparticles [8–11]. We are interested in how the static structure and flexibility of the DNA polymer influences its binding to a generic curved surface that is the size of a protein, such as a CdS nanoparticle. DNA can adopt quite different right-handed and left-handed double-helical forms depending on hydration level and salt [80], and at the base-pair level many local distortions have been characterized that can lead to sequence-dependent kinds and bends [81–88] (Fig. 7).

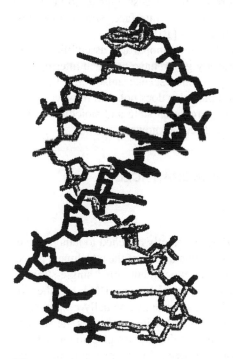

Figure 7 Crystal structure of an oligonucleotide having a 23° kink in its center, downloaded from the Protein Data Bank (file 126d) and visualized in RasMol [83].

In our work, CdS nanoparticles are synthesized that are surface-enriched in Cd(II) [8,31] or that have been capped with a thiolate so as to present an amine or alcoholic surface to the adsorbing DNA [9,10]. The wavelength maximum of the photoluminescence of the nanoparticles that are all 40–50 Å in diameter according to TEM varies significantly with the nature of the surface but in all cases is in the visible. We have performed photoluminescence titrations with straight and kinked double-stranded oligonucleotides and found that oligonucleotides containing intrinsic kinks adsorb better to the nanoparticle surface than straight ones [8–10]. Thus, the DNA provides shape-selective recognition, and the nanoparticle provides an optical signal in the experiment. Whether these kinks are statically curved or intrinsically flexible is a very interesting question that the protein–DNA community and the theoretical biophysics community are exploring vigorously; sequence-dependent structural distortions in DNA may be a mechanism of "indirect readout" in DNA function, damage, and repair [89–100]. In our experiments, an adsorption isotherm model is fit to the photoluminescence quenching data to extract relative binding constants [8–11]. More recent temperature- and salt-dependent experiments in our laboratory [11] support the notion that the binding of DNA to the Cd(II)-rich nanoparticle surface is entropically favorable and enthalpically unfavorable; the process appears to be driven by counterion release from the DNA–nanoparticle interface, similar to results found for nonspecific-protein–DNA interactions [101–105].

The properties of the DNA–semiconductor nanoparticle assemblies examined in both the Coffer group and the Murphy group have not been interrogated, for instance, by dynamic light scattering or fluorescence anisotropy. For both groups, the photoluminescence of the nanoparticles does not shift in wavelength, only in intensity, as DNA is added. The aggregation state and stoichiometry of the DNA–nanoparticle assemblies are not directly known.

V. COVALENT DNA–INORGANIC NANOPARTICLE ASSEMBLIES

A. Overall Strategies

The only examples in the literature of DNA covalently attached to nanoparticles have focused on gold nanoparticles (so far). Some details of the attachment procedures will be outlined below. In general, the experiments bind single-stranded DNAs to the nanoparticles and bring the nanoparticles and DNA together via Watson-Crick hydrogen bonding (Fig. 8). The strategies actually employed in the literature are more elegant. Alivisatos' group, for instance, uses the strategy outlined in Fig. 9. The Alivisatos strategy allows for precise placement of nanoparticles on the nanometer scale. Mirkin's strategy, in contrast, attaches many single-stranded oligonucleotides to the gold nanoparticles; hybridization with an

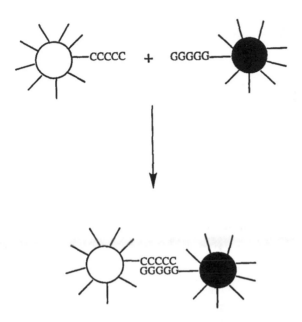

Figure 8 Generic strategy for hybridization of single-stranded complementary DNAs covalently attached to nanoparticles. The white particles are covered with 3'-CCCCC-5', which is attached through the 3' end to the nanoparticle; the black particles are covered with 5'-GGGGG-3', attached through the 3' end to the nanoparticle. In separate solutions the nanoparticles do not aggregate; upon mixing, GC base pairing mediates aggregation. Only two nanoparticles are shown, but in this strategy an entire weblike array would be produced.

added linker DNA (similar in concept to the Alivisatos technique) yields a three-dimensional array of nanoparticles (Fig. 10) [6,7,21,106].

Niemeyer et al. have used a more elaborate procedure to link DNA and metal nanoparticles [107]. In their strategy, most similar to that of Alivisatos, single-stranded DNA oligonucleotides are covalently attached to the protein streptavidin (STV) with a stoichiometry of one protein per DNA strand. Six distinct oligonucleotide sequences were used. With the addition of a 170mer linker RNA that has portions complementary to all six DNAs, DNA–RNA hybrids were produced, leading to the positioning of six STVs per long duplex. Meanwhile, monoamine-modified 1.4-nm gold nanoparticles were conjugated to biotin (one biotin molecule per nanoparticle); biotin and STV bind to each other with high affinity. Addition of the biotin-modified gold nanoparticles to the DNA-assembled STVs produces the nanoparticle–DNA–protein assembly. The work of Niemeyer et al. owes a debt to Schafer et al., who in 1991 attached biotin to DNA

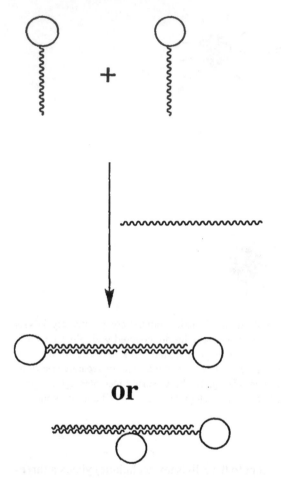

Figure 9 The Alivisatos strategy for making DNA–nanoparticle assemblies [22]. Commercially available 1.4-nm-diameter water-soluble gold nanoparticles are conjugated to single-stranded 18mers that are noncomplementary; there is on average one oligonucleotide per nanoparticle. A single-stranded 37mer that is complementary to both 18mers assembles two nanoparticles per duplex with differing orientations, depending on the orientation of the DNA attachment (5′ or 3′). Three nanoparticles can also be assembled with longer linker DNA [22].

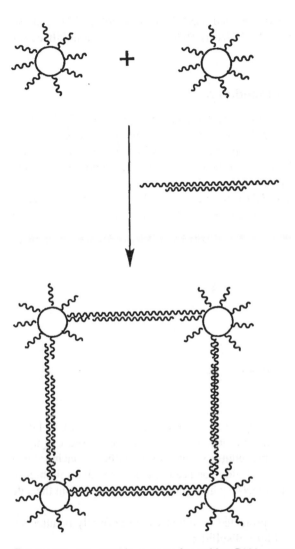

Figure 10 The Mirkin strategy for making DNA–nanoparticle assemblies [21]. Home-made 13-nm gold nanoparticles are conjugated to single-stranded noncomplementary 8mers via a 3'-thiol linkage. A linker duplex DNA that is 12 bp long in the middle and has single-stranded "sticky ends" complementary to both 8mers is added to the solution; complete hybridization results in a three-dimensional, crosslinked, nanoparticle–DNA net-work. Only a schematic two-dimensional assembly is illustrated for clarity.

and STV to 40-nm gold nanoparticles, and used the DNA–biotin–STV–Au assembly to image transcription of the DNA by RNA polymerase immobilized on a glass slide [5].

B. Covalent Attachment Procedures

One reason for the popularity of gold nanoparticles in these studies is that gold is readily covalently modified by thiols [108]. Many biological molecules can be derivatized to have a free thiol group (including DNA). The 1999 Sigma catalog, for example, offers 20 different modifications of gold nanoparticles; the surface groups include insulin, albumin, peroxidase, and STV.

Introduction of a thiol into DNA can be done at the 5′ and or 3′ end; such oligonucleotides can be obtained commercially. Shown below are the common 5′-thiol modifications:

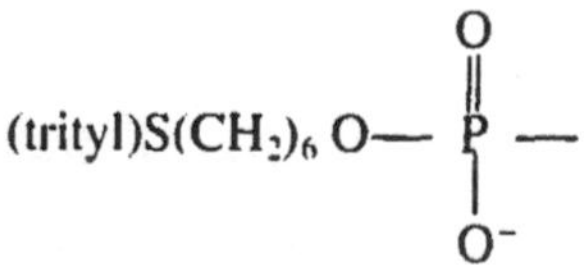

$$(trityl)S(CH_2)_6\ O—\overset{\overset{\textstyle O}{\|}}{\underset{\underset{\textstyle O^-}{|}}{P}}—\qquad \mathbf{1}$$

$$HO(CH_2)_6SS(CH_2)_6\ O—\overset{\overset{\textstyle O}{\|}}{\underset{\underset{\textstyle O^-}{|}}{P}}—\qquad \mathbf{2}$$

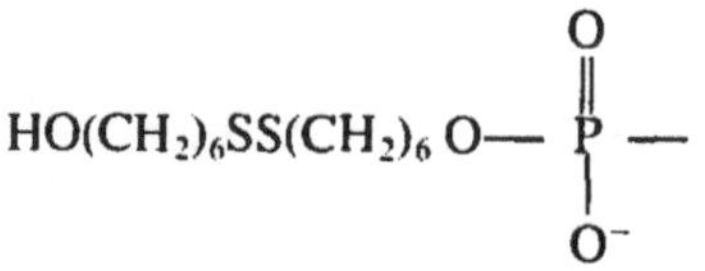

In **1** standard deprotection of the trityl group affords the free thiol, which should be immediately reacted with a maleiimide or iodoacetyl-derivatized conjugate (or gold). Standard deprotection procedures on the DNA synthesis machines can afford the disulfide **2** directly; the disulfide bond can be cleaved with dithiothreitol (DTT) to afford the free thiolated DNA, which should be separated from the side product $HO(CH_2)_6SH$.

In the 3′ modification, the protected base that would normally be attached to the solid support is replaced by a disulfide:

$$(DMTO)(CH_2)_3SS(CH_2)_3O—|$$

The *o*-dimethoxytrityl (DMTO) protecting group is a standard one in DNA phosphoramidite chemistry. DTT treatment again affords the free thiol, which can be coupled to gold. Mirkin and his group [6,7] have made use of both kinds of DNA modifications to develop hybridization assays using gold nanoparticles.

C. Optical Properties of Gold Nanoparticle–DNA Assemblies

The most striking optical data of nanoparticle–DNA conjugates has come from Mirkin's laboratory [6,7,21]. Aqueous solutions of 13-nm gold nanoparticles loaded with single-stranded oligonucleotides appear red. Upon hybridization with complementary linker DNA, the gold nanoparticles come within about 10 nm of each other, close enough to influence the plasmon absorption band, and the color of the solution changes to blue. The entire aggregate can precipitate [21], but optimization of salt and other conditions can keep the aggregate in solution for some time. The color change is reversible upon heating, which melts the duplexes back to single strands and thus separates the gold nanoparticles. Mirkin's group has optimized the system to detect 10 fmol of a target DNA single strand [6]. Mismatches are readily detected because they melt at a lower temperature than the perfect target. Mirkin's group has also developed a spot test in which the red or blue solutions at various temperatures are spotted onto a thin-layer chromatography plate; the red or blue color is retained permanently and serves as a record of the hybridization test [6,7].

D. Further Characterization of Nanoparticle–DNA Assemblies

Transmission electron and light microscopies are other tools to characterize nanoparticle–DNA assemblies [5,21–24,107]. In TEM measurements, the nanoparticle diameter can be measured, and interparticle spacing can be measured to confirm the length of the DNA linker [21,22]. With sufficiently large gold nanoparticles (40 nm), the brownian motion of the nanoparticles can be imaged under the light microscope; analysis of this kind of data has been used to measure the movement of individual RNA polymerase molecules along gold-tagged DNA [5]. The DNA–nanoparticle aggregates can also be examined by gel electrophoresis after fluorescent labeling, which permits estimates of molecular weight of the assembly [107].

VI. RELATED BIOLOGICAL APPLICATIONS OF INORGANIC NANOPARTICLES: NEW BIODETECTION SCHEMES AND NEW ROUTES TO BIOMATERIALS

There has been recent exciting work in which biological molecules other than DNA are attached or adsorbed to inorganic nanoparticles. Two different laboratories have simultaneously reported the use of CdSe nanoparticles that are highly photoluminescent as "fluorescent dyes" for biological cells [109,110]. In both cases CdSe quantum dots are surface-passivated with a thin layer of a larger bandgap material (CdS or ZnS), and a conjugatable group is attached to the outer surface. One key component in the syntheses was the production of water-soluble and stable nanoparticles. The Alivisatos group achieved this by coating their nanoparticles with a water-soluble silica shell [109]; the Nie group coordinated-—SCH_2COOH to their nanoparticles to render them water-soluble [110]. Subsequent attachment of small molecules, such as biotin, or large molecules, such as proteins, to the nanoparticle surface was achieved, and localization of the molecules that bound these nanoparticle-bound species in the cell could be imaged [109,110]. One major strength of the nanoparticle approach is that the color of the "dye" is a function of nanoparticle size; the same chemical strategies for attachment of groups to the outside of the particle can be used to achieve multicolor imaging and detection in vivo. For organic dyes that emit different colors, in contrast, new chemistry must be developed for each molecule to covalently attach it to a receptor of interest.

Surface plasmon resonance (SPR) is an emerging medical diagnostic technique. The method relies on the change in refractive index of a thin metal film upon adsorption of an analyte; specificity comes from chemical modification of the metal film. Natan and his group have found that gold nanoparticles enhance the ability of SPR to detect protein–protein interactions and have detected picomolar concentrations of human immunoglobin G [111].

Materials scientists, inspired by nature, are using the principles of biological recognition to assemble inorganic nanoparticles in new ways [112–116]. General strategies are similar to the DNA hybridization approach but use other ligand–receptor sets instead of complementary base pairing as the means to organize the nanoparticles. Other approaches use preorganized cavities in biological supramolecular structures as hosts for the synthesis of nanoparticle guests.

VII. SUMMARY AND OUTLOOK

The alliance of DNA with nanomaterials has already produced exciting results for both the biotechnological community and the materials community. The gold nanoparticle approach to detection of DNA hybridization, for example, does not require expensive and highly regulated radioactive labeling, or tedious gel electrophoresis, to determine whether the perfect complement to a target DNA strand is present. Semiconductor nanoparticles show promise as optical probes of distorted DNA structure and also as a new class of generic photoluminescent markers for biology. Medical diagnostics in which inorganic nanoparticles play an important role are already on the market [117]. Finally, DNA-based strategies are a new way to organize nanoparticles for molecular electronic or optoelectronic applications.

ACKNOWLEDGMENTS

I thank Samyah Hammami and Latha Gearheart for assistance with the preparation of this chapter. Work in our laboratory has been supported by the National Science Foundation, the National Institutes of Health, the Research Corporation, the Alfred P. Sloan Foundation, and the Dreyfus Foundation.

REFERENCES

1. Caruthers, M. H.; Beaton, G.; Wu, J. V.; Wiesler, W. *Meth. Enzymol.* **1992**, *211*, 3.
2. Brown, T.; Brown, D. J. S. *Meth. Enzymol.* **1992**, *211*, 20.
3. Goodman, S. L.; Hodges, G. M.; Trejdosiewicz, L. K.; Livingston, D. C. *J. Microscopy Pt 2* **1981**, *123*, 201.
4. Hayat, M. A., Ed. *Colloidal Gold: Principles, Methods, and Applications*, Vol. 1; Academic Press: San Diego, 1989.
5. Schafer, D. A.; Gelles, J.; Sheetz, M. P.; Landick, R. *Nature* **1991**, *352*, 444.
6. Elgahanian, R.; Storhoff, J. J.; Mucic, R. C.; Letsinger, R. L.; Mirkin, C. A. *Science* **1997**, *277*, 1078. An accompanying commentary: Service, R. F. *Science* **1997**, *277*, 1036.
7. Storhoff, J. J.; Elgahanian, R.; Mucic, R. C.; Mirkin, C. A.; Letsinger, R. L. *J. Am. Chem. Soc.* **1998**, *120*, 1959.
8. Mahtab, R.; Rogers, J. P.; Murphy, C. J. *J. Am. Chem. Soc.* **1995**, *117*, 9099.
9. Mahtab, R.; Rogers, J. P.; Singleton, C. P.; Murphy, C. J. *J. Am. Chem. Soc.* **1996**, *118*, 7028.
10. Murphy, C. J.; Brauns, E. B.; Gearheart, L. *Mater. Res. Soc. Symp. Proc.* **1997**, *452*, 597.
11. Mahtab, R.; Harden, H. H.; Murphy, C. J. *J. Am. Chem. Soc.*, **2000**, *122*, 14.
12. Cheng, L.; Ziegelhoffer, P. R.; Yang, N. S. *Proc. Natl. Acad. Sci. USA* **1993**, *90*, 4455.
13. Casas, A. M.; Kononowicz, A. K.; Zehr, U. B.; Tomes, D. T. Axtell, J. D.; Butler L. G. Bresson, R. A.; Hasegawa, P. M. *Proc. Natl. Acad. Sci. USA* **1993**, *90*, 11212.
14. Fynan, E. F.; Webster, R. G.; Fuller, D. H.; Haynes, J. R. Santoro, J. C.; Robinson, H. L. *Proc. Natl. Acad. Sci. USA* **1993**, *90*, 11478.
15. Kreuter, J. in *Microcapsules and Nanoparticles in Medicine and Pharmacy*, Donbrow, M., Ed.; CRC Press: Boca Raton, FL, 1992.
16. Sanford, J. C.; Smith, F. D.; Russell, J. A. *Meth. Enzymol.* **1993**, *217*, 483. A comment on the commercial potential of "gene gun" technology can be found in *Nature Biotechnol.* **1998**, *16*, 309.
17. Yaffee, M.; Walter, P.; Richter, C.; Muller, M. *Proc. Natl. Acad. Sci. USA* **1996**, *93*, 5341.
18. Nabiev, I.; Baranaov, A.; Chourpa, I.; Beljebbar, A.; Sockalingum, G. D.; Manfait, M. *J. Phys. Chem.* **1995**, *99*, 1608.
19. Sanchez-Cortez, S.; Miskovsky, P.; Jancurra, D.; Bertoluzza, A. *J. Phys. Chem.* **1996**, *100*, 1938.

20. Dou, X.; Takama, T.; Yamaguchi, Y.; Hirai, K.; Yamamoto, H.; Doi, S.; Ozaki, Y. *Appl. Optics* **1998**, *37*, 759.
21. Mirkin, C. A.; Letsinger, R. L.; Mucic, R. C.; Storhoff, J. J. *Nature* **1996**, *382*, 607.
22. Alivisatos, A. P.; Johnsson, K. P.; Peng, X.; Wilson, T. E.; Loweth, C. J.; Bruchez, Jr., M. P.; Schultz, P. G. *Nature* **1996**, *382*, 609.
23. Coffer, J. L.; Bigham, S. R.; Li, X.; Pinizzotto, R. F.; Rho, Y. G.; Pirtle, R. M.; Pirtle, I. L. *Appl. Phys. Lett.* **1996**, *69*, 3851.
24. Braun, E.; Eichen, Y.; Sivan, U.; Ben-Yoseph, G. *Nature* **1998**, *391*, 775.
25. Kim, C. W.; Rha, C. *Biotechnol. Bioeng.* **1989**, *33*, 1205.
26. Khanna, M.; Stotzky, G. *Appl. Environ. Microbiol.* **1992**, *58*, 1930.
27. Romanowski, G.; Lorenz, M. G.; Wackernagel, W. *Appl. Environ. Microbiol.* **1991**, *57* 1057.
28. Mahler, B. J.; Winkler, M.; Bennett, P.; Hillis, D. M. *Geology* **1998**, *26*, 831.
29. Day, P. J. R.; Flora, P. S.; Fox, J. E.; Walker, M. R. *Biochem. J.* **1991**, *278*, 735.
30. Fang, Y. *J. Chem. Phys.* **1998**, *108*, 4315.
31. Spanhel, L.; Haase, M.; Weller, H.; Henglein, A. *J. Am. Chem. Soc.* **1987**, *109*, 5649.
32. Henglein, A. *Chem. Rev.* **1989**, *89*, 1861.
33. Dameron, C.; Reese, R. N.; Mehra, R. K.; Kortan, A. R.; Carroll, P. J.; Steigerwald, M. L.; Brus, L. E.; Winge, D. R. *Nature* **1989**, *338*, 596.
34. Stucky, G. D.; MacDougall, J. E. *Science* **1990**, *247*, 669.
35. Herron, N.; Wang, Y.; Eckert, H. *J. Am. Chem. Soc.* **1990**, *112*, 1322.
36. Steigerwald, M. L.; Brus, L. E. *Acc. Chem. Res.* **1990**, *23*, 183.
37. Wang, Y.; Herron, N. *J. Phys. Chem.* **1991**, *95*, 525.
38. Pileni, M. P.; Motte, L.; Petit, C. *Chem. Mater.* **1992**, *4*, 338.
39. Weller, H. *Angew. Chem. Int. Ed. Engl.* **1993**, *32*, 41.
40. Murray, C. B.; Norris, D. B.; Bawendi, M. G. *J. Am. Chem. Soc.* **1993**, *115*, 8706.
41. Sooklal, K.; Cullum, B. S.; Angel, S. M.; Murphy, C. J. *J. Phys. Chem.* **1996**, *100*, 4551.
42. Sooklal, K.; Hanus, L. H.; Ploehn, H. J.; Murphy, C. J. *Adv. Mater.* **1998**, *10*, 1087.
43. Katari, J. E. B.; Colvin, V. L.; Alivisatos, A. P. *J. Phys. Chem.* **1994**, *98*, 4109.
44. Peng, X.; Wickham, P. J.; Alivisatos, A. P. *J. Am. Chem. Soc.* **1998**, *120*, 5343.
45. For a historical review, see: Kerker, M. *Appl. Optics* **1991**, *30*, 4699.
46. Creighton, J. A.; Blatchford, C. G.; Albrecht, M. G. *J. Chem. Soc. Faraday Trans. II* **1979**, *75*, 790.
47. Garrell R. L. *Anal. Chem.* **1989**, *61*, 401A.
48. Campion, A.; Kambhampati, P. *Chem. Soc. Rev.* **1998**, *27*, 241.
49. Keating, C. D.; Kovaleski, K. M.; Natan, M. J. *J. Phys. Chem. B* **1998**, *102*, 9404.
50. Andres, R. P.; Averback, R. S.; Brown, W. L.; Brus, L. E.; Goddard, III, W. A.; Kaldor, A.; Louie, S. G.; Moscovits, M.; Peercy, P. S.; Riley, S. J.; Siegel, R. W.; Spaepen, F.; Wang, Y. *J. Mater. Res.* **1989**, *4*, 704.
51. Weller, H. *Adv. Mater.* **1993**, *5*, 8.
52. Reed, M. A. *Sci. Amer.* **1993**, 113.
53. Kamat, P. *Chem. Rev.* **1993**, *93*, 267.
54. Wang, Y. *Adv. Photochem.* **1995**, *19*, 179.

55. Norris, D. J.; Bawendi, M. G. *Phys. Rev. B* **1996**, *53*, 16338.
56. Alivisatos, A. P. *Science* **1996**, *271*, 933.
57. Alivisatos, A. P. *J. Phys. Chem.* **1996**, *100*, 13226.
58. Murphy, C. J. *J. Cluster Sci.* **1996**, 7, 341.
59. Alivisatos, A. P.; Barbara, P. F.; Castleman, A. W.; Chang, J.; Dixon, D. A.; Klein, M. L.; McLendon, G. L.; Miller, J. S.; Ratner, M. A.; Rossky, P. J.; Stupp, S. I.; Thompson, M. E. *Adv. Mater.* **1998**, *10*, 1297.
60. Weller, H. *Curr. Opin. Colloid Surf. Sci.* **1998**, *3*, 194.
61. Brus, L. E. *J. Chem. Phys.* **1984**, *80*, 4403.
62. Fox, M. A.; Dulay, M. T. *Chem. Rev.* **1993**, *93*, 341.
63. Murphy, C. J.; Lisensky, G. C.; Leung, L. K.; Kowach, G. R.; Ellis, A. B. *J. Am. Chem. Soc.* **1990**, *112*, 8344.
64. Lisensky, G. C.; Penn, R. L.; Murphy, C. J.; Ellis, A. B. *Science* **1990**, *248*, 840.
65. Lewis, N. S. *Annu. Rev. Phys. Chem.* **1991**, *42*, 543.
66. Sequaris, J.-M.; Fritz, J.; Lewinsky, H.; Koglin, E. *J. Colloid Interfac. Sci.* **1985**, *105*, 417.
67. Kneipp, K.; Flemming, J. *J. Mol. Struct.* **1986**, *145*, 173.
68. Nabiev, I. R.; Sokolov, K. V.; Voloshin, O. N. *J. Raman Spectrosc.* **1990**, *21*, 333.
69. Kneipp, K.; Pohle, W.; Fabian, H. *J. Mol. Struct.* **1991**, *244*, 183.
70. Cotton, T. M.; Kim, J-H.; Chumanov, G. D. *J. Raman Spectrosc.* **1991**, 22, 729.
71. Sokolov, K.; Khodorchenko, P.; Petukhov, A.; Nabiev, I.; Chumanov, G.; Cotton, T. M. *Appl. Spectrosc.* **1993**, *47*, 515.
72. Garcia-Ramos, J. V.; Sanchez-Cortes, S. *J. Mol. Struct.* **1997**, *405*, 13.
73. Kneipp, K.; Kneipp, H.; Kartha, V. B.; Manoharan, R.; Deinum, G.; Itzkan, I.; Dasari, R. R.; Feld, M. S. *Phys. Rev. E* **1998**, *57*, R6281.
74. Lecomte, S.; Baron, M-H. *Biospectroscopy* **1997**, *3*, 31.
75. Coffer, J. L.; Chandler, R. R. *Mater. Res. Soc. Symp. Proc.* **1991**, *206*, 527.
76. Bigham, S. R.; Coffer, J. L. *J. Phys. Chem.* **1992**, *96*, 10581.
77. Coffer, J. L.; Bigham, S. R.; Pinizzotto, R. F.; Yang, H. *Nanotechnology* **1992**, *3*, 69.
78. Bigham, S. R.; Coffer, J. L. *Colloids Surfaces A: Physicochem. Eng. Aspects* **1995**, *95*, 211.
79. Coffer, J. L. *J. Cluster Sci.* **1997**, *8*, 159.
80. Saenger, W. *Principles of Nucleic Acid Structure*; Springer-Verlag: New York, 1984.
81. Crothers, D. M.; Haran, T. E.; Nadeau, J. G. *J. Biol. Chem.* **1990**, *265*, 7093.
82. Hagerman, P. J. *Annu. Rev. Biochem.* **1990**, *59*, 755.
83. Goodsell, D. S.; Kopka, M. L.; Cascio, D.; Dickerson, R. E. *Proc. Natl. Acad. Sci. USA* **1993**, *90*, 2930.
84. Wolffe, A. P.; Drew, H. R. in *Chromatin Structure and Gene Expression*, Elgin, S., Ed.; IRL Press: Oxford, 1995.
85. Grzeskowiak, K. *Chem. Biol.* **1996**, *3*, 785.
86. Dlakic, M.; Harrington, R. E. *Proc. Natl. Acad. Sci. USA* **1996**, *93*, 3847.
87. Allemann, R. K.; Egli, M. *Chem. Biol.* **1997**, *4*, 643.
88. Dickerson, R. E. *Nucl. Acids Res.* **1998**, *26*, 1906.
89. Goodman, S. D.; Nash, H. A. *Nature* **1989**, *341*, 251.

90. Gartenberg, M. R.; Crothers, D. M. *J. Mol. Biol.* **1991**, *219*, 217.
91. Schultz, S. C.; Shields, G. C.; Steitz, T. A. *Science* **1991**, *253*, 1001.
92. Kim, Y.; Geiger, J. H.; Hahn, S.; Sigler, P. B. *Nature* **1993**, *365*, 512.
93. Kim, J. L.; Nikolov, D. B.; Burley, S. K. *Nature* **1993**, *365*, 520.
94. Parvin, J. D.; McCormick, R. J.; Sharp, P. A.; Fisher, D. E. *Nature* **1995**, *373*, 724.
95. Grove, A.; Galeone, A.; Mayol, L. Geiduschek, E. P. *J. Mol. Biol.* **1996**, *260*, 120.
96. Rozenberg, H.; Rabinovich, D.; Frolow, F.; Hedge, R. S.; Shakked, Z. *Proc. Natl. Acad. Sci. USA* **1998**, *95*, 15194.
97. Young, M. A.; Ravishanker, G.; Beveridge, D. L.; Berman, H. M. *Biophys. J.* **1995**, *68*, 2454.
98. Olson, W. K.; Gorin, A. A.; Lu, X. J.; Hock, L. M.; Zhurkin, V. B. *Proc. Natl. Acad. Sci. USA* **1998**, *95*, 11163.
99. Flatters, D.; Lavery, R. *Biophys. J.* **1998**, *75*, 372.
100. Rouzina, I.; Bloomfield, V. A. *Biophys. J.* **1998**, *74*, 3152.
101. Manning, G. S. *Q. Rev. Biophys.* **1978**, *11*, 179.
102. Record, M. T., Jr.; Spolar, R. S. in *The Biology of Nonspecific DNA–Protein Interactions*, Rezvin, A., Ed.; CRC Press: Boca Raton, FL, 1990.
103. Takeda, T.; Ross, P. D.; Mudd, C. P. *Proc. Natl. Acad. Sci. USA* **1992**, *89*, 8180.
104. Fogoul, D.; Silva, J. L. *Proc. Natl. Acad. Sci. USA* **1994**, *91*, 8244.
105. Anderson, C. F.; Record, M. T., Jr. *Annu. Rev. Phys. Chem.* **1995**, *46*, 657.
106. Storhoff, J. J.; Mucic, R. C.; Mirkin, C. A. *J. Cluster Sci.* **1997**, *8*, 179.
107. Niemeyer, C. M.; Burger, W.; Peplies, J. *Angew. Chem. Intl. Ed. Engl.* **1998**, *37*, 2265.
108. Weisbecker, C. S.; Merritt, M. V.; Whitesides, G. M. *Langmuir* **1996**, *12*, 3763.
109. Bruchez, M., Jr.; Moronne, M.; Gin, P.; Weiss, S.; Alivisatos, A. P. *Science* **1998**, *281*, 2013.
110. Chan, W. C. W.; Nie, S. *Science* **1998**, *281*, 2016.
111. Lyon, L. A.; Musick, M. D.; Natan, M. *J. Anal. Chem.* **1998**, *70*, 5177.
112. Shenton, W.; Pum, D.; Sleytr, U. B.; Mann, S. *Nature* **1997**, *389*, 585.
113. Douglas, T.; Young, M. *Adv. Mater.* **1999**, *11*, 679.
114. Shenton, W.; Douglas, T.; Young, M.; Stubbs, G.; Mann, S. *Adv. Mater.* **1999**, *11*, 253.
115. Li, M.; Wong, K. W. K.; Mann, S. *Chem. Mater.* **1999**, *11*, 23.
116. Aherne, D.; Rao, S. N.; Fitzmaurice, D. *J. Phys. Chem. B* **1999**, *103*, 1821.
117. Martin, C. R.; Mitchell, D. T. *Anal. Chem.* **1998**, *70*, 322A.

8

Luminescence Quenching by Oxygen in Polymer Films

Xin Lu and Mitchell A. Winnik
University of Toronto, Toronto, Ontario, Canada

I. INTRODUCTION

This chapter describes the quenching by oxygen of the luminescence of dyes contained in a polymer matrix. This matrix is typically a polymer film. In most applications, one imagines that the dyes are molecularly dispersed (i.e., dissolved) in the film. There are, however, interesting examples where dyes that would normally be insoluble in the matrix are adsorbed on the surface of silica particles [1], which in turn are dispersed in the polymer matrix or covalently attached to the surface of controlled porous glass [2]. These various dye/polymer systems serve two general applications. The first involves oxygen sensors for monitoring the pressure or concentration of oxygen in a medium in contact with the dye-containing polymer film. Examples include the determination of oxygen in fluids such as blood [3a,b] in a tumor cell [3c], or in the gas phase in a variety of environments including one's breath [3d]. The second application involves the use of polymer films containing a luminescent dye to monitor the air–pressure profile across a surface, particularly in moving air. Created as a method for measuring air–pressure profiles for objects such as aviation models in wind tunnels [4], this technique is often referred to as "phosphorescence barometry"; and the polymer film applied to the substrate is called "pressure-sensitive paint" (PSP) [5]. These techniques are similar in that both rely on dyes with long excited-state lifetimes, typically with significant triplet character (phosphorescence).

311

In oxygen-sensing applications the average emission intensity or the emission decay profile is measured across the entire polymer film. The area monitored can be small. The dye-containing polymer is sometimes coated onto a portion of a fiberoptic [6] (see Fig. 1), which is inserted in the medium to be tested. Phosphorescence barometry is a two-dimensional extension of oxygen sensing. One illuminates a relatively large surface area and monitors the emission intensity with a digital camera. Each pixel serves as an individual sensor. The fundamental idea of this technology is that for thin polymer films (micrometers), the solubility

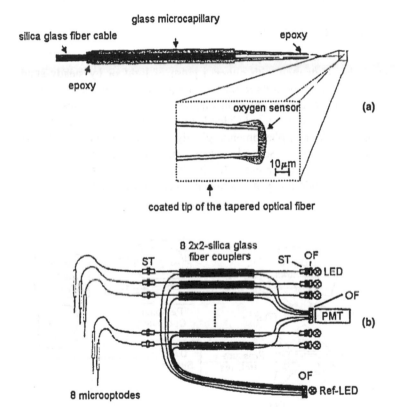

Figure 1 (a) Schematic drawing of an oxygen microsensor ("microoptode") built by Holst et al. The tapered silica glass fiber is fixed with epoxy in a glass microcapillary for better handling. The oxygen indicator layer (see insert) is applied by dipcoating. (b) The optical setup consists of several fused fiber couplers that have been especially assembled: PMT, photomultiplier tube assembly; OF, optical filters; LED, light-emitting diode; Ref-LED, reference light-emitting diode; ST, standard fiber connectors and receptacles. (From Ref. 6.)

of oxygen in the polymer film at each pixel will be related to the surface air pressure at that small area of the substrate. The higher the local concentration of O_2, the greater the extent of excited-state quenching. Thus regions of high surface air pressure will appear darker in the image than regions of lower air pressure. A schematic of the PSP data acquisition system is shown in Fig. 2.

Imaging of oxygen fluxes is also important in biology [7] as a means, for example, of studying local blood microcirculation in tissue. Activity in this area dates from the mid-1980s. Many of the publications describe imaging techniques based on spatially resolved luminescence decay measurements [8]. A very useful review of this field is provided in a recent paper by Hartmann et al. [9].

A. Luminescence Quenching

The kinetic scheme for a typical dynamic luminescence quenching reaction is shown below. A dye is excited with either a pulse of light or by steady-state illumination. In competition with its luminescence (either fluorescence or phosphorescence), the excited dye (dye*) can encounter a second species Q, the quencher. The interaction of dye* and Q leads to the deactivation of dye*. The term "quenching" is defined phenomenologically as any bimolecular process involving an excited dye and a second species that leads to a decrease in the luminescence intensity and an increase in the luminescence decay rate. The term says nothing about the underlying interaction mechanism between dye* and Q,

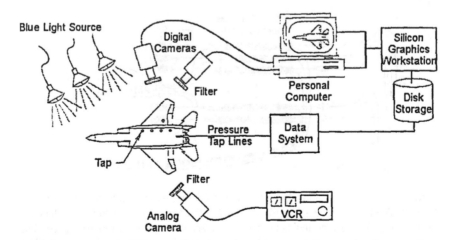

Figure 2 Schematic of a pressure-sensitive paint measurement system used by McDonnell Douglas Research Laboratories for testing air pressure profiles on an airplane model in a wind tunnel. (From Ref. 4b.)

which can, for example, be energy transfer, electron transfer, or complex formation (see scheme below).

$$\text{Dye} \xleftarrow{\ 1/\tau^0\ } \text{dye}^* + Q \xrightarrow{\ k_q[Q]\ } \text{dye} + Q^* \text{ or } Q$$

From this simple mechanism one derives the Stern–Volmer (SV) equation [eq. (1)] [10], which serves as the point of departure for all discussions of fluorescence and phosphorescence quenching involving mobile species in polymers. A different model, the Perrin model [10], is used to describe static quenching, the situation one encounters when neither dye* nor Q has significant mobility on the time scale of the excited state.

$$\frac{I^0}{I} = \frac{\tau^0}{\tau} = 1 + k_q \tau^0 [Q]_{eq} \tag{1}$$

In this expression, I is the emission intensity at some molar quencher concentration $[Q]$, and I^0 is the corresponding intensity in the absence of quencher. The decay of the electronically excited dye follows first-order kinetics in the absence of Q and pseudo-first-order kinetics in the presence of Q. In other words, dye* exhibits an exponential decay profile, with lifetime τ^0 in the absence of Q and τ in its presence. The concentration of Q does not change during the measurement. Since we are concerned with oxygen as a quencher, we use the subscript "eq" in Eq. (1) to emphasize that when the quencher itself is a gas, it must be in equilibrium with the condensed phase in which the quenching reaction occurs. This ensures a globally uniform concentration of Q throughout the medium. For dynamic quenching processes, I and τ are related. To determine the value of the second-order quenching rate constant k_q, one can carry out a series of lifetime measurements at different $[Q]$. Alternatively, one can determine k_q from steady-state intensity measurements at different $[Q]$ in conjunction with a measurement of τ^0.

The derivation of Eq. (1) implicitly assumes a "well-stirred reactor," which implies that the relaxation of the solvent around the excited dye and the redistribution of reactants in the solution are faster than the reaction itself. As a consequence, the concentration of Q in the solution is effectively uniform, and the radial distribution function of Q surrounding each dye* is uniform. We emphasize this point because many sensor experiments employ dyes which exhibit nonexponential decay profiles even in the absence of oxygen. In other systems, where the excited dye itself undergoes simple first-order decay in the absence of quencher, the decays become nonexponential in the presence of oxygen. Nonexponential decays in the absence of quencher point to a distribution of emitting species in the system. Deviations from simple pseudo-first-order kinetics in the presence of quencher are likely connected to spatial variations in quencher concentration which do not equilibrate on the time scale defined by τ^0. We will describe a number of these examples and, where possible, suggest reasons for this nonideal behavior.

Oxygen is a very effective quencher of both excited singlet and triplet states. In fluid solution, k_q approaches diffusion control ($k_q \approx k_{diff}$) [11,12]. In diffusion-controlled quenching reactions, individual dye* molecules that happen to have a nearby Q react very quickly, and the local concentration of Q in the vicinity of the remaining dye* becomes depleted. The system evolves in such a way that the rate becomes controlled by the flux of Q to the surviving dye*. The evolution of the distribution of Q contributes a nonexponential component at short times to the decay of dye*. In treating experimental data, the local concentration gradients of Q in system are reexpressed in terms of a time-dependent, diffusion-controlled rate coefficient. From the theory of partially diffusion controlled reactions [13], one derives Eq. (2), which describes the time evolution of k_{diff} for times not too close to $t = 0$:

$$k_{diff}(t) = \frac{4\pi N_A R_{eff} D}{1000} \left(1 + \frac{R_{eff}}{\sqrt{\pi\,Dt}} \right)$$ (2)

In this expression N_A is Avogadro's number and D is the sum of the diffusion coefficients of the interacting species. R_{eff} is an effective interaction distance at which quenching takes place. When quenching occurs on only a fraction of the encounters between dye* and Q, R_{eff} is smaller than the contact distance R at which orbital overlap occurs. If time-resolved quenching experiments can be carried out over sufficiently short times that the $t^{1/2}$ component (the "transient term") preceding the exponential decay can be monitored, the parameters D and R_{eff} can be obtained from the data [14]. If one can perturb the system to decrease D, the residence time of dye* in a dye*/Q encounter pair is prolonged, and the quenching efficiency increases. One can show that $1/R_{eff}$ approaches $1/R$ as D approaches zero [15].

For experiments in fluid solution, the transient term decays rapidly (a few nanoseconds at most) and is difficult to detect. Under these circumstances dye* decay profiles are exponential, and the second-order rate coefficient obtained from the data analysis is the steady-state value of k_{diff}:

$$k_{diff} = (4\pi N_A/1000)DR_{eff}$$ (3)

Under these circumstances, it is possible to relate R_{eff} to R using a theoretical model developed by Pilling and Rice [16].

B. Oxygen Quenching in Polymers

Studies of luminescence quenching in polymer matrices by mobile quenchers such as oxygen date from the early 1960s. In his review of this topic, Guillet

[17] cites Oster et al. [18] as among the first to study this phenomenon. They measured the quenching of a phosphorescent probe in poly(methyl methacrylate) (PMMA) by oxygen and showed that the emission intensity was sensitive to the diffusion of oxygen into the film. For experiments in polymers, one often has to contend with a system in which both the solubility and diffusivity of oxygen is low. Dyes with long-lived excited states are particularly useful. Another feature of polymers is that the diffusion constant of the dye (D_{dye}) is much smaller than that of oxygen (D_{O2}), particularly below the glass transition temperature (T_g). Under these circumstances, one can equate D in Eqs. (2) and (3) with D_{O2}.

II. GAS DIFFUSION IN POLYMERS

A. Diffusivity and Permeability

The diffusivity of oxygen and other gases in polymers is most commonly determined by measuring the rate of gas permeation across a membrane. At time zero, one exposes one side of a polymer film to a gas and measures the flux of the gas across the film as a function of time. The diffusion coefficient D is determined from the kinetics of the approach of the flux to its steady-state value. An example is shown in Fig. 3. The initial rate of transport is slow and is characterized by

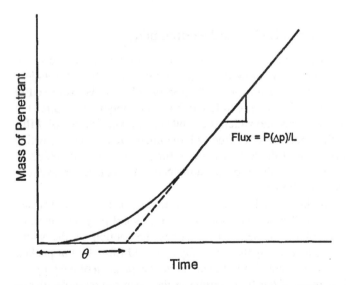

Figure 3 Typical membrane permeation response curve. Mass of gas that passes through a polymer film of thickness L as a function of time. Extrapolation of the steady-state flux line to the time axis gives the time lag θ as the intercept.

a lag time θ, after which the amount of gas transported increases linearly with time. The slope of this line is the steady-state flux F, related to permeability P of the gas by the expression

$$F = P(\Delta p)/L \qquad (4)$$

where L is the thickness of the polymer film and Δp is the difference in gas pressure across the film. P is related to D through the expression $P = DS$, where S is the solubility of the gas in the polymer at equilibrium [19]. The diffusion coefficient is determined from the lag time ($\theta = L^2/6D$), and S is calculated from the ratio P/D.

Gas diffusion in polymers normally satisfies Fick's laws of diffusion. D values for gases like oxygen in different polymers can vary by more than 10 orders of magnitude. For polymers above T_g, Henry's law is normally satisfied; S is proportional to the partial pressure of the gas in contact with the matrix. Below T_g, amorphous polymers are known to be inhomogeneous. Density fluctuations (free volume fluctuations), driven by brownian motion in simple liquids, are frozen in the glassy state [20]. The sorption of gases into glassy polymer films is more complex than that described by Henry's law. An additional term is needed to relate S to the partial gas pressure. In the dual-mode sorption model, one imagines that some of the sites of frozen free volume serve as sites for gas sorption ("Langmuir-type" adsorption) in addition to the Henry-type gas solubilization [21].

B. Mechanism of Gas Diffusion and Permeability

The diffusion of small gas molecules through solid amorphous polymers can occur even at temperatures well below the T_g of the polymer. The most widely accepted model for the mechanism of gas diffusion in polymers is expressed in terms of the free volume in the system. Solute diffusion in polymers requires the creation of molecule-size voids into which the solute can move. The rate of diffusion depends on the time necessary for free-volume fluctuations to occur, which in turn depends on the size of the hole necessary for diffusion to take place. A clear and concise description of this process in molecular terms is given by Guillet, [17] which we summarize here.

Whereas diffusion in simple liquids occurs primarily as a result of translational displacement of small molecules, the situation in solid polymers is more complex. In the free-volume model, one imagines the motion of the solute through the matrix to be assisted by the thermally activated formation of packets of free volume. As the temperature is increased, thermal expansion of the polymer leads to an overall increase of the free volume in the system. For molecules as small as nitrogen or oxygen, the sites of free volume that lead to diffusion are created or redistributed by the small-scale motion of polymer segments or the local movement of groups pendant to the polymer backbone. The rate of these processes increases with temperature in a manner consistent with an Arrhenius activation energy.

Local motions in polymers, such as phenyl group rotation in polystyrene or rotation of the ester group in PMMA, can be examined directly by solid-state nuclear magnetic resonance (NMR) or dielectric relaxation experiments. The consequences of this motion on the mechanical properties of the polymer can be studied by dynamic mechanical analysis (DMA) measurements. The transition temperature at which the rate of motion becomes faster than the characteristic frequency of the measurement is referred to as T_β (or T_β and T_γ, if there are two different types of motion detected; T_α refers to the glass transition temperature). If the activation energy for gas diffusion in a polymer at temperatures below T_g is close to that determined independently for the β or γ transition, one often concludes that the diffusion process is coupled in an important way to the pendant group motion associated with that transition. For example, Guillet and Andrews [12b] calculated $E_a \approx 30$ kJ/mol for oxygen diffusion in polystyrene (PS) over the temperature range $-100°C$ to $+23°C$, which falls in the same range as the activation energy for phenyl group rotation determined by NMR and by DMA. More precise D values for oxygen diffusion in PS determined by Wang and Ogilby [22] over a somewhat smaller temperature range also yield $E_a \approx 30$ kJ/mol.

We should note, however, that there are a number of examples in which gas permeability in a polymer appears to be coupled to the glass transition. Under these circumstances, an Arrhenius plot of gas permeability P vs. $1/T$ will show a change in slope in the vicinity of T_g [23].

Bazhin and co-workers [20,24] have shown that in PMMA oxygen diffusion is characterized by a single activation energy (32 kJ/mol) between 77 K and room temperature, which is likely related to ester group rotation. They have also argued that at low temperature the diffusivity of a gas solute in a glassy polymer increases exponentially with an increase in free volume, whereas its solubility increases linearly. Under these circumstances, measurements below T_g of the permeability of a gas in a polymer matrix as a function of temperature are dominated by changes in the diffusion coefficient.

The thermal expansion of a polymer is significantly larger above the glass transition temperature, and the free volume of the system permits large-amplitude motion of the polymer backbone to take place. Properties that are coupled to this motion, such as large-molecule diffusion or viscoelastic behavior, exhibit a temperature dependence above T_g that is more complex than that described by a single activation energy. Two parameters are required, as, for example, in the WLF (Williams-Landel-Ferry) expression [25]. For polymers in the glassy state, diffusion rates of simple solute molecules decrease exponentially with increasing size. For solute molecules as large as cyclohexane, pyrene, or acetophenone, one finds that their diffusion in an amorphous polymer matrix is accelerated by many orders of magnitude as the temperature of the system is raised above T_g. For example, Deppe et al. [26] have shown an almost linear increase in $\log D$ with $(T-T_g)$ over the range of T_g to $T_g + 20°C$ for various dyes in a series of methacrylate films. Over this narrow temperature range, D values increased by four orders of magnitude.

One consequence of the large amount of free volume polymer matrices above T_g is that the rate of diffusion of these molecules is often only one or two orders of magnitude slower than that in ordinary liquids. For high molecular weight polymers, the macroscopic viscosity η increases by many orders of magnitude with increasing chain length ($\eta \sim M_w^{3.5}$, where M_w is the weight-averaged molecular weight). In contrast, D values for typical gases is relatively insensitive to the molecular weight of the matrix polymer, except at lower M values. This type of result emphasizes a point often raised by Guillet [17,27] that there is a profound difference in the microscopic friction, experienced by a diffusing solute in a polymer matrix, and the bulk viscosity, which depends on long-range interactions such as entanglements.

Molecular dynamics calculations have now become so sophisticated that the processes of gas solubilization and gas diffusion can be modeled quantitatively. The molecular dynamics calculations are still limited by computer power and are capable of describing the motions of polymers in the bulk state on time scales up to nanoseconds [28]. On this time scale, one can observe the onset of fickian diffusion of gas molecules in the matrix. At shorter times, the diffusion-like motion is more complex because the solute has to wait for group motion in its vicinity in order to translate [28a]. This relaxation process, in the polymer physics community, is referred to as "anomalous diffusion." In the early 1990s, estimations based on these calculations often gave higher diffusion coefficient values and lower activation energies than those found experimentally [28b,c]; however, recent calculations show that the situation is now very much improved [28d].

III. OXYGEN DIFFUSION AND LUMINESCENCE QUENCHING

Membrane diffusion experiments require a free-standing film. This is often easy to achieve for glassy and semicrystalline polymers. Viscoelastic polymers that flow cannot be examined in this way. One often introduces crosslinks into the polymer to convert the polymer to an elastic film. These crosslinks have the potential to perturb both S and D. We will examine the implications of this statement on sensor and PSP applications in a later section of this chapter. Before examining the role of polymer structure on these applications, we consider in the next section the coupling between fickian diffusion of oxygen and the kinetics of oxygen quenching in an amorphous polymer matrix.

A. Quenching and Oxygen Permeability

For quenching experiments in systems at equilibrium, the Stern–Volmer expression often applies. We can rewrite Eq. (1) as

$$\frac{I^0}{I_{eq}} - 1 = A Q_{eq} = B \tag{5a}$$

$$B = 4\pi N_A R_{eff} \tau^0 (S_{O2} D_{O2}) p_{O2} = 4\pi N_A R_{eff} \tau^0 P_{O2} p_{O2} \tag{5b}$$

where $A = k_q \tau^0$, and we have for convenience set $Q_{eq} = [Q]_{eq}$. When the quenching rate is diffusion-controlled and Henry's law applies, A and B depend on the partial pressure of gas in contact with the polymer matrix. For the specific case of oxygen as a quencher, we write Eq. (5b), where p_{O2} is the partial pressure of oxygen, S_{O2} is its solubility in the film, D_{O2} is its diffusion coefficient, and P_{O2} is its permeability. The validity of Eq. (5) has been recognized for many years. For quenching experiments on polymer films, if the plot of I^0/I_{eq} vs. p_{O2} is linear, the permeability can be calculated from the slope. This calculation requires some assumptions about the magnitude of R_{eff}, which is often taken to be in the range of 7–10 Å.

In the gas sensor literature, Eq. (5) is often written as

$$\frac{I^0}{I} - 1 = K_{SV} \, p_{O2} \tag{5c}$$

where the proportionality constant K_{SV} is referred to as the Stern–Volmer constant. For processes in solution where the quencher concentration is known, the more traditional definition of the Stern–Volmer constant is $K_{SV} = A S_{O2} = k_q \tau^0 S_{O2}$.

B is a measure of the sensitivity of a particular dye–polymer combination to quenching. The magnitude of B obviously depends on whether the reference point for I_{eq} is pure oxygen ($p_{O2} = 1$) or air ($p_{O2} = 0.2$). Experimental values of B for air as a reference range from 1.01 to greater than 200 [29]. Meaningful measurements are possible only for B values (air reference) greater than about 1.5. In the field of phosphorescence barometry, one measures the change in intensity at high air pressure relative to that in still air. If we use p_1 and p_2 to represent two different oxygen partial pressures, the ratio of the emission intensities of the dye-containing films in equilibrium with O_2 at these pressures, I_1/I_2, is given by

$$\frac{I_1}{I_2} = \frac{1 + K p_2}{1 + K p_1} = \frac{1}{1 + K p_1} + \frac{K}{1 + K p_1} p_2; \qquad K = \frac{B}{p_{O2}} \tag{6}$$

When p_1 refers to a reference pressure, Eq. (6) can be rewritten as Eq. (7) by recognizing that $B = K p_{ref}$ and $K p_2 = B p_2 / p_{ref}$:

$$\frac{I_{ref}}{I_2} = \frac{1}{1+B} + \frac{B}{1+B}\left(\frac{p_2}{p_{ref}}\right) \tag{7}$$

In phosphorescence barometry, the air pressure P^{air} replaces the oxygen pressure p_{O2} in the above equations. Taking p_{ref} as P^{air} at 1 atm (P^{air}_{1atm}), the emission intensity (I) at P^{air}_{1atm} is related to that (I_{1atm}) at 1 atm air pressure, by the expression

$$\frac{I_{1atm}}{I} = \frac{1}{1+B} + \frac{B}{1+B}\left(\frac{P^{air}}{P^{air}_{1atm}}\right) \tag{8}$$

When we fit intensity data as a function of air pressure, we can rewrite this expression as

$$\frac{I_{1atm}}{I} = A' + Q_s\left(\frac{P^{air}}{P^{air}_{1atm}}\right) \tag{9}$$

where A' is a constant, and $Q_s = B/(1+B)$ is a measure of the sensitivity of the dye–film combination to quenching due to a change in air pressure. In this type of experiment, we refer to Q_s as the "quenching sensitivity." We note that while B can take a wide range of values, depending on the magnitude of $\tau^0 DS$, values of Q_s range from 0 to 1. When $B = 1$, $Q_s = 0.5$, whereas when B is very large, $Q_s \approx 1.0$.

B. Quenching and Fickian Diffusion

Many experiments reported in the literature involving luminescence detection of sorption of oxygen into a polymer film have been described in terms of a coupling of the Stern–Volmer expression and Fick's laws of diffusion. Until recently, however, these analyses of the data have not taken proper account of the spatial concentration profile created as a consequence of diffusion that satisfies Fick's laws. Instead, the authors of these papers assumed that one could model the quenching in terms of an average oxygen concentration in the film, so that an equation analogous to Eq. (5) could be employed. The differential equation describing fickian diffusion was solved to calculate the total moles of oxygen which had entered the film at time t, and then this amount was divided by the volume of the film to obtain the time-dependent spatially averaged quencher concentration in the film. This is an incorrect approach to relating the extent of quenching in a sorption experiment to the diffusivity of oxygen.

Ogilby and his group [22] overcame these problems by designing an experiment in which the total uptake of oxygen in a film can be measured. They use visible light to irradiate a sensitizer dye whose triplet state is quenched by oxygen to produce the singlet excited state of oxygen ($^1O_2$). The concentration of $^1O_2$ in the film is determined by monitoring the intensity of its phosphorescence at 1270 nm. For sufficiently low concentrations of oxygen in the film, the emission intensity at 1270 nm is proportional to the total concentration of O_2 in the film. From the growth in intensity over time as a film is exposed to oxygen, remarkably precise values of the oxygen diffusion coefficient D_{O2} can be calculated (Fig. 4).

Cox [30] took a different approach, using larger samples, a 1-cm² cell with a centimeter scale height. The cell was filled with polymer (a siloxane) containing a dye, and all oxygen was removed from the cell. A thin slice of the cell (about 1 mm thick) was illuminated with light absorbed by the dye, and at $t = 0$ the top of the cell (shown in Fig. 5) was exposed to oxygen. As the oxygen concentration profile propagates through the illuminated slice of the cell, the emission intensity is quenched. Because the slice illuminated is perpendicular to the propagation direction of the oxygen profile, the oxygen concentration in the illuminated region will be nearly uniform. For this geometry, rigorous application of Fick's laws of diffusion is possible.

A numerical analysis of the contribution of oxygen diffusion to oxygen quenching experiments in thin polymer film was first reported by Mills [31]. A

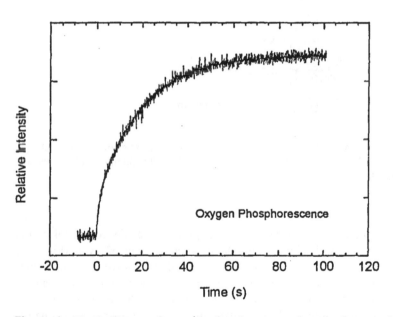

Figure 4 The Ogilby experiment. $^1O_2$ phosphorescence intensity from a polycarbonate film as a function of elapsed exposure time to an oxygen atmosphere of 30 Torr. The fitting function was obtained by incorporating the solution of Fick's law into a nonlinear least-squares routine. (From Ref. 22b.)

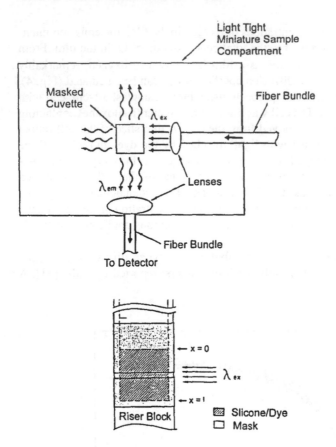

Figure 5 The Cox experiment. Schematic diagram of the sample compartment. Oxygen penetrates from the top of the cell. A cross-section of the cell is illuminated from the side. (From Ref. 30.)

derivation-based fundamental physical principles was developed by Yekta et al. [32]. This analysis begins by dividing the film into a series of layers (Fig. 6). If these layers are sufficiently thin, one can assume that the *local concentration* of oxygen in the layer located between x and $x + dx$ at time t [$Q(x,t)$] is essentially uniform. Under these circumstances, the instantaneous elemental luminescence intensity emanating from this layer is given by a Stern–Volmer type of equation.

$$\frac{\delta I^0}{\delta I(x,t)} = 1 + AQ(x,t); \quad \delta I^0 = I^0 dx/L \tag{10}$$

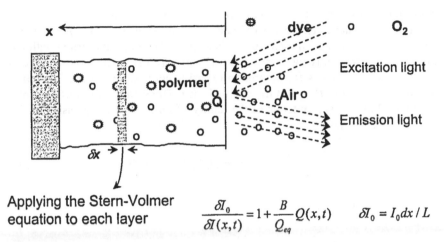

Applying the Stern-Volmer
equation to each layer

$$\frac{\delta I_0}{\delta I(x,t)} = 1 + \frac{B}{Q_{eq}} Q(x,t) \qquad \delta I_0 = I_0 dx / L$$

Figure 6 The Yekta Experiment. A polymer film containing a uniform concentration of dye is supported on a substrate. The film is illuminated with light as the film is exposed to an atmosphere of air or nitrogen. One monitors luminescence intensity as a function of exposure time.

What is measured is the sum total of the elemental intensities. When one compares the emission intensity between two equilibrium states of the system, in which Q is uniform, Eqs. (1) and (5) are recovered.

Diffusion that satisfies Fick's second law is given by the expression:

$$\frac{\partial Q(x,t)}{\partial t} = D \frac{\partial^2 Q(x,t)}{\partial x^2} \qquad (11)$$

For simplicity of notation, we define a dimensionless concentration profile and obtain

$$\rho(x,t) \equiv \frac{Q(x,t)}{Q_{eq}} \qquad (12a)$$

$$\frac{\partial \rho(x,t)}{\partial t} = D \frac{\partial^2 \rho(x,t)}{\partial x^2} \qquad (12b)$$

Diffusion generates a concentration profile of quencher in the direction normal to the film surface. Because of quenching, the emission intensity, $I(x,t)$ varies as a function of x and t as described in Eq. (10). To describe the total intensity monitored in a measurement, one has to solve Fick's law under the appropriate boundary conditions. These boundary conditions and the solution of the equations have been described in detail by Yekta et al. [32]. For a free-standing film of thickness L, the film has two faces of exposure at $x = 0$ and $x = L$. For a film on a substrate, there is only one exposed face at $x = 0$. One assumes that a thin layer at the exposed surface is in instantaneous equilibrium with the external pressure. Thus for sorption experiments, $\rho(x,t)$ at the exposed face is unity at $t = 0$, whereas for desorption experiments into an oxygen-free atmosphere, at $t = 0$, $\rho(x,t) = 0$ at the exposed face. The solution to $\rho(x,t)$ for free-standing films is given in terms of the family of terms in the expression [33]:

Oxygen sorption:

$$\rho(x,t) = 1 - \frac{4}{\pi} \sum_{n=\text{odd}}^{\infty} \frac{1}{n} \exp\{-n^2\pi^2 Dt/L^2\} \sin(n\pi x/L) \tag{13a}$$

Oxygen desorption:

$$\rho(x,t) = \frac{4}{\pi} \sum_{n=\text{odd}}^{\infty} \frac{1}{n} \exp\{-n^2\pi^2 Dt/L^2\} \sin(n\pi x/L) \tag{13b}$$

In the case of a film with only one side of exposure, thickness L in above equations would be replaced by $2L$.

IV. EXPERIMENTAL DATA

A. Response and Recovery

Many authors who work in the oxygen sensor field have noticed that the time response for the luminescent sensor when exposed to oxygen is significantly faster than the recovery time when oxygen is removed from the external environment (Fig. 7). In many of these papers, the data are reported without comment, whereas in others, explanations are attempted in terms of special or unusual properties of polymers as matrices. Until very recently it had escaped notice that this type of behavior is a direct consequence of a diffusion process that satisfies Fick's laws of diffusion. Mills [31] describes the first example of this kind of analysis, based on data obtained from the literature. As an example, in Fig. 7 we present plots of data obtained in our laboratory for oxygen diffusion in a poly(n-butylthio-nylphosphazene) (C_4PATP) matrix containing platinum octaethylporphine (PtOEP) as the luminescent dye [34]. The experimental data are compared with data calculated according to Eq. (13) in which D_{O2}/L^2 is the only adjustable parameter. Although the characteristic half-times for response and recovery to exposure to oxygen are very different, both curves can be fitted with the same value of D_{O2}/L^2. From nine separate experiments on this film of $L = 0.30$ mm, $D_{O2} = (1.4 \pm 0.4) \times 10^{-5}$ cm^2 s^{-1}.

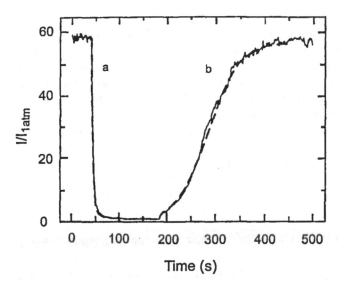

Figure 7   The Yekta experiment. Time scan of the phosphorescence of 10 ppm PtOEP in a film ($L = 300$ μm) of C₄PATP supported on quartz plate. The emission is observed at 644 nm during steady-state excitation at 535 nm. The continuous trace represents the experimental data; the dashed line shows the theoretical fit to Eq. (13). (a) The sample is initially equilibrated in an atmosphere of pure N_2 ($p_{O2} = 0.0$ atm) and the chamber is immediately flushed with air ($p_{O2} = 0.2$ atm). (b) The sample is initially equilibrated with air and the chamber is immediately flushed with pure N_2. Note that the response and recovery times are very different. (From Ref. 34.)

B.   Fluorescence Decay Profiles in the Absence of Quencher

1. Nonexponential Decays

Dyes dissolved in a polymer film, particularly in a glassy polymer film, often exhibit nonexponential fluorescence or phosphorescence decay profiles. In this section we focus on examples of true solutions of the dye in an amorphous matrix, meaning that the dye molecules are molecularly dispersed. If one were to take a snapshot (femtoseconds) of an ensemble of excited dye molecules in solution, one would find that different individual dye molecules experience somewhat different solvation environments. These differences may refer to tiny differences in the local density, which effects the local polarizability of the surrounding solvent molecules. In fluid solution, these differences in solvation normally relax on a time scale faster than the emission decay time of dye*. In viscous media and in rigid polymer films, this relaxation can be slow compared to the rate of emission.

The different solvation fields experienced by different dye* molecules can affect the radiative (k_r) and nonradiative (k_{nr}) decay rates of subsets of the ensemble of molecules that find themselves in different environments. Some dyes are sensitive to these differences and others are not. In our laboratory we have observed that naphthalene and some of its derivatives exhibit nonexponential fluorescence decay profiles in glassy polymer matrices such as PMMA or poly(butyl methacrylate) (PBMA), whereas the fluorescence decay profile of phenanthrene and its derivatives is almost always a simple exponential. Some dyes that exhibit exponential emission decay profiles in polymers above T_g exhibit nonexponential decays below T_g [35]. This type of result suggests that above T_g, the local environment can relax much faster than emission from the dye.

In sensor applications, the photostability of the dye is of paramount importance. Leading research groups in this field often build their sensors around ionic organometallic dyes, in which the heavy metal promotes rapid formation of triplet-like states and relatively long-lived emission. Many of these dyes are based on Ru(II) salts, such as tris(2,2'-bipryidyl)Ru(II) [Ru(bpy)$_3$] and tris(1,10'-phenanthroline)Ru(II) [Ru(phen)$_3$] salts, which have decay times on the order of a few microseconds. The solubility of the dye in the matrix depends on the ligands and the counterions. Demas and DeGraff [36] examined a number of RuL$_3^{2+}$ salts (here L is the ligand) in a crosslinked silicone polymer matrix. Because these salts have poor solubility in the rubbery polymer matrix, they are adsorbed onto nonporous silica particles dispersed in the matrix. Tris(4,7-diphenyl-1,10'-phenanthroline)Ru(II) [Ru(dpp)$_3$(II)] salts are more soluble in many polymers. (For the chemical structures of the various ligands of luminescent dyes referred to in this chapter, see Fig. 8.) Li et al. [37] looked at the optical properties as well as the oxygen-quenching response of the perchlorate salt in a silicone rubber film and found that the performance of the system was very sensitive to the sample preparation conditions. This is suggestive of dye aggregation in the film. Hartman et al. examined this dye as an oxygen sensor in polystyrene matrices [38].

None of these dyes exhibit exponential decay profiles, even in the absence of oxygen. While some experiments are carried out in matrices of high-T_g polymers, where frozen density fluctuations and differential dye solubilization can lead to nonexponential decays, most of the experiments referred to in the preceding paragraph were carried out in silicon rubber matrices at temperatures well above T_g. We have to seek alternative explanations for this type of emission decay for these dye–polymer systems.

Limited solubility can lead to aggregation of the dye molecules. Aggregation can affect the luminescence of dyes in fluid solution, but the problem is more pronounced in a polymer phase. It is well understood that the entropy of mixing decreases with increasing molecular size [39]. This effect is particularly strong in polymers. As the molecular weight increases, the range of solvents in which a polymer will dissolve diminishes, and the polymer becomes less soluble

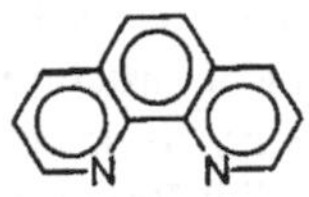

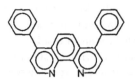

phen (1,10'-phenanthroline)

dpp (4,7-diphenyl-1,10'-phenanthroline)

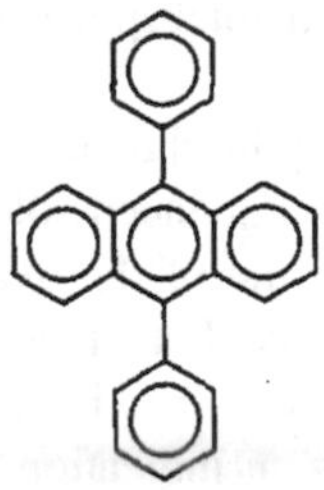

bpy (2,2'-bipryidy)

dpa (9,10-diphenylanthracene)

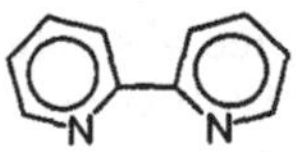

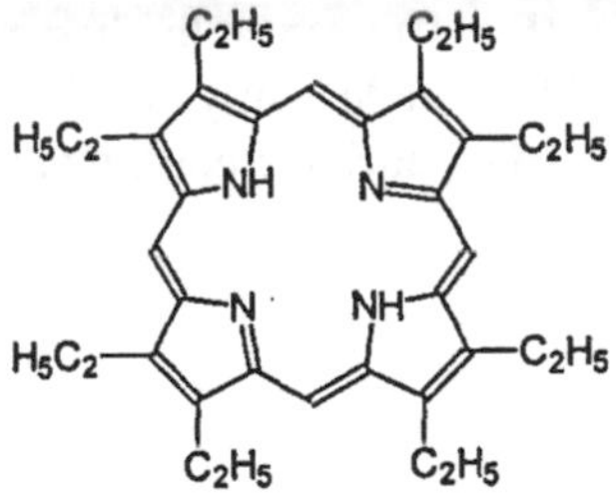

OEP (octaethylporphine)

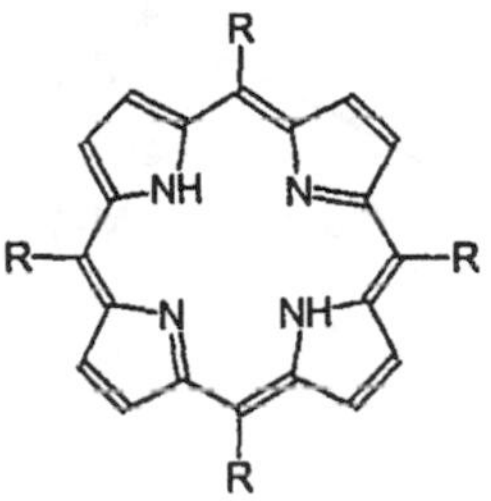

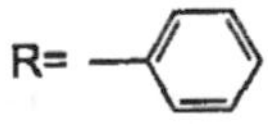

R=

TPP (tetraphenylporphine)

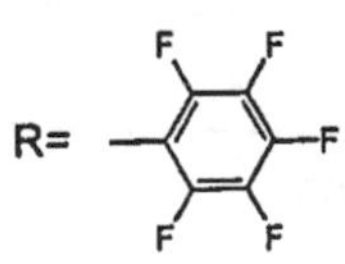

R=

TFPP (tetrafluorophenyl porphine)

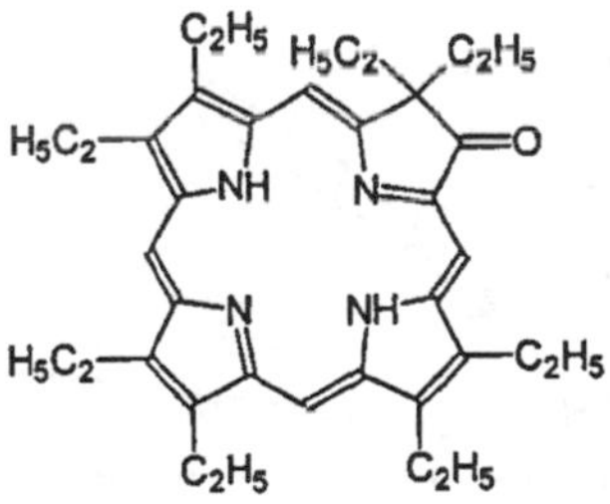

OEPK (octaethylporphine ketone)

Figure 8 Structures of ligands for luminescent dyes and their abbreviations used in this chapter.

in "poor" solvents. At the same time, the solubility of solutes characterized by an unfavorable enthalpy of mixing decreases [39]. Most dyes are less soluble in a polymer matrix than in a solvent medium of similar chemical structure. Dye aggregates (dimers, trimers, small clusters) are not always easy to detect. Dye aggregates are different from microcrystals, which would exhibit Bragg diffraction when exposed to x-rays. These species are sometimes nonfluorescent due to self-quenching and the aggregates can act as quenching traps for the fraction of unassociated dyes in the system. Some dye aggregates have distinct spectroscopic properties with shifted absorption and emission spectra.

Klimant and Wolfbeis [3a] found that increasing the hydrophobicity of the counterion (dodecyl sulfate, trimethylsilylpropanesulfonate) enhanced solubility of RuL_3^{2+} dyes in a variety of silicone rubber matrices. Some of the dye-containing polymer films were turbid, indicating poor solubility, but others were transparent. They interpret the nonlinear Stern–Volmer plots obtained in the presence of O_2 (Fig. 9) even in the best performing system, as an indication of the presence of both molecularly dissolved dyes and noncrystalline aggregates. Hartmann et al. [38a] attribute nonexponential decays for $Ru(dpp)_3(II)$ perchlorate in

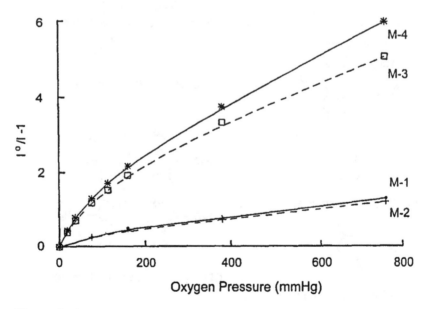

Figure 9 Nonlinear Stern–Volmer plots for the quenching by O_2 of the luminescence Ru(II) dyes in silicone rubber membranes E-4 (Wacker Co.) for two different ligands and two different counter ions. M-1: $Ru(phen)_3(DS)_2$; M-2: $Ru(phen)_3(TSPS)_2$; M-3: $Ru(dpp)_3(DS)_2$; M-4: $Ru(dpp)_3(TSPS)_2$, where phen is [1,10-phenanthroline)], DS is dodecyl sulfate, and TSPS is trimethylsilylpropansulfonate. (From Ref. 3a.)

polystyrene to the formation of noncrystalline aggregates, whereas in a second paper [38b] they explain these decays in terms of "spatial disorder in the polymer structure." Their x-ray diffraction measurements indicate a continuous distribution of scattering centers in the matrix rather than the presence of crystalline aggregates.

Another explanation for nonexponential decays for ionic dyes in polymer films is related to the distribution of counterions in the medium. If the counterion mobility in the matrix is slow compared to the excited-state lifetime, ion pairs with different spatial arrangements will emit light or decay nonradiatively at different rates. One very good model for the influence of the microenvironment on the rate of emission decay has been presented by Draxler and Lippitsch [40]. They develop a Förster-like expression [10] for the nonradiative transfer of energy from the excited dye to dipolar trap sites in the matrix. This model predicts a stretched exponential form for the emission decay profile rather than a sum of exponentials.

$$I(t) = I(0) \exp\left[-(1 - c)(t/\tau) - a(t/\tau)^{1/2}\right] \qquad (14)$$

In this expression, a is a parameter proportional to the number of quenching sites, and c, the "dynamic quenching parameter," is predicted to be proportional to the concentration of quencher (oxygen) in the matrix. The stretched exponential is a powerful function for fitting decay profiles, but the meaningfulness of the fit must be assessed independently. One of the impressive features of this treatment is that for $Ru(dpp)_3(II)$ perchlorate in polystyrene, over a wide range of external oxygen pressures, the luminescence decay profiles fit to Eq. (14), and the fitting parameter c is proportional to p_{O2} (Fig. 10).

2. Exponential Decays

In striking contrast to the behavior described above, some nonionic dyes in the absence of oxygen give simple exponential decays in polymer films. It is common for nonionic dyes to exhibit exponential decays in polymers above T_g, but some of the dyes described below show exponential decays upon pulsed excitation in polymers below T_g. Important examples include platinum octaethylporphine (PtOEP) [4a,32,34], platinum tetraphenylporphine (PtTPP), and the corresponding palladium derivatives PdOEP and PdTPP. The TPP derivatives have a lower solubility in polymer matrices than the OEP derivatives. The lifetimes of the two OEPs in polymer matrices are very different; in a polystyrene matrix, PtOEP has a lifetime of 91 µs and PdOEP has a lifetime of 1 ms [41]! Jayarajah [42] found that PtOEP in a series of aminothiolphosphazene matrices exhibited exponential decays in the absence of oxygen, with lifetimes ranging from 70 to 100 µs, depending on the chemical structure of the matrix.

Perhaps some of the resistance to using such well-behaved porphyrin derivatives as oxygen sensors is related to their lack of photostability [43]. Papkovsky et al. [44] have shown that when PtOEP and PdOEP are treated with a strong

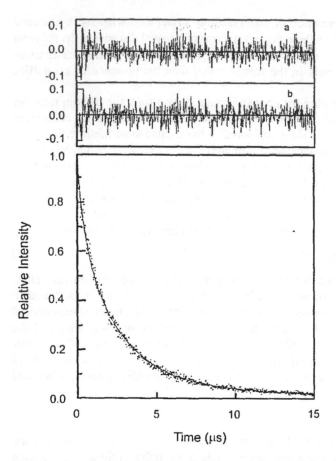

Figure 10 Luminescence decay of Ru(dpp)$_3$(ClO$_4$)$_2$ in polystyrene, at an oxygen partial pressure of 369 Torr. The decay curve is nonexponential. The authors fit the decay curve to two different equations. Experimental decay, fitted curve according to Eq. (14) (a) and a conventional multiexponential fit (b). The weighted residuals for these fits are shown at the top of the figure. (From Ref. 40.)

oxidizing agent (OsO$_4$), they are converted cleanly to ketone derivatives (PtOEPK and PdOEPK) that have a number of very desirable features. The absorption bands are red-shifted as is the emission, which appears in the near-infrared. The dyes are soluble in polystyrene films where the luminescence decays are exponential ($\tau_{PtOEPK} = 61.4$ μs; $\tau_{PdOEPK} = 480$ μs). PtOEPK is reported to be 10 times more photostable than PtOEP [44]. Hartmann and Trettnak reported that the quenching property of PtOEPK and PdOEPK is affected by the structure of polymer matrix. In poly(vinyl chloride), PtOEPK and PdOEPK exhibit perfectly linear Stern–Volmer intensity plots and almost single-exponential excited-state decays. In polystyrene, the slightly nonlinear Stern–Volmer plots and the nonexponential decays of PdOEPK show the significance of matrix effect of polystyrene, which is in the glassy state at the measurement temperature [45].

C. Response to Oxygen Concentration

One of the most interesting and challenging aspects to understand in the quenching of dye* luminescence by oxygen is the response of the measured intensity to increases in oxygen concentration or external oxygen partial pressure. In some systems the behavior is simple. In fluid solution, virtually all dyes with sufficiently long-lived excited states are quenched by oxygen in accord with the Stern–Volmer equation, irrespective of whether one examines intensities or lifetimes. The quenching kinetics can be explained in terms of the theory of partially diffusion-controlled reactions. The platinum and palladium porphyrin derivatives described above exhibit linear Stern–Volmer plots [I^0/I vs. p_{O2}, cf. Eq. (1)] in most amorphous polymer matrices. Two organogold derivatives dissolved in polystyrene films were reported by Mills et al. [46] to give linear Stern–Volmer plots over a range of p_{O2} from 10 to 750 Torr.

In stark contrast to this simple behavior, many systems incorporating Ru(II) dyes exhibit Stern–Volmer plots with pronounced downward curvature. Important examples are provided in the papers from the group of Demas [1,36,47] and that of Wolfbeis [3a,b]. A recent publication by Mills and Thomas [48], who investigate plasticizer effects on oxygen quenching of Ru(II) dyes in PMMA and cellulose acetate butyrate (CAB) films, provides an up-to-date commentary on this situation. These authors fit their intensity data to the expression

$$\frac{I^0}{I} = \frac{1}{f_1/(1 + K_1 \cdot p_{O2}) + (1 - f_1)/(1 + K_2 \cdot p_{O2})} \tag{15}$$

This expression is based on the idea that there are different quenching sites in the matrix and that each site has its own Stern–Volmer constant K_{sv} [47]. Strictly speaking, this expression is derived for a two-site model in which a fraction f_1 of the quenching is described by K_{sv1} and the remaining fraction is described by K_{sv2}, but in actual practice, Eq. (15) is used as a phenomenological expression to parameterize curved Stern–Volmer plots.

A decade ago, Krasnansky et al. [49] examined the quenching by oxygen of the fluorescence of pyrene and 9,10-diphenylanthracene adsorbed to the surface of nonporous silica, employing both steady-state and time-resolved fluorescence measurements. They found nonexponential fluorescence decays and downward curvature in the steady-state Stern–Volmer plots. The data could be described quantitatively in terms of a gaussian distribution of reaction rates, but only in terms of a model in which the oxygen molecules adsorbed to the surface of the silica in a rapid fast step. Quenching occurred subsequently through surface diffusion of the two reactants (dye* + O_2). Demas and DeGraff [50] are aware of this work. They have commented that Eq. (5) might mask a very complex gaussian or multiple gaussian distribution of sites in the (dye + polymer) or (dye + polymer + silica) mix.

In most of the experiments described above, oxygen quenching was determined through measurements of the steady-state luminescence intensity. As Eq. (1) makes clear, a useful alternative is the direct determination of luminescence decay times. These can be measured in pulsed experiments in which one monitors the intensity decay after excitation, or in phase shift experiments in which the intensity of the excitation beam is modulated sinusoidally at a frequency close to the decay rate of the chromophore [10b]. As a consequence of the excited-state lifetime, the emission detected is out of phase with the excitation. From the phase shift and intensity change, the value of τ can be determined. Both methods have been applied extensively to systems of biological interest [6,7], both to simple sensor applications and to imaging of oxygen fluxes. Papers are beginning to appear describing application of time-resolved and phase shift methods for imaging air pressure profiles for objects in wind tunnels [44,51–53] (Fig. 11).

There are two major advantages of luminescence decay measurements over intensity measurements. There are fewer artifacts, arising, for example, from scattering, competitive light absorption, emission reabsorption, and photodegradation of the sensor dye. In addition, decay measurements give absolute values of the luminescence decay time (or mean decay time if decays are not exponential). As a consequence, it is no longer necessary to compare results to an intensity reference.

D. Pressure-Sensitive Paint

In 1980, Peterson and Fitzgerald [54] demonstrated a surface pattern flow visualization technique based on the principle of oxygen quenching of dye fluorescence. The sensor used in their experiments was a chromatography plate, consisting of silica gel powder with a dye (fluorescein yellow) adsorbed on it. Their initial experimental results revealed the possibility of using this technique for studies of surface pressure profile measurements. This pressure-sensitive paint (PSP) concept has been developed into a sophisticated technology for monitoring air pressure profiles in wind tunnels. Its major application is in the design of aircraft, where it complements or replaces traditional pressure sensors [5]. Its advantages include high spatial resolution, large dynamic range, and direct visualization of

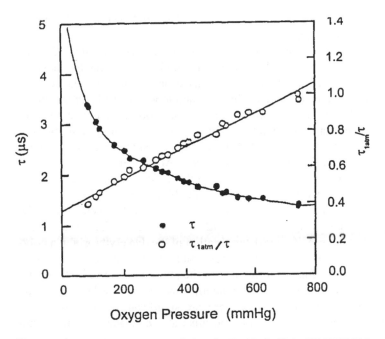

Figure 11 Lifetime–pressure relation for Ru(dpp)$_3$Cl$_2$ in GE-RTV 118 at 22°C, where τ_{1atm} is the lifetime at ambient pressure. Note that none of the individual decay profiles are exponential. The values reported are all mean decay times. (From Ref. 51a.)

the pressure profiles. With further refinements, one can image other lower wind velocity applications for wind profile testing, including the design of automobiles, buildings, and bridges. Because the technology commonly employs dyes that emit from triplet-like excited states, the PSP technique is also referred to as "phosphorescence barometry." Like the more traditional oxygen sensor described above, the PSP technique employs oxygen-sensitive lumophores dissolved or dispersed in a polymer binder. In order to capture images and analyze pressure profiles, and to map the pressure field over aerodynamic surfaces, PSP technology has a strong dependence on quantitative digital video and image-processing techniques.

Excellent reviews have been published describing the principles and applications of luminescent paints for aerodynamic pressure sensing. These include a very accessible introduction to the topic by Gouterman and co-workers [52], a comprehensive review of temperature- and pressure-sensitive luminescent paints by Sullivan and co-workers [51a], and a recent book by Mosharov et al. [5] stressing the measurement technology. Since many descriptions of the technology, including image capture and data reduction, are available elsewhere, we will focus on materials. We will examine issues related to the choice of polymer matrix and luminescent dye.

Many of the pioneering experiments on PSP were carried out at the Central Aero-Hydrodynamic Institute (TsAGI) in Russia [5,55]. Their initial luminescent pressure sensors involved a special cement based on starch and sugar to glue silica gel powder to the model surface, in conjunction with trypaflavin or amino-anthraquinone as luminophore. Unfortunately, these coatings exhibited a strong light intensity dependence of the delayed fluorescence when they were excited. A relatively short lifetime of around 0.2 ms was observed at low excitation light intensity, and this emission was sensitive to p_{O_2}. At high light intensities, the lifetime increased to 300 ms. Another group at the TsAGI developed a lumines-cent pressure sensor consisting of a polyurethane binder and acridine orange as the dye [5]. This system was not very successful because the quenching by O_2 in their polyurethane was highly time-dependent. More successful was a method based on phosphorescence lifetime determination of the dye erythrosin in a poly-vinylbutyral (PVB) matrix. PVB is more permeable to oxygen than the polyure-thane resin. The authors found a difference in the phosphorescence lifetimes ob-tained at the same temperature during the heating and cooling cycles. This group reached the important conclusion that there are significant advantages to em-ploying as the PSP matrix an amorphous polymer above its glass transition tem-perature [5].

For PSP applications, many of the same dyes are employed as in oxygen sensor applications. These include platinum porphyrins (PtOEP, platinum *meso*-tetra(pentafluorophenyl)porphine PtTFPP) and ruthenium complexes ([Ru(bpy)$_3$], [Ru(phen)$_3$], and [Ru(dpp)$_3$] salts). Most of the polymeric binders are silicone rubber resins [29,56,57], which have high oxygen permeability and also have the advantage of being commercially available. Other binders include silica gel [58] and sol-gel-derived coatings [59]. The choice of a binder for PSP is in many ways more demanding than for oxygen sensor applications. In addition to concern for the oxygen permeability of the resin and humidity effects on the dye–polymer combination, which are common to both technologies, temperature effects, adhesion, and mechanical stability are of particular importance for PSP applications. For example, a surface exposed to high-velocity wind is subjected to considerable abrasion by dust in the air. The toughness and abrasion resistance of the coating is an important consideration. The light intensity and time of expo-sure are longer in PSP applications than oxygen sensor applications. As a conse-quence, photodegradation of the dye is a more serious problem. For these reasons, one of the major directions in PSP materials research involves the search for new polymeric binders, of which the poly(aminothionylphosphazenes) (PATP) [29,34,42,60] and polyfluoroacrylate copolymers such as poly(*n*-heptafluorobutyl methacrylate-*co*-hexafluoroisopropyl methacrylate) (PFIB) [52,61] have shown promise. The chemical structures of these various binders are shown below:

$R = -NH - C_mH_{2m+1}$, $m = 1-7$ $CH_2(CF_2)_2CF_3$ $CH(CF_3)_2$

PATP   **PFIB**

The fundamental principles of phosphorescence barometry and some of the important technical problems in the application of this technique are described in the classic paper by Kavandi et al. [56a]. They used PtOEP as the dye, dissolved in a commercial curable silicone rubber (GP-197, Genesee Polymer Co.) as the binder. This coating is sensitive to air pressure. It exhibits a linear Stern–Volmer behavior in the pressure range of 0–1 atm of air. In terms of Eq. (9), this coating is characterized by values of $A' = 0.32$ and $Q_s = 0.70$. In wind tunnel testing applications, they obtained good agreement between pressures calculated from the luminescence intensity and those measured with conventional pressure transducers. There was poor agreement between the two techniques near the leading edge, where the surface curvature was high. The authors expressed their concern about temperature effects on the measurement and the sensitivity of the dye to photodegradation. They report a temperature dependence of the response near room temperature of $-1.7\%/°C$ [56b]. This group attempted to quantify the effects of photobleaching of the dye by monitoring the relative intensity I_{1atm}/I for films subjected to continuous illumination at an intensity equal to that used for making air pressure profile measurements [62]. They found that this ratio was reduced by approximately 45% when the film was illuminated for 1 h at $\lambda = 380$ nm and by 40% for illumination at $\lambda = 540$ nm. In later experiments, the dye PtOEP was replaced by the fluorinated derivative platinum *meso*-tetra (pentafluorophenyl)porphine (PtTFPP)] [52,63], which is less susceptible to photodegradation but appears to be more susceptible to aggregation in the polymer matrix.

Masoumi et al. [29] examined a series of commercially available curable silicone resins as potential binders for PSP. All of these binders have a polymer backbone based on poly(dimethylsiloxane) (PDMS) with different pendant functionalities for crosslinking. For PtOEP as the sensor dye, linear Stern–Volmer plots were obtained when intensities were measured as a function of p_{O2}. When decay profiles were measured, they report the curious situation that the decay profiles in the absence of oxygen were clearly exponential, but deviations occurred that became more severe at elevated oxygen pressure. From each decay curve, one could calculate a mean decay time $\langle \tau \rangle$. Plots of $\tau^0/\langle \tau \rangle$ vs. p_{O2} (e.g., Fig. 12) are linear.

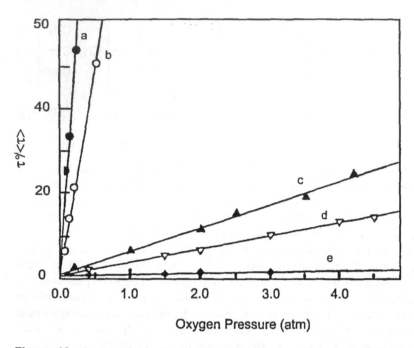

Figure 12 Stern–Volmer plots for PtOEP in a series of commercial silicone rubber polymers. Each decay trace is fit to a sum of three to four exponentials and the mean decay time $\langle\tau\rangle$ was calculated from the expression $\langle\tau\rangle = \sum_i A_i\tau_i^2 / \sum_i A_i\tau_i$, where A_i is the preexponential factor and τ_i is the corresponding lifetime of one term. τ^0 refers to the lifetime of the exponential decay of PtOEP in the absence of oxygen. From the top to bottom the silicone resins (from Genesee Polymer Co.) are (a) GP-277; (b) GP-187; (c) EXP36-X-20; (d) GP-197; (e) EXP38-X-20.

These authors [29] also report B values [I^0/I_{eq}, Eq. (5)] from steady-state fluorescence measurements as well as D_{O_2} values from time scan experiments [Eq. (13)]. They found that the different resins had very different sensitivities to oxygen quenching, with B values ranging from 2 to 200. It is not widely appreciated that different commercial silicone rubber compositions, all based on a PDMS backbone, can have very different oxygen permeability and diffusivity properties.

More disturbingly, the B values obtained by this group changed as the films aged, suggesting that cure was incomplete and that further reaction proceeded as the coated substrate was allowed to age. In addition, even the unquenched lifetime of the dye, PtOEP, changed as the samples aged. In one case, τ^0 increased from 70 μs in the freshly cast film to 92 μs in the same film aged 2 months at room temperature. For resins that depend on crosslinking chemistry, the response to oxygen can change with time if additional crosslinking occurs after the initial application of the coating to the substrate.

As an approach to avoid the problems of crosslinking, Manners and co-workers [34,64] synthesized the linear elastomer poly(n-butylaminothionylphosphazene) (C₄PATP). They prepared a sample of sufficient molecular weight that its films on a coated substrate did not flow at room temperature. This polymer is very useful for PSP applications. One of the attractive features of this polymer is that the polar functions in the backbone assist dye solubility. In this resin, PtOEP gives exponential decays in the absence of oxygen, and its Stern–Volmer plots are strictly linear. Stern–Volmer plots for this polymer, containing $Ru(dpp)_3Cl_2$, exhibit only small deviations from linearity. Nevertheless, under pulsed excitation, the emission decays are nonexponential, indicating that dye aggregation has not been eliminated. D_{O2} and p_{O2} values have been determined for a homologous series of C_nPATP polymers [65,66]. The pendant group has a small but important influence on the T_g of the polymer as well as on the permeability and diffusivity of oxygen in the matrix (Table 1).

When the size of the pendant groups is increased, a decrease of T_g can be observed, accompanied by an increase of both D_{O2} and P_{O2}, whereas S_{O2} remains almost constant. Since the polymer with lower T_g is considered to have larger free volume, those results are consistent with the idea that high free volume favors oxygen diffusion in the polymer matrix [65].

For C₄PATP in the absence of oxygen, the emission decay of $Ru(dpp)_3Cl_2$ was nearly exponential. Pronounced deviations from an exponential profile appeared when oxygen was introduced into the system. Plots of $\tau^0/\langle\tau\rangle$ vs. P_{O2} were linear [29], whereas plots of I^0/I vs. P_{O2} exhibited a slight downward curvature [64]. In a specific application of C₄PATP to phosphorescence barometry, plots of I_{1atm}/I vs. P^{air}/P^{air}_{1atm} were linear for PtOEP in C₄PATP, whereas the corresponding plot for $Ru(dpp)_3Cl_2$ in C₄PATP exhibited a very slight curvature, as shown in Fig. 13 [64]. For these films, the response and recovery upon exposure to

Table 1 Average Values Calculated from Time-Scan Experiments[a]

Polymer	T_g, (°C)	B	D	P	S	τ^0
C₂PATP	4	9.7 ± 0.2	0.50 ± 0.04	0.62 ± 0.01	1.2 ± 0.1	105
C₃PATP	6	28 ± 1	1.2 ± 0.2	1.8 ± 0.1	1.5 ± 0.3	NA
C₄PATP	−16	74 ± 6	4.0 ± 0.6	4.8 ± 0.4	1.2 ± 0.2	103
C₆PATP	−18	80 ± 1	6.2 ± 0.8	5.4 ± 0.1	0.87 ± 0.11	98
C₇PATP	NA	74 ± 4	6.0 ± 0.5	5.1 ± 0.3	0.85 ± 0.08	96

[a]D is given in 10^{-6} cm² s⁻¹, P in 10^{-12} mol cm⁻¹ s⁻¹ atm⁻¹, S in 10^{-3} M atm⁻¹, and τ^0 in μs.

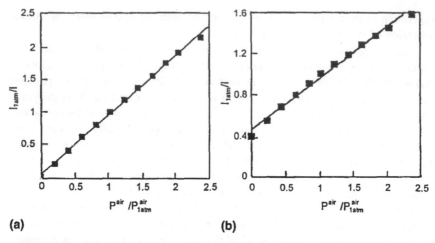

Figure 13 Air-quenching Stern–Volmer plots for two dyes in C₄PATP polymer films. The solid line represents the best straight line fit to the data. (a) PtOEP; (b) Ru(dpp)₃Cl₂. (From Ref. 64.)

oxygen were completely consistent with the model of fickian diffusion described above.

The Gouterman group [61] examined the fluoroacrylic copolymer poly(*n*-heptafluorobutyl methacrylate-*co*-hexafluoroisopropyl methacrylate) (PFIB) as a matrix for PSP measurements, using the fluorinated dye PtTFPP. PFIB is in the glassy state at room temperature, with T_g reported to be 70°C. This resin is reported to form a smooth, robust coating that displays a higher pressure sensitivity than the silicone resin GP-197. It has a better response time, better photostability, and lower temperature dependence (−0.6%/°C; see below) than the silicone resin. The intensity Stern–Volmer plots for this polymer/dye combination show a small downward curvature. The data cannot be fitted to Eq. (9). Rather, the authors employ an empirical equation [Eq. (17)], which adds a quadratic term to Eq. (9), and obtain good fits to their data with the parameters ($A = 0.09587$, $B = 1.081$, $C = −0.1867$, at 25°C).

$$\frac{I_{1atm}(T)}{I(p,T)} = A + B(P^{air}/P^{air}_{1atm}) + C(P^{air}/P^{air}_{1atm})^2 \qquad (17)$$

E. Temperature Effects

Unlike oxygen sensors used for biological and medical applications, where changes in temperature are normally not a serious problem, the use of pressure-sensitive coatings for air pressure profile determination requires that temperature effects be properly taken into account. In wind tunnel testing, a reference intensity at a reference pressure is required to calculate the pressure [Eq. (9)] at each point on the surface of the object. The reference point is commonly taken as air at 1 atm, i.e., static air. In the PSP literature, this condition is called "wind-off." "Wind-on" refers to intensity measurements made after the fans are turned on and air flows in the wind tunnel. The temperature in the wind tunnel might increase from 10°C to 50°C during the course of a long experiment [52]. As a consequence, I_{1atm} and I in Eq. (9) are often measured at different temperatures, and a temperature correction is required. A second problem arises due to temperature gradients that exist over the surface of the aerodynamic model. The first problem requires a correction for the effect of temperature on the luminescence intensity. The second problem requires that the correction be made almost pixel by pixel.

Changes in temperature operate on the binder, where the solubility of O_2 and its diffusivity are affected, and they operate on the dye. We first consider the influence of temperature on the decay rates of the excited dye. One can write the expression for the fluorescence quantum yield Φ_e of a dye as Eq. (18a), where the subscript e (emission) refers to either fluorescence or phosphorescence. The rate constants k_r and k_{nr} refer respectively to the radiative and nonradiative decay rates of the excited state. For triplets, the expression has to be modified to take account of the quantum efficiency (Φ_{ST}) of triplet formation. For the Ru and Pt dyes described in this chapter, Φ_{ST} is likely to be unity.

$$^1\Phi_e = k_r\tau = \frac{k_r}{k_r + k_{nr} + k_q[Q]} \tag{18a}$$

$$^3\Phi_e = \Phi_{ST}k_r\tau = \frac{k_r\Phi_{ST}}{k_r + k_{nr} + k_q[Q]} \tag{18b}$$

Equation (1) can now be reexpressed as

$$\frac{\Phi_e^0}{\Phi_e} = \frac{I^0}{I} = \frac{k_r + k_{nr} + k_q[Q]}{k_r + k_{nr}} = 1 + \frac{k_q[Q]}{k_r + k_{nr}}$$

The radiative rate is weakly sensitive to temperature, varying only with index of refraction changes in the medium. The nonradiative rate can have a significant temperature sensitivity. Deactivation of an electronically excited dye most commonly involves coupling of the excited state with vibrational levels of the dye in its ground state and with vibrational levels of the matrix. For dyes containing metal atoms, particularly RuL_3^{2+} dyes, temperature changes affect the rate of radiationless decay through coupling of the emissive triplet metal-to-ligand charge transfer state (MLCT) to a nearby metal-centered (MC) triplet state [67]. In a Stern–Volmer quenching experiment, temperature affects the unquenched lifetime τ^0 through its influence on k_{nr}. It affects k_q through its influence on D_{O2}, and it affects $[Q]$ through its influence on S_{O2}.

From this perspective, Eq. (9) can be rewritten as

$$\frac{I_{1atm}}{I} = \frac{k_r + k_{nr}}{k_r + k_{nr} + k_q P_{1atm}^{air}} + \frac{k_q P_{1atm}^{air}}{k_r + k_{nr} + k_q P_{1atm}^{air}} \left(\frac{P^{air}}{P_{1atm}^{air}}\right) = A' + Q_s \left(\frac{P^{air}}{P_{1atm}^{air}}\right) \quad (19)$$

Experimental evidence indicates that the temperature dependence of the nonradiative decay rate follows an Arrhenius form:

$$k_{nr} = A_{nr} e^{-E_{nr}/RT} \quad (20)$$

where A_{nr} is a preexponential factor and E_{nr} the activation energy. The solubility of diatomic gases in condensed liquids or polymers is only weakly dependent on temperature [68]. The major influence on the $k_q[Q]$ term in Eq. (1) is the change in temperature of D_{O2}. Diffusion in polymers depends upon free volume, which increases with the thermal expansion of the matrix. Polymer properties, including processes such as gas diffusion, which are coupled to the β transition of the polymer, satisfy an Arrhenius expression over a large temperature range.

$$D = A_D e^{-E_D/RT} \quad (21)$$

In Eq. (21), A_D is the preexponential term and E_D is the activation energy for oxygen diffusion in the matrix. Since the quenching rate constant k_q depends on diffusion, one anticipates that E_D for oxygen diffusion will be equal to the activation energy for quenching, E_q, determined from the temperature dependence of k_q [69]. In Eq. (19), temperature-dependent terms enter into the two fitting parameters A' and Q_s. Thus, both would be expected to vary as the temperature is changed.

Schanze and co-workers [70] investigated the temperature-dependent properties of Ru(dpp)$_3$Cl$_2$ in a diacetoxy-end-capped PDMS matrix. They calculate a value of $E_{nr} = 5.1$ kcal/mol (21 kJ/mol) for this system from an Arrhenius plot of the unquenched lifetime of Ru(dpp)$_3$Cl$_2$ as a function of temperature. From an examination of the quenched and unquenched lifetimes they obtained a value of $E_q = 2.7$ kcal/mol). This compares to a value of $E_D = 4.7$ kcal/mol determined by Cox [71] from direct examination of oxygen diffusion in an amine-cured silicone rubber. Although the silicone resins used in these two sets of experiments are not the same, the results taken together suggest that E_D and E_q are somewhat smaller than E_{nr}. When this is the case, one expects to find a larger influence of temperature on Q_s than on A' in Eq. (19). One example is presented in Fig. 14.

Schanze et al. [70] compared the temperature dependence of the Ru(dpp)$_3$Cl$_2$ lifetime in the silicone rubber matrix with its behavior as a solution in ethanol both in the presence and in the absence of oxygen. The ethanol system is characterized by a value of $E_{nr} = 7.3$ kcal/mol, which is significantly different

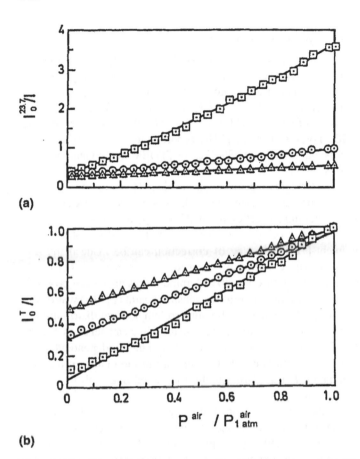

(a)

(b)

Figure 14 Effect of temperature on the calibration curve of PSP coatings based on PtOEP in a Genesee GP-197 silicone matrix. (a) The I_0 for each curve was measured at 23.7°C; (b) the I_0 for each curve was taken at its respective temperature. (From Ref. 56b.)

from that ($E = 5.1$ kcal/mol) found in the polymer matrix. The value found for E_q was close to zero ($E_q < 0.2$ kcal/mol). This seemingly surprising result is not uncommon for quenching processes that occur at rates close to, but less than, the diffusion-controlled limit. When the quenching mechanism involves a (dye*/Q) complex with a finite lifetime, an increase in temperature can promote dissociation of the complex to (dye* + Q) if $\Delta H < 0$ for formation of the complex. As shown in the scheme below, the experimentally determined "activation energy" is equal to ($E_a + \Delta H$). This sum can be positive, zero, or negative.

$$\text{Dye* + Q} \xrightleftharpoons{\Delta H} \text{dye*/Q} \xrightarrow{E_a} \text{quenching}$$

A plot of I vs T provides a useful measure of the sensitivity of a dye/matrix combination to a change in temperature. This type of plot is normally linear if the variation in temperature is not too large. Thus, the Gouterman group [56b] reported a temperature sensitivity of $-1.7\%/°C$ for PtOEP in Genesee GP-197, which compares to a value of $-0.6\%/°C$ for the dye PtTFPP in the fluoroacrylate polymer matrix PFIB. Torgerson et al. [51b] report values of $-1.8\%/°C$ for PtTFPP in airplane dope paint, and $-0.78\%/°C$ for PtTFPP in the General Electric silicone resin GE RTV 118.

There are various strategies that one can consider to overcome problems caused by variation in temperature. One can seek dyes or dye matrix combinations for which the temperature coefficient of A' and Q_s in Eq. (19) is small. Alternatively, one can try to develop strategies to correct the effects of temperature on the measurement [72]. This type of correction can be made if the film contains a second sensor whose emission is sensitive to temperature but not to the presence of oxygen. Temperature-sensitive luminophores have been proposed for other applications. The idea of temperature-sensitive paints (TSPs) was developed by Kolodner et al. [73] to measure the surface temperature distribution of an operating integrated circuit. A temperature-sensitive dye can serve as a reference: if E_{nr} for this reference dye in the matrix is determined independently, the intensity of its fluorescence reports on the temperature of the matrix. It then becomes possible to compare I [Eq. (18)] for the long-lived dye with that of the reference dye I_{ref} instead of I_{1atm}.

There are three strategies for incorporating temperature-sensitive sensors into a PSP formulation. First, one could use an undercoat in which a dye is dissolved in polymer matrix with low oxygen permeability. Irrespective of the surface air pressure, no significant oxygen quenching would occur in the oxygen-poor polymer layer. Second, one could disperse a solid pigment into the PSP coating. For example, microcrystalline dyes and phosphors are insensitive to oxygen quenching. Alternatively, one could take advantage of a principle expressed in Eq. (3), that dyes with short lifetimes (a few nanoseconds) are insensitive to quenching at oxygen concentrations that easily quench phosphorescence of dyes with 10^3 or 10^4 longer lifetimes.

The Gouterman group has examined the latter two alternatives. They introduced a solid phosphor, $BaMg_2Al_{16}O_{27}:Eu^{2+}$, as a reference to the PtTFPP–PFIB system [72]. This approach suffered from technical problems related to competitive absorption of the excitation light by the PtTFPP, the absorption of light emitted by the phosphor by PtTFPP, and inhomogeneous distribution of the phosphor particles in the matrix. More successful was the use of rhodamine B dissolved in the GP-197 matrix [56c]. Rhodamine B has a lifetime less than 10 ns and has a thermally activated radiationless decay pathway involving an internal rotation in the molecule. Under PSP measurement conditions, its intensity changes by $-2.3\%/°C$. More recently, the Gouterman group has employed silicon oetaethylporphine (SiOEP) as the reference, along with PtTFPP in PFIB as the polymer matrix [63]. The absorption spectrum of SiOEP is similar to that of PtTFPP. Both dyes can be excited at 400 nm. SiOEP has two maxima in its fluorescence spectrum, at 580 and 630 nm, whereas the phosphorescence maximum of PtTFPP appears at 650 nm. One can excite both dyes simultaneously but detect the two emissions independently.

F. Additives, Plasticizers, and Copenetrants

Plasticized poly(vinyl chloride) (PVC) is a common polymer matrix for oxygen sensor applications. A plasticizer is an additive to a polymer matrix that lowers the glass transition temperature and decreases the modulus [74]. PVC itself is a brittle glassy polymer at room temperature, whereas heavily plasticized PVC is a transparent rubber. The effect of plasticizers on polymers films is normally explained in terms of their contribution to the total free volume of the system. Plasticizers are liquids with a low glass transition temperature. The glass transition temperature of a solution of a plasticizer in a polymer matrix can be estimated very roughly via the Flory–Fox equation [39]. To have a significant effect on T_g, significant amounts of plasticizer must be present.

Ferry devotes a chapter to the topic of the influence of plasticizers on the viscoelastic behavior of polymers at temperatures above T_g. He comments that diluting a polymer with a low molecular weight solvent with which it forms a true solution normally leads to a sharp reduction in the local friction coefficient [75]. From this point of view, plasticizers should act to increase diffusion rates in polymer films. Below T_g, the effects are smaller. Ferry cites an example (see figure 15-8 in reference 75) where the presence of 9.1% and 16.7% dibutylphthalate in PMMA leads to a "slight loosening" of the local structure, facilitating the side group motions to which the β mechanism is attributed.

One of the most important contributions from the group of Ogilby has been an investigation of the effect of small concentrations of plasticizers and other species on the rate of oxygen diffusion in polymer films [22]. We recall that this group uses a spectroscopic method involving $^1O_2$ detection to monitor the uptake of O_2 into polymer films. A rigorous analysis of the data is possible, and precise values of D_{O2} over a wide temperature range could be determined. They examined the influence on oxygen diffusion of traditional plasticizers such as dimethylphthalate (DMP) and diphenylphthalate (DPP) as additives in polystyrene and polycarbonate films [22b]. The additives have a significant effect on D_{O2}, leading in some instances to a decrease in oxygen diffusivity and in other instances to an increase. Plots of $\ln(D_{O2})$ vs. $1/T$ were linear, and E_a values were 3 kJ/mol higher for 4.9% DMP and 1% DPP in polycarbonate than in the additive-free polymer. Two lines with different slopes will cross. At temperatures lower than

30°C. 1.0 DPP acts to lower oxygen diffusivity in polycarbonate, but at higher temperatures D_{O_2} is larger than in the neat polymer. For DMP, D_{O_2} values were decreased over the entire range of temperatures studied. Similar results were found for polystyrene. E_a was higher in the presence of 7.2% DMP, and the Arrhenius plot crosses that for polystyrene at about 50°C. Here 1.0% DPP had little effect. All of these experiments were carried out well below T_g for the two polymers [22d].

One of the most remarkable results on oxygen permeation in polymer films is the effect of the presence of a second gas on the oxygen diffusion rate. This effect was first reported by Chu and Thomas [76], who investigated oxygen quenching of triplet bromopyrene in a solvent-cast epoxy polymer SU8 (Intrex Co.) at various oxygen pressures below 300 mm Hg. They found that in the presence of nitrogen the quenching efficiency by oxygen is increased. We show their data in the Fig. 15, where we plot k_{obsd} (1/τ) vs. external oxygen pressure.

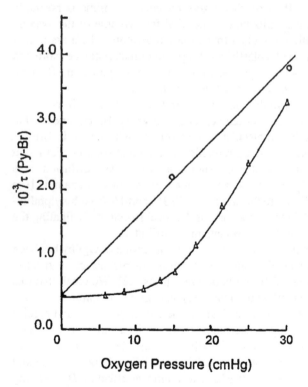

Figure 15 Effect of oxygen on the triplet decay rate of excited bromopyrene. ($\triangle$) Quenching by oxygen alone; ($\bigcirc$) quenching by oxygen in the presence of nitrogen with $P_{O_2} + P_{N_2} = 1$ atm. The polymer film (SU8 epoxy resin from Interez, Inc.) was dried at 40–50°C for 8 h. (From Ref. 76.)

In term of Stern–Volmer kinetics, $k_{obsd} = k_0 + k_q[O_2]$, where k_{obsd} and k_0 are the first-order decay rate constants of the excited dye in the presence and absence of quencher oxygen, respectively; and k_q is the quenching rate constant.

Ogilby and co-workers have investigated this effect in detail, directly examining oxygen diffusion in polymer films. In Fig. 16 we reproduce as an example the influence of 110 Torr of nitrogen on a $\ln D_{O2}$ vs. $1/T$ plot for a polystyrene film exposed to 30 Torr O_2 [22b]. The E_a values are essentially identical, but the presence of N_2 increases the value of D_{O2}. In polycarbonate D_{O2} is higher in the presence of 41 Torr of N_2 between 0 and 65° C, but the E_a is higher in the presence of N_2. In a second paper [22a], they examine the influence of other gases on the rate of oxygen diffusion in polystyrene and polycarbonate. The Ogilby group found that helium, argon, and nitrogen accelerate oxygen diffusion (E_a values are indeed not much affected for all gases except for nitrogen in polycarbonate). The effect is more pronounced in the presence of higher partial pressures of the second gas but appears to level off as this partial pressure exceeds 300 Torr. The authors consider a number of possible explanations for this unexpected behavior.

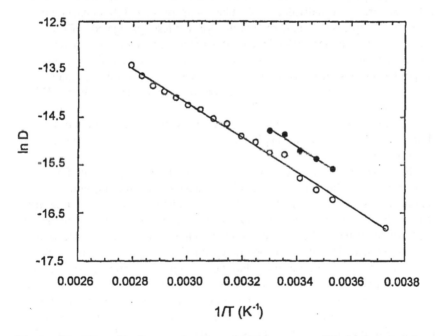

Figure 16 Plots of $\ln D_{O2}$ as a function of $1/T$ for oxygen diffusion in polycarbonate films in the presence (●) and absence (○) of nitrogen. Filled circles refer to a film exposed to an atmosphere (30 Torr of oxygen + 110 Torr of nitrogen). Open squares refer to the polymer film exposed to an atmosphere of oxygen only, at 30 Torr. (From Ref. 22b.)

One explanation is expressed in terms of the Langmuir component of the dual-mode sorption model. If some of the sites of oxygen adsorption bind or trap oxygen transport more effectively than others, adsorption of the copenetrant to sites that hinder the diffusion of oxygen would accelerate the overall rate of oxygen transport in the system. An alternative explanation distinguishes between the static and dynamic free-volume sites in the matrix.

V. CONCLUSIONS AND FUTURE DIRECTIONS

Solutions of luminescent dyes in polymer films are sensitive and effective sensors for oxygen. In the best-case scenarios, the luminescence is bright in the absence of oxygen and decreases in the presence of oxygen as described by the Stern–Volmer expression. Stern–Volmer plots with downward curvature is a common feature of ionic dyes, particularly dyes which are derivatives of Ru(II). Deviations from linear Stern–Volmer behavior may be associated with poor solubility of the dye in the polymer matrix. Relatively little attention has been paid to ascertaining if dye–polymer combinations form true solutions and, if they do, whether the dyes are present as individual molecules, or as associated (but noncrystalline) aggregates. It appears that ionic ruthenium dyes are less aggregated in PATP than in silicone rubber matrices. PATP is a modestly polar polymer rich in nonbonding electron pairs which may help dissolve ionic dyes. Even so, the solubility of $Ru(dpp)_3Cl_2$ in these films is limited. Dye miscibility in a polymer matrix can be enhanced by covalently attaching the dye to the polymer backbone. This is an active area of current research. Covalent attachment to the polymer does not guarantee the absence of aggregates. If the chemical structure of the dye and polymer are very different, such polymer-bound dyes are still susceptible to association in the matrix.

Effective sensors should be based on dyes with high quantum efficiencies of fluorescence or phosphorescence. It is not easy to measure absolute emission quantum yields for dyes in polymer films (but radiationless yields can be measured by photoacoustic spectroscopy [77]). When dyes form dimers or other associated structures, these often provide new radiationless pathways for deactivation of an excited dye. The luminescence quantum yield drops as a consequence.

In the sensor literature, there has been a lot of attention paid to finding highly sensitive dye–polymer combinations. While the search has not been hit-or-miss, attention has been focused on the Stern–Volmer constant K_{sv} itself, without giving due attention to its component parts. For systems involving a true solution of a dye in a polymer, the sensitivity of the excited dye to quenching depends on the product $(\tau^0 S_{O_2} D_{O_2}) \equiv (\tau^0 P_{O_2})$. It is not sufficient to consider only the diffusion coefficient for oxygen D_{O_2} in the matrix (or k_q for quenching). One must also take into account the solubility S_{O_2} of oxygen in the polymer. For many polymers, values of the oxygen permeability ($P_{O2} = D_{O2}\, S_{O2}$) can be found in the literature [68].

For detecting low concentrations of oxygen in fluids or small partial pressures of oxygen in gas mixtures, dyes with long-lived excited states have a particular advantage. For polymer films above T_g, where Henry's law is satisfied, the optimum choice may be PdOEP or PdOEPK with 1-ms and 0.5-ms lifetimes, respectively. For phosphorescence barometry, the choice is different. Most applications compare the luminescence intensity in air at 1.0 atm with air at elevated pressure. If both p_{O2} and S_{O2} are large, the extensive quenching that occurs under ambient conditions in still air places strenuous demands on the illumination and detection system, and also promotes photo-oxidation of the dye. One has to balance the permeability of the polymer to oxygen against the lifetime of the dye to optimize the response of the system. For C_4PATP as a polymer matrix, PtOEP has a lifetime (100 µs) that is almost too long for convenient monitoring of air pressure profiles. The shorter-lived Ru(dpp)$_3$Cl$_2$, with its 5-µs lifetime, is easier to detect.

In summary, there are two types of strategies that one can pursue to develop reliable oxygen sensors. One can develop mathematical expressions with good predictive power that allow one to describe the response of a formulated system to oxygen concentration or oxygen partial pressure irrespective of the composition of the system. Alternatively, one can try to understand structural features of the system that lead to complex behavior. Armed with this kind of knowledge, one may in the future be able to design systems which exhibit a simpler behavior. At the moment, this goal remains elusive, although we believe that much of the complexity observed for ionic dyes in silicone rubber or polystyrene films and for PtTFPP in PFIB films is related to the limited solubility of the dyes in the matrix.

REFERENCES

1. Xu, W.; McDonough, R. C. III; Langsdorf, B.; Demas, J. N.; DeGraff, B. A. *Anal. Chem.*, **1994**, *66*, 4133–4141.
2. Xavier, M. P.; Garcia-Frenadillo, D.; Moreno-Bondi, M. C.; Orellana, G. *Anal. Chem.*, **1998**, *70*, 5184–5189.
3. (a) Klimant, I.; Wolfbeis, O. S. *Anal. Chem.*, **1995**, *67*, 3160–3166; (b) Preininger, C.; Klimant, I.; Wolfbeis, O. S. *Anal. Chem.*, **1994**, *66*, 1841–1846; (c) Young, W. K.; Vojnovic, B.; Wardman, P. *Br. J. Cancer*, **1996**, *74* (Supp. 27), 256–259; (d) Bacon, J. R.; Demas, J. N. *Anal. Chem.* **1987**, *59*, 2780–2785; (e) Demas, J. N.; DeGraff, B. A. *J. Chem. Ed.*, **1997**, *74*(6), 690–695.
4. (a) Kavandi, J.; Callis, J.; Gouterman, M.; Khali, G.; Wright, D.; Green, E.; Burns, D; McLachlan, B. *Rev. Sci. Instrum.*, **1990**, *61*(11), 3340–3347; (b) Dowgwillo, R. M.; Morris, M. J.; Donovan, J. F.; Benne, M. E. *J. Aircraft*, **1996**, *33*(1), 109–116.

5. Moshasrov, V.; Radchenko, V.; Fonov, S. *Luminescent Pressure Sensors in Aerodynamic Experiments*, Central Aerohydrodynamic Institute (TsAGI), Moscow, 1998.

6. Holst, G.; Glud, R. N.; Kühl, M.; Klimant, I. *Sensors and Actuators B* 1997, *38–39*, 122–129.

7. (a) Rumsey, W. L.; Vanderkooi, J. M.; Wilson, D. F. *Science*, 1988, *241*, 1649; (b) Lübbers, D. W.; Köster, T.; Holst, G. A. *Proc. SPIE*, 1995, *2388*, 507–518.

8. (a) Wang, X. F.; Uchida, T.; Coleman, D. M.; Minami, S., *Appl. Spectrosc.*, 1991, *45*, 360–366; (b) Marriott, G.; Clegg, R. M.; Arndt-Jovin, T. M.; Jovin, T. M., *Biophys. J.*, 1991, *60*, 1374–1387; (c) Lakowicz, J. R.; Berndt, K. W., *Rev. Sci. Instrum.*, 1991, *62*, 1727–1734; (d) Clegg, R. M.; Feddersen, B.; Gratton, E.; Jovin, T. M., *Proc. SPIE*, Vol. 1640, 1992, pp. 448–460.

9. Hartmann, P.; Ziegler, W.; Holst, G.; Lübbers, D. W. *Sensors and Actuators B*, 1997, 38–39, 110–115.

10. (a) Birks, J. B. *Photophysics of Aromatic Molecules*, Wiley-Interscience, New York, 1971; (b) Lakowicz, J. R., *Principles of Fluorescence Spectroscopy*, Plenum, New York, 1983.

11. For reasons associated with spin statistics, the quenching of excited triplet states (dye[*3]) by O_2 is often slower than the quenching of singlet excited states (dye[*1]).[12] For studies in solution see (a) Gijzeman, O. L. J.; Kaufman, F.; Porter, G. *J. Chem. Soc., Faraday Trans. II*, 1973, *69*, 708–720; (b) Patterson, L. K.; Porter, G.; Topp, M. R. *Chem. Phys. Lett.*, 1970, *7*, 612–614.

12. For a more recent discussion of spin statistics, particularly for oxygen quenching in polymer films, see (a) Charlesworth, J. M.; Gan, T. H. *J. Phys. Chem.* 1996, *100*(36), 14922–14927; (b) Guillet, J. E.; Andrews, M. *Macromolecules*, 1992, *25*, 2752–2756.

13. Rice, S. A., In *Chemical Kinetics*, vol. 25, Bamford C. H., Tipper, C. F. H., Compton, R. G., Eds. Elsevier: New York, 1985.

14. Nemzek, T. L.; Ware, W. R. *J. Chem. Phys.*, 1975, *62*(2), 477–489.

15. Martinho, J. M. G.; Winnik, M. A. *J. Phys. Chem.*, 1987, *91*(13), 3640–3644.

16. Pilling, M. J.; Rice, S. A. *J. Chem. Soc., Faraday Trans. II*, 1975, *71*, 1563–1571.

17. Guillet, J. E., in *Photophysical and Photochemical Tools in Polymer Science*, Winnik, M. A. ed., NATO ASI C182, D. Reidel, Dordrecht, 1986, 467.

18. Oster, G.; Geacintov, N.; Khan, A. V. *Nature (London)*, 1962, *196*, 1089–1090.

19. The units for S are usually expressed as [cm^3 (gas at STP)]/[cm^3 (polymer)]/[Pa (of the external pressure)]. P = [(quantity of permeant)(film thickness)]/[(film area) (time)(pressure drop across the film)] = [cm^3 (gas at STP) cm]/[cm^2.s.Pa] = DS.

20. Bagryansky, V. A.; Korolev, V. V.; Tolkatchev, V. A.; Bazhin, N. M., *J. Polym. Sci. Part B: Polym. Phys.*, 1992, *30*, 951–958, and references cited therein.

21. (a) Koros, W. J. *J. Polym. Sci. Polym. Part B: Phys. Ed.*, 1980, *18*, 981–992; (b) Tshudy, J. A.; Frankenberg, C. *J. Polym. Sci. A2*, 1973, *11*, 2027–2037. (dual model sorption)

22. (a) Wang, B.; Ogilby, P. R. *Can. J. Chem.*, 1995, *73*, 1831–1840; (b) Gao, Y.; Baca, A. M.; Wang, B.; Ogilby, P. R. *Macromolecules*, 1994, *27*(24), 7041–7048; (c) Gao, Y.; Ogilby, P. R. *Macromolecules*, 1992, *25*(19), 4962–4966; (d) The authors also report an interesting change in slope of the ln D vs 1/T plot for polystyrene in the presence of 7.2% dimethyl phthalate.

23. Laguna, M.; Guzman, J.; Riande, E. *Macromolecules,* **1998**, *31,* 7488–7494.
24. Korolev, V. V.; Mamaev, A. L.; Bol'shakov, B. V.; Bazhin, N. M. *J. Polym. Sci. Part B: Polym. Phys.,* **1998**, *36,* 127–131.
25. Ferry, J. D. *Viscoelastic Properties of Polymers,* Wiley, New York, 1980, Chapter 11.
26. Deppe, D. D.; Dhinojwala, A.; Torkelson, J. M. *Macromolecules,* **1996**, *29,* 3898–3908.
27. Guillet, J. E. *Polymer Photophysics and Photochemistry,* Cambridge University Press, Cambridge, 1985.
28. (a) Müller-Plathe, F. *Comput. Polym. Sci.,* **1995**, *5,* 89–98; (b) Müller-Plathe, F. *Acta Polymer.,* **1994**, *45,* 259–293; (c) Müller-Plathe, F. *J. Chem. Phys.,* **1991**, *94,* 3192–3199; (d) Kim, W-K.; Mattice, W. L. *Macromolecules,* **1998**, *31,* 9337–9344.
29. Masoumi, Z.; Stoeva, V.; Yekta, A.; Winnik, M. A.; Manners, I. in *Polymers and Organic Solids,* Shi, L.; Zhu, D. Eds., Science Press, Beijing, 1997, pp. 157–168.
30. Cox, M. E., Dunn, B. *J. Polym. Sci. Part A: Polym. Chem.,* **1986**, *24,* 621–636.
31. Mills, A.; Chang, Q. *Analyst,* **1992**, *117,* 1461–1466.
32. Yekta, A.; Masoumi, Z.; Winnik, M. A. *Can. J. Chem.,* **1995**, *73,* 2021–2029.
33. Crank, J. *The Mathematics of Diffusion,* Clarender Press, Oxford, 1975, Chapter 1.
34. Masoumi, Z.; Stoeva, V.; Yekta, A.; Pang, Z. Manners, I.; Winnik, M. A. *Chem. Phys. Lett.,* **1996**, *261,* 551–557.
35. Horie, K; Mita, I; *Chem. Phys. Lett.* **1982**, *93,* 61–65.
36. (a) Carraway, E. R.; Demas, J. N.; DeGraff, B. A. *Anal. Chem.,* **1991**, *63,* 332–336; (b) Carraway, E. R.; Demas, J. N.; DeGraff, B. A.; Bacon J. R. *Anal. Chem.,* **1991**, *63,* 337–342.
37. Li, X.; Ruan, F.; Wong, K. *Analyst,* **1993**, *118,* 289–292.
38. (a) Hartmann, P.; Leiner, M. J. P.; Lippitsch, M. E. *Anal. Chem.,* **1995**, *67*(1), 88–93; (b) *Sensors and Actuators B,* **1995**, *29,* 251–257.
39. Flory, P. J., *Principles of Polymer Chemistry,* Cornell University Press, Ithaca NY, 1953.
40. Draxler, S.; Lippitsch, M. E. *Anal. Chem.,* **1996**, *68,* 753–757.
41. (a) Papkovsky, D. B. *Sensor and Actuators B,* **1995**, *29,* 213–218; (b) Papkovsky, D. B. *Sensors and Actuators B,* **1993**, *11,* 293–300.
42. Jayarajah, C. M. Sc. Thesis, University of Toronto, 1998.
43. All of these dyes are very stable in the absence of light; see for example Lee, S-K.; Okura, I. *Spectrochimica Acta Part A,* **1998**, *54,* 91–100.
44. Papkovsky, D. M.; Ponomarev, G. V.; Trettnak, W.; O'Leary, P. *Anal. Chem.,* **1995**, *67,* 4112–4117.
45. Hartmann, P.; Trettnak, W. *Anal. Chem.,* **1996**, *68,* 2615–2620.
46. Mills, A.; Lepre, A.; Theobald, B. R. C.; Slade, E.; Murrer, B. A. *Anal. Chem.,* **1997**, *69,* 2842–2847.
47. Demas, J. N., DeGraff, B. A., Xu, W. *Anal. Chem.,* **1995**, *67,* 1377–1380.
48. Mills, A.; Thomas, M. D. *Analyst,* **1998**, *123,* 1135–1140.
49. Krasnansky, R.; Koike, K.; Thomas, J. K. *J. Phys. Chem.,* **1990**, *94,* 4521–4528.
50. (a) Demas, J. N.; DeGraff, B. A. *Sens. Actuators B,* **1993**, *11,* 35–41; (b) Kneas, K. A., Xu, W.; Demas, J. N.; DeGraff, B. A. *Appl. Spectr.,* **1997**, *51,* 1346–1351.

51. (a) Liu, T.; Campbell, B. T.; Burns, S. P.; Sullivan, J. P. *Appl. Mech. Rev.*, **1997**, *50*, 227–246.; (b) Torgerson, S. D.; Liu, T.; Sullivan, J. P. *AIAA Paper 96-2184*.

52. Gouterman, M. *J. Chem. Edu.*, **1997**, *74*(6), 697–702.

53. Lachendro, N.; Crafton, J.; Guille, M.; Sullivan, J.; Carney,T.; Stirm, B. *Sixth Annual Pressure Sensitive Paint Workshop*, Boeing Co., Seattle, WA, **1998**, 23.1–23.31.

54. Peterson, J. I.; Fitzgerald, R. V. *Rev. Sci. Instrum.*, **1980**, *51*(5), 670–671.

55. (a) Ardrasheva, M. M.; Nevskii, L. B.; Pervushin, G. E., *J. Appl. Mech. Tech. Phys.*, **1985**, *26*, n 4, 24–33; (b) Troyanovsky, I.; Sadovskii, N.; Kuzmin, M.; Mosharov, V., Orlov, A.; Radchenko, V., Phonov, S. *Sens. Actuators B*, **1993**, *11*, 201–206.

56. (a) Kavandi, J.; Callis, J.; Gouterman, M.; Khalil, G.; Wright, D.; Green, E.; Burns, D.; McLachlan, B. *Rev. Sci. Instrum.*, **1990**, *61*(11), 3340–3347; (b) McLachlan, B. G.; Kavandi, J. L.; Callis, J. B.; Gouterman, M.; Green, E.; Khalil, G.; Burns, D. *Experiments in Fluids*, **1993**, *14*, 33–41; (c) Gallery, J.; Gouterman, M.; Callis, J.; Khalil, G.; McLachlan, B.; Bell, J. *Rev. Sci. Instrum.* **1994**, *65*(3), 712–720.

57. Carroll, B. F.; Abbitt, J. D.; Lukas, E. W.; Morris, M. J. *AIAA Journal*, **1996**, *34*(3), 521–526

58. Baron, A. E.; Danielson, J. D. S.; Gouterman, M.; Wan, J. R., Callis, J. B.; McLachlan, B. *Rev. Sci. Instrum.*, **1993**, *64*(12), 3394–3402.

59. Jordan, J. D.; Watkins, A. N.; Weaver, W. L.; Trump, D. D.; Goss, L. P.; Bell, J. H.; Dale, G. A.; Navarra, K. R. *Sixth Annual Pressure Sensitive Paint Workshop*, Boeing Co., Seattle, WA, **1998**, 29.1–29.15.

60. Evans, C.; Ruffolo, R.; Liu, X.; McWilliams, A.; Lu, X.; Winnik, M. A.; Manners, I. *Sixth Annual Pressure Sensitive Paint Workshop*, Boeing Co., Seattle, WA, **1998**, 30.1–30.30 and 32.1–32.21.

61. Puklin, E.; Carlson, B.; Gouin, S.; Costin, C.; Green, E.; Ponomarev, S.; Tanji, H.; Gouterman, M. *Sixth Annual Pressure Sensitive Paint Workshop*, Boeing Co., Seattle, WA, **1998**, 8.1–8.22.

62. Morris, M. J.; Donovan, J. F.; Kegelman, J. T.; Schwab, S. D.; Levy, R. L.; Crites, R. C. *AIAA J.*, **1993**, *31*(3), 419–425.

63. Coyle, L. M.; Chapman, D.; Khalil, G.; Schibi, E.; Gouterman, M. *Sixth Annual Pressure Sensitive Paint Workshop*, Boeing Co., Seattle, WA, **1998**, 7.1–7.12.

64. Pang, Z.; Gu, X.; Yekta, A.; Masoumi, Z.; Coll, J. B.; Winnik, M. A.; Manners, I. *Adv. Mater.*, **1996**, *8*, 768–771.

65. Jayarajah, C. N., M. Sc. Thesis, University of Toronto, 1998.

66. Jayarajah, C. N.; Yekta, A.; Winnik, M. A.; Manners, I. Manuscript in preparation.

67. Mongey, K. F.; Vos, J. G.; MacGeraith, B. D.; McDonagh, C. M.; Coates, C.; McGarvey, J. J., *J. Mater. Chem.*, **1997**, *7*, 1473–1479.

68. (a) Pauly, S., in *Polymer Handbook*, 3rd Ed., Brandrup, J.; Immergut, E. H., Eds., Wiley-Interscience, New York, 1989, pp. VI435–VI449; (b) Naylor, T.; in *Comprehensive Polymer Science*, Vol 2, Allen, G.; Bevington, J. C., Eds., Pergamon, Oxford UK, 1989, pp. 643–668.

69. From a more rigorous perspective, these two values should be different. According to eq. 3, kq depends upon the product (D R_{eff}) and $R_{eff}^{-1} \propto$ D. Thus a change in temperature that leads to a decrease in D is accompanied by an increase in R_{eff}.

70. Schanze, K. S.; Carroll, B. F.; Korotkevitch, S. *AIAA J.*, **1997**, *35*(2), 306–310.

71. Cox, M. E. *J. Polym. Sci. B: Polym. Chem.*, **1986**, *24*, 621–636.

72. Harris, J.; Gouterman, M. *Sixth Annual Pressure Sensitive Paint Workshop*, Boeing Co., Seattle, WA, **1998**, 19.1–19.13.
73. (a) Kolodner, P.; Tyson, A. *Appl. Phys. Lett.*, **1982**, *40*(9), 782–784; (b) Kolodner, P.; Tyson, A. *Appl. Phys. Lett.*, **1983**, *42*(1), 117–119.
74. Antiplasticizers are more rare. These substances act to raise T_g and the polymer modulus.
75. Ferry, J. D., *Viscoelastic Properties of Polymers*, Wiley, New York, **1980**, Chapter 17.
76. Chu, D. Y.; Thomas, J. K. *Macromolecules*, **1988**, *21*, 2094–2100.
77. Zharov, V. P.; Letokhov, V. S. *Laser Optoacoustic Spectroscopy*, Springer-Verlag, New York, **1986**.

Index

Milton Keynes UK
Ingram Content Group UK Ltd.
UKHW021818071024
449327UK00021B/1340

9 780367 398255